Hydrothermal Synthesis of Nanoparticles

Hydrothermal Synthesis of Nanoparticles

Editors

Gimyeong Seong
Juan Carlos Rendón-Angeles

MDPI • Basel • Beijing • Wuhan • Barcelona • Belgrade • Manchester • Tokyo • Cluj • Tianjin

Editors
Gimyeong Seong
The University of Suwon
Hwaseong-si, Republic of
Korea

Juan Carlos Rendón-Angeles
Center for Research and
Advanced Studies of the
National Polytechnic Institute
Mexico City, Mexico

Editorial Office
MDPI
St. Alban-Anlage 66
4052 Basel, Switzerland

This is a reprint of articles from the Special Issue published online in the open access journal *Nanomaterials* (ISSN 2079-4991) (available at: https://www.mdpi.com/journal/nanomaterials/special_issues/hydrotherm_nano).

For citation purposes, cite each article independently as indicated on the article page online and as indicated below:

LastName, A.A.; LastName, B.B.; LastName, C.C. Article Title. *Journal Name* **Year**, *Volume Number*, Page Range.

ISBN 978-3-0365-8064-7 (Hbk)
ISBN 978-3-0365-8065-4 (PDF)

Cover image courtesy of Juan Carlos Rendón-Angeles

© 2023 by the authors. Articles in this book are Open Access and distributed under the Creative Commons Attribution (CC BY) license, which allows users to download, copy and build upon published articles, as long as the author and publisher are properly credited, which ensures maximum dissemination and a wider impact of our publications.
The book as a whole is distributed by MDPI under the terms and conditions of the Creative Commons license CC BY-NC-ND.

Contents

About the Editors vii

Juan Carlos Rendón-Angeles and Gimyeong Seong
Editorial for the Special Issue: "Hydrothermal Synthesis of Nanoparticles"
Reprinted from: *Nanomaterials* 2023, 13, 1463, doi:10.3390/nano13091463 1

Md. Shah Alam, Bodrun Nahar, Md. Abdul Gafur, Gimyeong Seong and Muhammad Zamir Hossain
Forced Convective Heat Transfer Coefficient Measurement of Low Concentration Nanorods ZnO–Ethylene Glycol Nanofluids in Laminar Flow
Reprinted from: *Nanomaterials* 2022, 12, 1568, doi:10.3390/nano12091568 7

Zully Matamoros-Veloza, Juan Carlos Rendon-Angeles, Kazumichi Yanagisawa, Tadaharu Ueda, Kongjun Zhu and Benjamin Moreno-Perez
Preparation of Silicon Hydroxyapatite Nanopowders under Microwave-Assisted Hydrothermal Method
Reprinted from: *Nanomaterials* 2021, 11, 1548, doi:10.3390/nano11061548 23

David L. Burnett, Christopher D. Vincent, Jasmine A. Clayton, Reza J. Kashtiban and Richard I. Walton
Hydrothermal Synthesis of Iridium-Substituted $NaTaO_3$ Perovskites
Reprinted from: *Nanomaterials* 2021, 11, 1537, doi:10.3390/nano11061537 39

Akito Watanabe, Arisa Magi, Akira Yoko, Gimyeong Seong, Takaaki Tomai, Tadafumi Adschiri, et al.
Fabrication of Liquid Scintillators Loaded with 6-Phenylhexanoic Acid-Modified ZrO_2 Nanoparticles for Observation of Neutrinoless Double Beta Decay
Reprinted from: *Nanomaterials* 2021, 11, 1124, doi:10.3390/nano11051124 55

Natsuko Asano, Jinfeng Lu, Shunsuke Asahina and Seiichi Takami
Direct Observation Techniques Using Scanning Electron Microscope for Hydrothermally Synthesized Nanocrystals and Nanoclusters
Reprinted from: *Nanomaterials* 2021, 11, 908, doi:10.3390/nano11040908 67

Asato Nakagiri, Kazuya Imamura, Kazumichi Yanagisawa and Ayumu Onda
The Role of the Surface Acid–Base Nature of Nanocrystalline Hydroxyapatite Catalysts in the 1,6-Hexanediol Conversion
Reprinted from: *Nanomaterials* 2021, 11, 659, doi:10.3390/nano11030659 77

Hongjuan Zheng, Kongjun Zhu, Ayumu Onda and Kazumichi Yanagisawa
Hydrothermal Synthesis of Various Shape-Controlled Europium Hydroxides
Reprinted from: *Nanomaterials* 2021, 11, 529, doi:10.3390/nano11020529 93

Juan Carlos Rendón-Angeles, Zully Matamoros-Veloza, Jose Luis Rodríguez-Galicia, Gimyeong Seong, Kazumichi Yanagisawa, Aitana Tamayo, et al.
One-Pot Hydrothermal Synthesis of Victoria Green ($Ca_3Cr_2Si_3O_{12}$) Nanoparticles in Alkaline Fluids and Its Colour Hue Characterisation
Reprinted from: *Nanomaterials* 2021, 11, 521, doi:10.3390/nano11020521 103

Wenpo Luo and Abdelhafed Taleb
Large-Scale Synthesis Route of TiO_2 Nanomaterials with Controlled Morphologies Using Hydrothermal Method and TiO_2 Aggregates as Precursor
Reprinted from: *Nanomaterials* 2021, 11, 365, doi:10.3390/nano11020365 119

Nicolas Biscay, Lucile Henry, Tadafumi Adschiri, Masahiro Yoshimura and Cyril Aymonier
Behavior of Silicon Carbide Materials under Dry to Hydrothermal Conditions
Reprinted from: *Nanomaterials* **2021**, *11*, 1351, doi:10.3390/nano11051351 . . . . **137**

About the Editors

Gimyeong Seong

Gimyeong Seong is currently a full-time professor in the Environmental and Energy Engineering Department at Suwon University. He is actively engaged in environmental and energy-related research, both domestically and internationally. His research primarily focuses on hydrogen and renewable energy; low-carbon process development; and carbon capture, utilization, and storage (CCUS). Through his research, he aims to contribute to sustainable energy and environmental issues. He graduated from Sogang University, a prestigious private university in South Korea, with a degree in Chemical Engineering in February 2006. He obtained his Master's degree in Thermodynamics and Phase Equilibrium from the same university in 2008. He has had a strong interest in environmental issues, such as pollution and greenhouse gases, from the early stages of his career, which he initiated by developing alternative refrigerants to replace HFCs. Following that, he pursued his studies abroad at Tohoku University in Japan, where he received his Ph.D. in Engineering in March 2012 under Prof. Tadafumi Adschiri, a leading authority in the supercritical fluid synthesis and reactions field. After completing his Ph.D., he worked as an assistant professor, and later an associate professor at Tohoku University's New Industry Creation Hatchery Centre (NICHe). During this period, he researched various fields, including the development of nanocatalysts and energy conversion systems. He has published numerous research papers in reputable domestic and international academic journals, and presented his research findings at international conferences. His research in supercritical nanoparticle synthesis has gained recognition, and since joining Suwon University in April 2023, he has continued actively contributing to research in his field.

Juan Carlos Rendón-Angeles

Juan Carlos Rendón-Angeles is a full-time researcher and lecturer at the Centre for Research and Advanced Studies' Department of Ceramic Engineering. He has extensive research and academic experience in hydrothermal and inorganic compound processing, and particle crystal growth in aqueous solutions. He completed his undergraduate studies at the University College of Applied Chemistry of the National Polytechnic Institute in 1990, and obtained his Master's degree from the same university in 1993. In 1997, he earned a Ph.D. in Engineering from the Faculty of Mechanical Engineering at Tohoku University in Japan. He held a Postdoctoral position at Kochi University's Research Laboratory of Hydrothermal Chemistry from 1997 to 2000, where he conducted research and taught laboratory courses, allowing him to delve deeply into the fundamental chemistry aspects of hydrothermal reaction mechanisms. In 2000, he joined CINVESTAV Saltillo Campus as a research assistant, and later advanced to the positions of lecturer and associate professor in 2009, a position he currently holds. His research interests in hydrothermal chemistry revolve around developing environmentally friendly hydrothermal processing methods. He specialises in preparing powder nanoparticles of catalysts, electro-ceramics, and cold pigments using raw mineral precursors, as well as particle growth, waste management, and microwave-assisted powder processing.

Editorial

Editorial for the Special Issue: "Hydrothermal Synthesis of Nanoparticles"

Juan Carlos Rendón-Angeles [1,*] and Gimyeong Seong [2,3,*]

1. Center for Research and Advanced Studies of the National Polytechnic Institute, Mexico City 25900, Mexico
2. Department of Environmental and Energy Engineering, The University of Suwon, 17, Wauan-Gil, Hwaseong-si 18323, Gyeonggi-do, Republic of Korea
3. New Industry Creation Hatchery Center, Tohoku University, 6-6-10 Aoba, Aramaki, Aoba-Ku, Sendai 980-8579, Japan
* Correspondence: jcarlos.rendon@cinvestav.edu.mx (J.C.R.-A.); soppua4@suwon.ac.kr (G.S.)

Research and development in materials science has improved tremendously over the past few decades, resulting in benefits to the quality of life of people worldwide. In parallel, nanoparticle synthesis has emerged as a rapidly growing field of nanotechnology research. Nanoparticles (NPs), including organic and inorganic materials (such as metals, metal oxides, polymers, etc.), are tiny finite solids with at least one geometrical dimension below 100 nm. According to their size, NPs possess specific physical, chemical, optical, catalytic, and electronic properties compared to their large-sized parent bulk materials. Therefore, nanoparticles have various applications in catalysis, medicine, pharmaceutical, agriculture, and electronics, among other specific research fields. These applications are microstructurally dependent on nanoparticles' large surface-to-volume ratio, which allows them to interact rapidly with other particles of larger size.

Since natural NPs were discovered, the synthesis of NPs has become a primary research subject. Hitherto, the techniques proposed for the preparation of NPs are based on physical and chemical principles. Physical methods are founded on a top-down methodology, grinding being the most representative among these. It involves the dimensional reduction of solid matter to a nanoscale level. Other advanced techniques based on the same principle involve sputtering, laser ablation, electrospraying, and electron beam evaporation. On the contrary, the chemical methods developed for NP preparation differ in the technical philosophy applied; these methods are established on a bottom-up methodology that comprises the controlled manoeuvre of the atoms or molecules employed to form the NPs. Chemical processing techniques provide suitable reaction conditions for the production of NPs with specific microstructural aspects (e.g., size and morphology), compositions, and crystalline structures, which are the significant technological advantages of these processing methods. Nowadays, solution-based techniques such as coprecipitation, sol–gel, membrane-based, sonochemical, pyrolysis, vapour deposition, microwave-assisted, ion exchange, and combustion techniques have been widely employed to produce a wide variety of NPs.

Hydrothermal synthesis is a well-known technique used by various research groups worldwide to prepare NPs under subcritical or supercritical water conditions. This technique follows the bottom-up approach and has recently been used as a potential route for preparing a broad type of inorganic NPs. In particular, the heterogeneous chemical reactions that trigger the manipulation of hydrolysed atomic species in water can be achieved in a temperature range of 100–500 °C and high pressures varying from 0.1 to 22.5 MPa or more. However, some researchers also correlate the "hydrothermal" term with aqueous systems under reaction below 100 °C. In the last two decades, particular efforts have been made to provide faster nucleation kinetics and limited particle growth, resulting in the preparation of particles with specific morphologies and nanosized dimensions.

Citation: Rendón-Angeles, J.C.; Seong, G. Editorial for the Special Issue: "Hydrothermal Synthesis of Nanoparticles". *Nanomaterials* **2023**, *13*, 1463. https://doi.org/10.3390/nano13091463

Received: 14 April 2023
Accepted: 24 April 2023
Published: 25 April 2023

Copyright: © 2023 by the authors. Licensee MDPI, Basel, Switzerland. This article is an open access article distributed under the terms and conditions of the Creative Commons Attribution (CC BY) license (https://creativecommons.org/licenses/by/4.0/).

Additionally, the excellent properties of water have made it the research subject of new research groups for use in the production of nanoparticles and microparticles in a wide area of investigation encompassing various production scenarios, from laboratory-scale (basic chemical reaction studies) to pilot plant-scale production. Furthermore, the prepared nanomaterials' chemical compositions are controlled in the hydrothermal reaction media through a liquid phase or multiphase chemical reactions. Hence, these advantages provide an efficient processing technology to produce metal, oxide, and hybrid NPs.

This Special Issue, titled "Hydrothermal Synthesis of Nanoparticles", compiles the efforts of various groups of researchers worldwide and aims to communicate the recent advances in this research field. Nine original research papers focusing on synthesising different inorganic materials for heat transfer, biomedical, photocatalysis, scintillator, and photoluminescence applications are included. Furthermore, one review presents the state-of-the-art on SiC oxidation under dry and hydrothermal conditions. The following paragraphs give an overview highlighting the contributions of each published article.

The effort to control heat transfer is essential in the industry because of the high cost implied with the waste of heating energy. Hence, heat management has become a subject of concern in the equipment used in the industry, such as heat exchangers, automobiles, high-voltage transformers, refrigerators, electronic circuits, nuclear systems, solar energy harvesting plants, and desalinisation machines. Alam et al. [1] studied an innovative methodology for preparing nanofluids with thermal insulating properties. Zinc oxide nanorods particles were proposed as an insulating material, and the preparation of ZnO nanorods was successfully achieved at a low temperature of 170 °C for 6 h. A surfactant agent (polyvinylpyrrolidone, PVP) controlled the particle morphology and size. Heat transfer coefficient measurements were conducted to determine the nanofluid's efficiency. The results revealed that the laminar nanofluids constituted by 0.3 vol% of ZnO nanorods and ethylene glycol had a good heat transfer coefficient. These results open a new possibility for expanding this technology for heat management applications at the industrial level.

Various studies have been carried out in the last two decades to prepare biomaterials with similar chemical properties to biological calcium hydroxyapatite (HAp). The incorporation of silicon in the hexagonal structure of HAp has become a research subject, in which the veracity of the Si^{4+} substitution is under discussion to some extent. Matamoros-Veloza et al. [2] investigated the preparation of substituted silicon hydroxyapatite (Si-HAp) nanoparticles via the microwave-assisted hydrothermal process and employing $C_4H_{13}NO_5Si_2$ (TMAS) as a Si^{4+} precursor. Si^{4+} uptake improved by saturating the hydrothermal media with TMAS. The highest content of Si^{4+} obtained at 150 °C for 1 h was 12.16 mol%. Excess Si^{4+} also triggered the formation of Si-HAp agglomerates, constituted by fine nanorod particles of 32 nm.

Furthermore, the fine particles exhibited a 3D hierarchical self-assembly process favoured by the Si^{4+} ions in aqueous media coupled with rapid kinetic reaction conditions assisted by microwave heating. Additionally, the functionality of HAp NPs as potential catalyser agents was proposed by Nakagiri et al. [3]. The catalytic properties of calcium hydroxyapatite NPs were investigated as a function of the Ca/P molar ratio of the HAp. HAp NPs with various Ca/P ratios, between 1.54 and 1.72, were prepared under hydrothermal conditions at 110 °C for 16 h. The catalyst containing nonstoichiometric HAp NPs and a mixture of SiO_2/P_2O_5 exhibited remarkable catalytic behaviour for converting 1,6 hexanediol. The dependence on the Ca/P molar ratio in preparing unsaturated and cycloalkane alcohols was demonstrated for the first time. The particle size and the compositional variation inherited from the HAp NPs resulted in higher yields and sufficient selectivity than that reported in previous studies related to the catalytic behaviour of stoichiometric HAp particles.

Perovskite materials have been prepared under a wide range of hydrothermal conditions; thermodynamically, oxides are highly stable under alkaline hydrothermal media. In this research field, Burnett et al. [4] have presented an exciting study to optimise the chemical reactivity of the raw materials Ta_2O_5 and $IrCl_3$ to establish chemical equilibrium to produce iridium containing NaTaO$_3$ in a single step. A non-common alkaline medium constituted by 10 M NaOH in 40 vol% H$_2$O$_2$ heated at 240 °C was used. This aqueous medium is highly reactive to produce cube-shaped NaTa$_{1-x}$Ir$_x$O$_3$ crystallites around 100 nm in size. Interestingly, the perovskite nanoparticles underwent a heat treatment that caused the reduction of Ir^{4+}, partially substituting Ta^{5+} in the perovskite structure.

Consequently, the oxide particles were covered by the Ir^0. A ten-fold capability on hydrogen production, determined using a water–methanol solution, was observed on the NaTa$_{1-x}$Ir$_x$O$_3$ crystallites rather than the parent NaTaO$_3$. A significant proportion of H$_2$ was ascribed to visible light absorption; therefore, these materials have the potential to be employed in the water-splitting catalyst medium for H$_2$ generation.

Nanoparticle dispersibility is an essential property for preparing organic solvent scintillators that can be improved via the morphological modification of the particles. In this regard, the subcritical and supercritical hydrothermal conditions provide an adequate reaction environment to maximise the crystallisation of highly monodispersed nanoparticles. The study conducted by Watanabe et al. [5] aimed to improve the dispersibility of ZrO$_2$ NPs in toluene because it is well-known that the double-beta decay of ^{96}Zr has a high Q value of 3350 keV; therefore, this oxide can play the role of an isotope for $0\nu\beta\beta$ detection via radioluminescence spectroscopy. According to the research results, the organic 6-phenyl hexanoic acid part controlled the NP's growth via surface modification and was established as a function of the temperature. At a ZrO$_2$ NPs content of 0.33 wt%, prepared under subcritical conditions (300 °C for 10 min) and dispersed in toluene with 1,4-bis(5-phenyl-2-oxazolyl) benzene and 2,5-diphenyl oxazole, the as-prepared liquid scintillator had the maximum X-ray-induced radioluminescence response. It was concluded that only a small amount (0.0092 wt%) of ^{96}Zr isotope is required to prepare a liquid scintillator for potential usage in radiotherapy.

Fundamental studies based on parametric optimisation analyses correlating hydrothermal media selection, solvent pH, the chemical stability of the target crystalline phase, and precursor reactivity, among other things, are essential to determine the feasible morphological tuning of NPs to be prepared under hydrothermal conditions. Zheng et al. [6] applied this methodology for synthesising Eu(OH)$_3$ submicron particles and nanoparticles. Eu(OH)$_3$ was chemically stable in the pH range of 7.26–12; in acidic conditions, the dominating phase was Eu$_2$(OH)$_5$NO$_3\bullet$2H$_2$O in the Eu$_2$O$_3$-NO$_3$-NH$_4$OH-H$_2$O system. A remarkable difference in the particle morphologies occurred in the experiments conducted between 80 and 240 °C for 24 h. Eu(OH)$_3$ produced at pH values over 9 exhibited nanorods, nanotubes, and euhedral shapes. The differences in the solute's saturation level, which is temperature-dependent, trigger the formation of a large molar volume of the nucleus that limits the growth of Eu(OH)$_3$ crystals at temperatures over 200 °C. Likewise, the low solubility of Eu(OH)$_3$ also hinders the dissolution–recrystallisation mechanism that commonly triggers NP agglomeration, particle coarsening, and faceted growth.

An analogous investigation was carried out by Rendon-Angeles et al. [7], which focused on the one-step synthesis of Uvarovite NPs (Ca$_3$Cr$_2$Si$_3$O$_{12}$), in which the single-phase stability depends on the alkaline concentration of the hydrothermal media. The single-step chemical reaction associated with the Ca$_3$Cr$_2$Si$_3$O$_{12}$ formation proceeded in a 5 M KOH solution under hydrothermal conditions at 200 °C for 12 h, resulting in the preparation of fine agglomerates with a popcorn-like shape and sizes varying between 66 and 156 nm. A peculiar self-assembly process involving primary irregular anhedral crystals (8.05–12.25 nm) led to the formation of the Ca$_3$Cr$_2$Si$_3$O$_{12}$ agglomerates. The Ca(OH)$_z$$^{n+}$ and Cr(OH)$_m$$^{n+}$ formed in the reaction media due to a deficient Si^{4+} nominal content, resulting in a size decrease in the 3D popcorn-shaped agglomerates. The particle size variation led to a variation in the colour hue. At the same time, the faceted surface texture inherited from the tiny primary NPs achieved a high-reflectance near-infrared spectrum, which is vital in designing cold pigments for industrial insulating applications.

Differences in the shapes and growth mechanisms of NPs likely depend on the chemical reaction pathway between the solid precursor and the aqueous media, which was demonstrated by the study conducted by Luo and Taleb [8]. Their analysis considers the usage of TiO_2 aggregates constituted by NPs and aims to elucidate the mechanism that controls the morphology of TiO_2 under hydrothermal conditions in a highly concentrated NaOH medium (10 M). An increase in the reaction temperature from 100 to 200 °C caused the consumption of aggregated particles of nano-urchin-shaped TiO_2 anatase to produce TiO_2 nanotubes and nanobelts. The preparation of the ultimate TiO_2 nanobelts occurred by a particular assembly mechanism promoted by the Na^+ ions in the hydrothermal media; these ions caused the exfoliation of TiO_2 rolled nanosheets self-assembled to produce the TiO_2 nanotubes, and with the increase in both the temperature and reaction interval, the production of TiO_2 nanobelts. According to their morphological features, TiO_2 nanobelts can be used to prepare anodes for Li-ion batteries.

In addition, the characterisation of microstructural NPs is essential for the morphological study of their features (particle size, surface area, geometry, and chemical compositional analysis). The details of the NPs' microstructures are currently revealed by transmission electron microscopy (TEM). However, additional information, such as overlapping particles, faceted particle growth, and 3D hierarchical self-assembly architectures, are complicated to analyse using TEM observations. Hence, Asano et al. [9] proposed an innovative technological approach based on scanning electron microscopy (SEM) to overcome the deficiencies of TEM observation. A peculiar holder constituted by an ultrathin Si_3N_4 window allowed the authors to observe CeO_2 NPs in water in the SEM machine. This tool makes it possible to directly observe the architectural features of CeO_2 nanoclusters synthesised under hydrothermal conditions. Improvements in new SEM devices boost their performance and versatility, allowing them to obtain high-resolution images of hydrothermally synthesised nanoparticles. Various examples, including details of particle preparation and SEM operating conditions, are discussed in detail in this paper. The methodologies derived from this work will increase the motivation of researchers to characterise NPs using this technique.

Finally, a review of the fundamentals of the oxidation process of SiC materials is published by the research group of Biscay et al. [10]. The review provides a detailed overview of the state-of-the-art in this specific research field. SiC ceramics have excellent thermomechanical and corrosion resistance and are thus suitable for high-performance applications in the aerospace and nuclear industries. Therefore, the conscientious study of the corrosion behaviour of SiC in dry and aqueous environments is required. The review presents a critical analysis of the oxidation pathways for each case, which are correlated to thermodynamic and chemical kinetics. A detailed evaluation of the models proposed for SiC oxidation under hydrothermal conditions is discussed. The last section of the review provides a comprehensive examination from an application point of view of the hydrothermal corrosion process of SiC materials. This methodology can be applied to designing SiC devices for aerospace applications; however, a dangerous scenario can arise for some critical applications, such as employing SiC to generate new nuclear reactors.

Nanotechnology is developing quickly, and significant progress is expected in developing efficient, clean, and sustainable materials. This Special Issue presents high-quality original research works and a review of advanced nanomaterials obtained via "hydrothermal synthesis" for various applications. We hope these articles will provide valuable information that motivates a generation of young scientists to continue the growth of this research field. The research in this field is booming, and significant advances in developing efficient, clean, and sustainable materials and technologies are expected.

Funding: No funding was received to accomplish the Editorial corresponding to the Special Issue "Hydrothermal Synthesis of Nanoparticles"; individual research funding information is provided in each published paper in the Funding section.

Data Availability Statement: All papers comprising the Special Issue "Hydrothermal Synthesis of Nanoparticles", including the supporting information data files, are available at https://www.mdpi.com/journal/nanomaterials/special_issues/hydrotherm_nano (accessed on 23 April 2023).

Acknowledgments: Special acknowledge are offered to all the authors for their state-of-the-art contributions to hydrothermal nanoparticle preparation. They consider the Special Issue an academic medium to spread their research globally. We particularly acknowledge all the academic editors, reviewers, and editorial assistant MDPI staff, who ensured that this Special Issue's preparation was a very efficient process.

Conflicts of Interest: The authors declare no conflict of interest.

References

1. Alam, M.S.; Nahar, B.; Gafur, M.A.; Seong, G.; Hossain, M.Z. Forced convective heat transfer coefficient measurement of low concentration nanorods ZnO-ethylene glycol nanofluids in laminar flow. *Nanomaterials* **2022**, *12*, 1568. [CrossRef] [PubMed]
2. Matamoros-Veloza, Z.; Rendon-Angeles, J.C.; Yanagisawa, K.; Ueda, T.; Zhu, K.; Moreno-Perez, B. Preparation of silicon hydroxyapatite nanopowders under microwave-assisted hydrothermal method. *Nanomaterials* **2021**, *11*, 1548. [CrossRef] [PubMed]
3. Nakagiri, A.; Imamura, K.; Yanagisawa, K.; Onda, A. The role of the surface acid-base nature of nanocrystalline hydroxyapatite catalysts in the 1,6-hexanediol conversion. *Nanomaterials* **2021**, *11*, 659. [CrossRef] [PubMed]
4. Burnett, D.L.; Vincent, C.D.; Clayton, J.A.; Kashtiban, R.J.; Walton, R.I. Hydrothermal synthesis of iridium substituted $NaTaO_3$ perovskites. *Nanomaterials* **2021**, *11*, 1537. [CrossRef] [PubMed]
5. Watanabe, A.; Magi, A.; Yoko, A.; Seong, G.; Tomai, T.; Adschiri, T.; Hayashi, Y.; Koshimizu, M.; Fujimoto, Y.; Asai, K. Fabrication of liquid scintillators loaded with 6-phenyl hexanoic acid-modified ZrO_2 nanoparticles for observation of neutrinoless double beta decay. *Nanomaterials* **2021**, *11*, 1124. [CrossRef] [PubMed]
6. Zheng, H.; Zhu, K.; Onda, A.; Yanagisawa, K. Hydrothermal synthesis of various shape-controlled Europium hydroxides. *Nanomaterials* **2021**, *11*, 529. [CrossRef] [PubMed]
7. Rendón-Angeles, J.C.; Matamoros-Veloza, Z.; Rodríguez-Galicia, J.L.; Seong, G.; Yanagisawa, K.; Tamayo, A.; Rubio, J.; Anaya-Chavira, L.A. One-pot hydrothermal synthesis of Victoria green ($Ca_3Cr_2Si_3O_{12}$) nanoparticles in alkaline fluids and its colour hue characterisation. *Nanomaterials* **2021**, *11*, 521. [CrossRef] [PubMed]
8. Luo, W.; Taleb, A. Large-scale synthesis route of TiO_2 nanomaterials with controlled morphologies using the hydrothermal method and TiO_2 aggregates as a precursor. *Nanomaterials* **2021**, *11*, 365. [CrossRef] [PubMed]
9. Asano, N.; Lu, J.F.; Asahina, S.; Takami, S. Direct observation techniques using scanning electron microscope for hydrothermally synthesised nanocrystals and nanoclusters. *Nanomaterials* **2021**, *11*, 908. [CrossRef] [PubMed]
10. Biscay, N.; Henry, L.; Adschiri, T.; Yoshimura, M.; Aymonier, C. Behavior of silicon carbide materials under dry to hydrothermal conditions. *Nanomaterials* **2021**, *11*, 1351. [CrossRef] [PubMed]

Disclaimer/Publisher's Note: The statements, opinions and data contained in all publications are solely those of the individual author(s) and contributor(s) and not of MDPI and/or the editor(s). MDPI and/or the editor(s) disclaim responsibility for any injury to people or property resulting from any ideas, methods, instructions or products referred to in the content.

Article

Forced Convective Heat Transfer Coefficient Measurement of Low Concentration Nanorods ZnO–Ethylene Glycol Nanofluids in Laminar Flow

Md. Shah Alam [1], Bodrun Nahar [1], Md. Abdul Gafur [2], Gimyeong Seong [3,*] and Muhammad Zamir Hossain [1,*]

1. Department of Chemistry, Jagannath University, Dhaka 1100, Bangladesh; m170301033@chem.jnu.ac.bd (M.S.A.); bodrunnahar@chem.jnu.ac.bd (B.N.)
2. Pilot Plant & Process Development Centre, Bangladesh Council of Scientific and Industrial Research (BCSIR), Dhaka 1205, Bangladesh; d_r_magafur@bcsir.gov.bd
3. New Industry Creation Hatchery Center, Tohoku University, 6-6-10 Aoba, Aramaki, Aoba-ku, Sendai 980-8579, Japan
* Correspondence: kimei.sei.c6@tohoku.ac.jp (G.S.); zamir@chem.jnu.ac.bd (M.Z.H.)

Abstract: This paper presents the experimental forced convective heat transfer coefficient (HTC) of nanorods (NRs) zinc oxide–ethylene glycol nanofluids (ZnO–EG NFs) in laminar flow. First, ZnO NRs were synthesized using a hydrothermal method that uses zinc acetate dihydrate [$Zn(CH_3COO)_2 \cdot 2H_2O$] as a precursor, sodium hydroxide as a reducing agent, and polyvinylpyrrolidone (PVP) as a surfactant. The hydrothermal reaction was performed at 170 °C for 6 h in a Teflon-lined stainless-steel tube autoclave. The sample's X-ray diffraction (XRD) pattern confirmed the formation of the hexagonal wurtzite phase of ZnO, and transmission electron microscopy (TEM) analysis revealed the NRs of the products with an average aspect ratio (length/diameter) of 2.25. Then, 0.1, 0.2, and 0.3 vol% of ZnO–EG NFs were prepared by adding the required ZnO NRs to 100 mL of EG. After that, time-lapse sedimentation observation, zeta potential (ζ), and ultraviolet-visible (UV–vis) spectroscopy was used to assess the stability of the NFs. Furthermore, the viscosity (μ) and density (ρ) of NFs were measured experimentally as a function of vol% from ambient temperature to 60 °C. Finally, the HTC of NFs was evaluated utilizing a vertical shell and tube heat transfer apparatus and a computer-based data recorder to quantify the forced convective HTC of NFs in laminar flow at Reynolds numbers (Re) of 400, 500, and 600. The obtained results indicate that adding only small amounts of ZnO NRs to EG can significantly increase the HTC, encouraging industrial and other heat management applications.

Keywords: hydrothermal; zinc oxide nanorods; zinc oxide–ethylene glycol NF; zeta potential; ultraviolet-visible spectroscopy; shell and tube heat exchanger; convective HTC; laminar flow

1. Introduction

Heat transfer enhancement is a significant aspect of engineering research since it is associated with heat management in industries: heat exchangers, automobiles, high-voltage transformers, refrigerators, electronics, nuclear systems, solar energy harvesters, and desalinization [1]. Water, EG, engine oil, silicone oil, gear oil, kerosene, and other conventional fluids have been used for heat transfer purposes for a long time. The poor heat transfer of conventional heat transfer fluids (e.g., the thermal conductivity of liquids lies between that of the insulators and non-metallic solids ($0.1 < k_L < 10$ W/mK)) has been causing problems in many sectors, including industries [2]. That heat transfer needs to be increased. Increased heat transfer can intensify some cooling processes, saving time, energy, and the amount of heat transfer equipment. Generally, adding a small number of solid particles to conventional fluids can increase heat transmission capacity remarkably [3]. However, the rapid sedimentation of bulk solids in conventional fluids is a problem in this case [4].

Nanomaterials (NMs—at least one of the dimensions is between 1 and 100 nm) are being considered to solve or mitigate the rapid sedimentation problem. NFs, the dispersion of NMs in base fluids (BFs), is one of the alternatives to conventional heat transfer fluids [5]. NFs are more stable than micron-sized materials dispersed fluids since NMs remain dispersed in a BF for a longer time [6]. Usually, metals, metal oxides, carbon nanotubes, graphene, and other NMs are used to prepare NFs to enhance the thermal conductivity by increasing the conduction and convection coefficient [7]. Especially, nano ZnO can be used to prepare NFs in conventional fluids such as EG because the thermal conductivity of ZnO (29 WmK^{-1}) is more than a hundred times higher than that of ethylene glycol (0.26 WmK^{-1}) [2]. Additionally, experimental results revealed that ZnO-based NFs exhibited enhanced thermal conductivity compared to other NFs [8–10].

BF is a key issue to consider for NFs preparation, and HCT results of NFs often depend on BFs. The literature survey provides information on using water-based [11] and oil-based NFs [12], their thermophysical properties, and their advantages and disadvantages. For instance, inside a vertical circular enclosure heated from above, the influence of an alumina-water NF on natural convection heat transfer reported that the Nusselt number of the alumina–water NF is lower than that of the BF. This suggests that, as compared to pure water, utilizing alumina–water NFs harms the HTC [13]. Therefore, choosing either an aqueous or non-aqueous BF may be worthy.

Many efforts have been made to improve the thermal conductivity of ZnO-based NFs further. Experiments were done considering the affecting factors related to the convection heat transfer of NFs. For instance, ZnO–water NFs were used in an experiment to improve convective heat transfer in a car radiator at a higher Reynolds number [14]. Heat transfer was evaluated at the laminar flow condition, and the results revealed that the heat transfer behavior of ZnO–EG-based NFs depends on the flow condition [15]. The geometric shape effect coupled with the heat exchanger of NFs in heat transfer applications was investigated and evaluated [16]. Shell and tube heat exchangers were used to experiment with the convective heat transfer enhancement of graphene NFs and the results revealed that the HTC of graphene NFs rose by 23.9% at 38 °C when the graphene concentration was increased from 0.025 to 0.1 wt% [17]. The literature review paper reveals that employing NF to improve natural convection heat transfer is still controversial, and there is an ongoing discussion about the role of NPs in heat transfer enhancement [18]. Few studies focused on the factors that affect the thermal conductivity of NFs and summarized that the dispersion stability of NFs [19], NMs' volume%, NF preparation method, sonication time, and the morphology (size and shape) of NMs are the essential factors that affect the heat transfer properties of NFs [19,20].

The morphology, i.e., size and shape of NMs [21,22], and preparation method significantly influence the heat transfer properties of NFs. The result of a thermal conductivity dependence on ZnO-based NFs shows that the thermal conductivity increases linearly with increasing particle size [23]. An experimental investigation of the ZnO particle shape-dependent thermal conductivity of ZnO–EG NFs showed a 23% enhancement in the thermal conductivity of 0.1 mg/1 mL of ZnO NF compared to the BF (water). Again, the NF with non-spherical NPs exhibits greater thermal diffusivity than an NF with spherical NPs [24]. Moreover, instability in NFs results in a decrease in thermophysical properties as well as a reduction in heat transfer efficiency [25]. The thermal conductivity of NFs is demonstrated to be strongly influenced by the suspension stability of boehmite alumina in various shapes [26]. According to a molecular dynamic simulation, the heat transfer following the collision of cylindrical NPs with a heat source under the most favorable conditions can be substantially higher than that of spherical NPs of equal volume. As evidenced by experimental data, this can potentially result in a more significant enhancement in the thermal conductivity of an NF, including elongated particles, compared to one having spherical NPs [27]. In addition, the thermal conductivity of ZnO NFs with roughly rectangular and sphere NP morphologies was also examined experimentally at varied NP volume concentrations ranging from 0.05 to 5.0 vol% in a different study. Compared to the BF (water), the thermal conductivity of the ZnO NFs rose by 12% and 18% at 5.0 vol% for the spherical and almost rectangular forms of NPs, respectively. The shape of the particles is found to impact the increases in thermal conductivity substantially [28]. Molecular dynamics simulations were also done to investigate the effect of NPs' aggregation morphology on the thermal conductivity of NF [29]. Moreover, the effect of the friction factor of NFs containing cylindrical nanoparticles in laminar flow was evaluated by simulation work, and the results concluded that the friction factor decreases with the increase in Reynolds numbers [30,31]. However, these papers do not include ZnO NPs. Additionally, the layering phenomenon at the liquid-solid interface in Cu- and CuO-based NFs [32], thermal conductivity based on the phonon theory of liquid [33], and clustering phenomenon in Al_2O_3 NFs were investigated. Results revealed that clustering reduces the thermal conductivity of NFs; however, the sedimentation phenomenon, which increases with cluster size, can cause results to alter [34]. The findings of heat transfer coefficient measurement inside Al_2O_3–water-filled square cuboid enclosures reveal that inclination angle is only a significant component in natural convection for enclosures with a large aspect ratio [35].

In another research, thermal conductivity measurement as a function of temperatures (10–70 °C) and ZnO NPs' concentration (0.5–3.75 vol%) reported that, at 30 °C, a maximum thermal conductivity improvement of 40% (3.75 vol% ZnO) is found [36]. Various shapes of ZnO NMs have been used to test the thermal conductivity of ZnO–EG NFs so far. A few reports are found on the effect of NPs' shape on the heat transfer of a shell and tube heat exchanger, and that does not include ZnO NRs [37]. According to our survey, no report is available for the study of ZnO–EG NFs with rod-shaped ZnO NPs in the laminar flow condition at low volume concentrations.

The shape of NMs has a significant impact on heat transfer performance in many applications because the specific surface area of NMs changes with the change in sizes and shapes. Moreover, NMs of the same diameter with various shapes may have different surface areas. As a result, synthesizing NMs with the desired shape is significant for improving heat transfer performance. Therefore, this study aimed to synthesize ZnO NRs and forced the convective HTC measurement of low concentration ZnO–EG NFs at laminar flow. Hence, the first task of this research is the preparation of ZnO–EG NFs.

The NF preparation procedure is significant, because it involves something beyond simply mixing the NPs with the BF. NFs can be prepared either by a two-step or a one-step method [12,38]. In the one-step method, the synthesis of NMs and the preparation of NFs are performed in a single step. In the two-step method, however, NMs are synthesized, and then the NFs are prepared by dispersing the required amount of NMs in BFs. Due to its simplicity, the two-step approach is frequently used to prepare the desired concentration. Surfactants or capping agents are often used to prepare stable NFs, and PVP surfactant has the highest degree of stability over other surfactants [39]. ZnO-based dispersion stability and thermophysical properties were assessed to evaluate the heat transfer earlier [40]. However, some limitations of water-based NFs for HTC measurement have been reported [35]. Once the NFs are prepared, stability can be assessed by observing the sedimentation and UV–vis analysis [41]. Before ZnO NRs-based NFs' preparations and stability assessment, ZnO NRs need to be synthesized.

Therefore, synthesizing ZnO NRs by a simple and cost-effective method is essential. Physical [42], chemical [43], biological [44], and green [45] approaches are used to produce ZnO NMs. Chemical synthesis is popular because it converts a large percentage of precursors to products in a short period. Chemical precipitation, chemical vapor deposition, sol-gel, spray pyrolysis, sputtering, microwave-assisted, hydrothermal, and other synthesis methods include examples [46]. Hydrothermal synthesis has several advantages over other synthetic methods, including inexpensive water as the solvent, one-pot synthesis, low aggregation, high purity, and great control of particle size and morphology. ZnO may be synthesized at the nanoscale in a variety of shapes, including spheres [47], needles [47], flowers [48], flakes [49], tablets [50], pencils [50], and rods [51].

As the convective HTC of rod-shaped, ZnO-based NFs is not reported earlier, in this study, firstly, ZnO NRs were synthesized using a hydrothermal route for HTC measurement. Next, using ultrasonication, 0.1, 0.2, and 0.3 vol% rod-shaped ZnO–EG NFs are formulated by dispersing the required amount of ZnO NRs in EG, and their dispersion stability was assessed by time-lapse sedimentation observation, zeta potential measurement, and UV–vis analysis. Finally, employing a vertical shell and tube heat exchanger, the forced convective HTC of ZnO–EG NFs at 0.1, 0.2, and 0.3 vol% was investigated under a constant heat flux at different temperatures and laminar flow conditions of Re between 400 and 600.

2. Experimental Section

2.1. Materials

Zinc acetate dihydrate [$Zn(CH_3COO)_2 \cdot 2H_2O$] was supplied by Scharlau, Barcelona, Spain. Sodium hydroxide (NaOH), polyvinylpyrrolidone (PVP) as a surfactant, and ethylene glycol as a base fluid were supplied by Merck, India. All the materials were analytical grade and used without further purification. Distilled water supplied by Active Fine Chemicals. Dhaka, Bangladesh was used in all experiments.

2.2. Synthesis of ZnO NRs

ZnO NRs were synthesized using a slightly modified hydrothermal procedure described previously [51]. A 50 mL aqueous solution of zinc acetate dihydrate [$Zn(CH_3COO)_2 \cdot 2H_2O$] was prepared in a beaker. A 50 mL aqueous NaOH solution was prepared in another beaker by dissolving the requisite amounts of NaOH pellets into the distilled water. The 50 mL aqueous solution of NaOH was then slowly added dropwise into the precursor solution while stirring continuously. $Zn(CH_3COO)_2 \cdot 2H_2O$ and NaOH had a molar ratio of 1:2. Next, 0.25 g of PVP was added to the solution mixture. After 2 h of stirring at room temperature, the mixture became milky. The solution mixtures were then transferred to a stainless-steel autoclave with a volume of 120 mL and heated in an oven at 170 °C for 6 h. The autoclave was then allowed to cool. The product was collected and rinsed three times with distilled water through centrifugation (10,100 rpm at an ambient temperature) and decantation. Finally, the product is dried in the oven at 120 °C for 3 h.

2.3. Characterization of ZnO NRs

An X-ray diffractometer (Phillips X'Pert PRO PW 3040, The Netherlands) using CuK radiation in a 2θ-configuration was used to analyze the crystal structure of the produced products. At a scanning rate of 0.02°/0.6 s, the scanned value of 2θ angles was between 30° and 70°. Data from the Joint Committee for Powder Diffraction Studies (JCPDS) file for ZnO were compared to the measured data (Card No. 01-070-8072). The particle size and shape of the produced ZnO NRs were examined using a transmission electron microscope (TEM, TALOS F200X, 200 KeV, The Netherlands) with an acceleration voltage of 200 kV.

2.4. ZnO–EG NFs Preparation

A two-step approach was used to make ZnO–EG NFs [1]. ZnO NRs were synthesized and then dispersed in EG. The loaded amount of ZnO NRs was 0.1, 0.2, and 0.3 vol% in EG. Afterward, the ZnO–EG NFs were sonicated at ambient conditions for 1 h at a frequency of 40 kHz using an ultrasonic cleaner (Model: VGT-1860QTD, China).

2.5. Stability Assessment of ZnO–EG NFs

The stability of the prepared 0.1, 0.2, and 0.3 vol% of ZnO–EG NFs was assessed using sedimentation observation with time interval image capturing using a digital camera, zeta potential measurement (Zeta sizer, Nano ZS, Malvern, UK), and UV–vis spectroscopy with quartz cuvettes and a path length of 10 mm using a spectrophotometer (UV-1800, Shimadzu, Japan). The measurement was performed in the wavelength range from 200 to 800 nm. Absorbance was taken at 1, 2, 3, and 4 h for 0.1, 0.2, and 0.3 vol% ZnO–EG NFs and plotted as a function of wavelength. UV–vis spectrums were compared with the stability change of the NFs with time.

2.6. Viscosity Measurements of ZnO–EG NFs

Viscosity (μ) of the BF, i.e., 0% and 0.1, 0.2, and 0.3 vol% ZnO–EG NFs, was measured using a DV-II + Pro EXTRA, BROOKFIELD VISCOMETER, the USA, at ambient temperature and 40, 50, and 60 °C, and 100 rpm following the user's manual. A total of 250 mL NFs was needed for viscosity measurement using a spindle. The uncertainty of the temperature control is ± 5 °C.

2.7. Density Measurements of ZnO–EG NFs

The density (ρ) of the BF and 0.1, 0.2, and 0.3 vol% ZnO–EG NFs were measured using a Pycnometer. The 50 mL samples were used for measurements. The weight of the blank Pycnometer was taken at ambient temperature, 40, 50, and 60 °C. Similarly, the weights of the pycnometer with BF, 0.1, 0.2, and 0.3 vol% ZnO–EG NFs were taken at ambient temperature, 40, 50 and 60 °C. Next, the density was calculated by subtracting the weight and dividing it by mass. The uncertainty of the balancing is 0.001 g.

2.8. HTC Measurement of ZnO–EG NFs

An experimental setup was established to determine the convective HTC of the NFs. Figure 1 schematically depicts the experimental setup. Previously, this kind of setup was used for HTC measurements using CuO–PVA NFs [52]. The experimental setup consisted of a 350 mL reservoir shell in which hot water was kept at a constant temperature of 100 °C. Moreover, hot water was kept at a constant temperature of 40, 50, and 60 °C for some measurements. The water is heated by 2 kW immiscible heaters (Watlow ref. L14JX8B) installed in the shell. The heat flux was maintained at a constant rate. Incorporated inside the shell was a spiral tube constructed of smooth copper tubing with a 0.44×10^{-3} m outer diameter, 0.15×10^{-3} m thickness, 0.29×10^{-3} m inner diameter, and a heat exchange length of 0.6858 m. The NFs and BF were picked up into a reservoir using a centrifugal pump (Espa Ref. XVM8 03F15T). The Reynolds number of the flow was between 400 and 600. Then, the NFs and BFs were passed through the test area. Laminar flow was confirmed by trialing the flow of ZnO–EG NFs. Some pigments were added in the ZnO–EG, and the flow was observed. No vertical flow of pigments was observed or no layer change of pigments was observed. In this way, it was characterized that the flow was laminar. The fluids absorbed the heat from the hot water and then exited the tube. Hot fluids were collected in a beaker as they traveled through the tube. The test area was appropriately insulated with glass wool to reduce heat loss from the system. The temperature of the shell and outflow side tube is measured using K-type thermocouples.

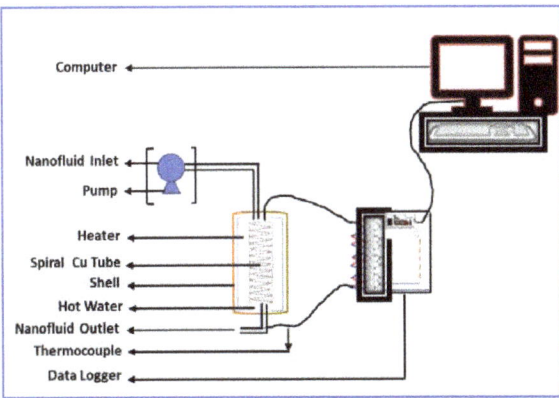

Figure 1. Experimental setup for the measurement of forced convective HTC of ZnO–EG NFs.

The heat flux determines the HTC via a single wall surface and heat conductivity across the boundary layer over the wall surface is required to complete heat transfer. With a change in wall temperature, the HTC is proportionality constant between the heat supplied and the thermodynamic driving force of heat flow through the per unit area. The spatial molecular diffusion of heat throughout the fluid is related to the thermal conductivity of the fluid [52]. The NFs were used within the stable period of 2 h from their preparation. The experiments were repeated at least three times and the average value was taken as the HTC coefficient. Standard deviation was included by adding an error bar in the column graph.

3. Results and Discussion

3.1. Characterization of Synthesized ZnO NRs

Analysis of the XRD peak pattern of the product can reveal the crystalline nature and phase structure of the product. Figure 2 depicts the XRD pattern of powder-type ZnO. The corresponding diffractogram in the 2-theta range of 30–70° indicated ZnO's hexagonal wurtzite phase (b), corresponding to the standard card [Card No. 01-070-8072] (a). No other peaks due to contamination were observed. Thus, the single-phase structure of ZnO has successfully synthesized with a {100} plane and the results were similar to other previous reports [15,16].

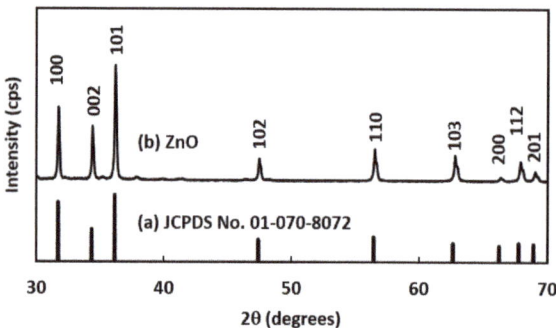

Figure 2. XRD peak pattern of the products: (**a**) JCPDS and (**b**) Synthesized ZnO NRs.

The morphology of ZnO was examined using TEM analysis. Figure 3 shows the TEM image of the synthesized ZnO, which shows the rod-shaped particles. As shown in Figure 3a, different sizes of ZnO rods were created with different diameters and lengths. The smallest diameter of a single ZnO NR is estimated at 50 nm, indicating the production of nanoscale ZnO NRs. Figure 3b shows the absence of rod branching, indicating that the ZnO NRs were developed from spontaneous nucleation with excellent crystal perfection [53]. Of course, their wide size distribution is seen in the images. The ZnO NR's aspect ratio was calculated by dividing the length by the diameter of the ZnO NR. In many applications, the aspect ratio of rod-shaped particles is significant, as the NRs with a higher aspect ratio provide larger specific surface areas than the NRs with a smaller aspect ratio. The NRs' average length and diameter are 270 and 120 nm, respectively, and the mean aspect ratio of ZnO NRs was found to be 2.25.

3.2. Stability Assessment of ZnO–EG NFs

First, the sedimentation observation method was performed for the stability test of NFs [54]. All the prepared NFs appeared to be white or whitish initially, and a slight color change occurred as sedimentation increased with increasing time intervals. Figure 4 shows the time-lapse observation during the sedimentation procedure of NFs. As seen in Figure 4a,b, no sedimentation is observed for all three samples. As the time increased, however, ZnO NRs tended to settle down to the bottom of the container due to sedimentation. Figure 4c,d shows that some sedimentation was deposited at the bottom of the bottles at 3 and 4 h, especially for the 0.2 vol% and 0.3 vol% NF samples. In the case of 4 h (Figure 4d), the upper dispersion phase becomes slightly transparent, showing more sedimentation at the bottom of the bottles. Therefore, it is considered that the prepared NFs are stable for at least 2 h. Considering the instability and sedimentation, HTC experiments were conducted within the stability period of NFs.

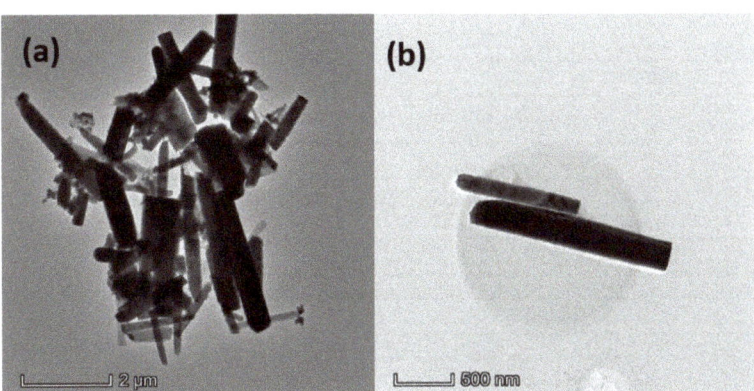

Figure 3. TEM images of the synthesized ZnO NRs using the hydrothermal method: (**a**) images at 2 µm scale with the magnification of 7000 times and (**b**) images at 500 nm scale with the magnification of 17,500 times.

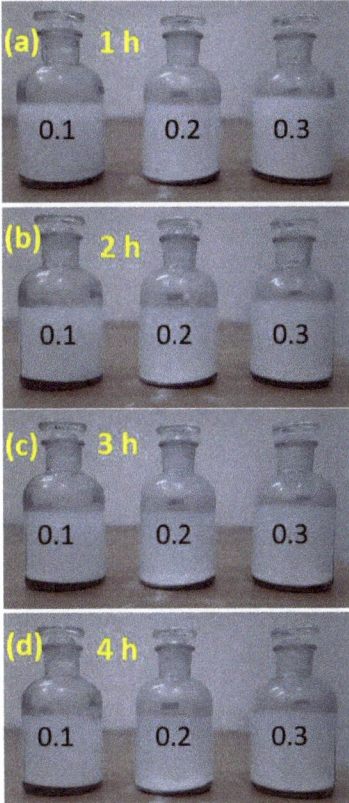

Figure 4. Time-dependence sedimentation photo images of ZnO–EG NFs (0.1, 0.2, 0.3 vol%) after the NFs preparation: (**a**) 1 h; (**b**) 2 h; (**c**) 3 h; (**d**) 4 h.

Agglomeration and settlement of NPs in BFs greatly contribute to channel clogging and affect thermal conductivity. Thus, stability is one of the most important factors affecting NF quality. Zeta potential is an important tool to assess the stability of NFs. In NF stability, the zeta potential of (+60 mV) to (−60 mV) is generally used as a border value. NFs having zeta potential values of more than +60 mV or less than −60 mV are often stable [55]. However, zeta potentials of between +10 mV and −10 mV or close to zero indicate the instability of NFs [56]. Table 1 shows the zeta potential of prepared NFs. The zeta potentials of 0.1, 0.2, and 0.3 vol% ZnO–EG NFs at 2 h of their preparation were −26.3 mV, 13.5 mV, and 4.1 mV, respectively. According to zeta potential measurements, the stability of the prepared NFs decreases with the increase in concentration. Moreover, 0.1 and 0.2 vol% ZnO–EG NFs are within the stability range, indicating a homogenous dispersion and good stability. However, 0.3 vol% ZnO–EG NF is out of the stability range.

Table 1. Zeta potential of 0.1, 0.2, and 0.3 vol% ZnO–EG NFs at 2 h of their preparation.

Concentration (vol%)	Zeta Potential, ζ (mV)
0.1	−26.3
0.2	13.5
0.3	4.1

Next, UV–vis analysis was performed on the NFs samples to assess the stability further. This is because the state and stability of the NFs can be evaluated by the difference in UV–vis absorption wavelength [57]. Here, the UV–vis spectra of ZnO–EG NFs were measured at 1, 2, 3, and 4 h of their preparation. Figure 5 shows the UV–vis spectrum of ZnO–EG NFs at 0.1 vol% (a), 0.2 vol% (b), and 0.3 vol% (c) at different times between 1 and 4 h. Figure 5a shows that the absorbance of NFs at a wavelength of 800 nm decreases harmonically with the increase in time. The spectrum pattern for 1, 2, and 3 h is similar, except for that of 4 h. The absorbance at 4 h significantly decreased from the initial state (1 h), which indicates that the 0.1 vol% ZnO–EG NFs is stable up to 3 h after their preparation. The drastic change in the absorbance pattern indicates that agglomeration and sedimentation occurred rapidly after 3 h of its preparation.

Figure 5b shows the UV–vis spectra of 0.2 vol% ZnO–EG NFs. The absorbance of 0.2 vol% ZnO–EG NFs at the wavelength of 800 nm decreases with the increase in time. Generally, absorbance values of 0.2 vol% ZnO–EG NFs at 800 nm are lower than that of 0.1 vol% ZnO–EG NFs, indicating a relatively low concentration of the dispersed phase. Although this phenomenon could not be accurately confirmed with a photograph, it can be considered reasonable because it coincides with the change in the zeta-potential value. It can be seen that absorbance after 2 h becomes lower fast; this is probably because unstable particles were removed by precipitation due to the time-lapse of 0.2% ZnO–EG NFs. Thus, Figure 5b indicates that the 0.2 vol% ZnO–EG NFs are stable until 2 h of its preparation.

Figure 5c shows the UV–vis spectrum of 0.3 vol% ZnO–EG NFs. Similar to Figure 5b, the initial absorbance value decreased with an increase in time. Here, the absorbance at 800 nm decreased quickly when the 2 h time had passed. Finally, it can be said that 0.1 vol% ZnO–EG NF is stable for up to 3 h, and 0.2 and 0.3 vol% ZnO–EG NFs are stable for at least 2 h.

3.3. Viscosity of ZnO–EG NFs

Average viscosity of BF (EG) and 0.1, 0.2, and 0.3 vol% ZnO–EG NFs at ambient temperature and 40, 50, and 60 °C at 100 rpm were found to be 7.40, 7.60, 7.85, and 8.46 cP, respectively, as shown in Table 2. Relative viscosity was also calculated, which indicates no significant differences at different concentrations of NFs. It was found that adding NPs to NFs increased their viscosity by a tiny amount, making them suitable for heat transfer applications with minimal pressure drop in the flow channels. Viscosity values are important to understand the stability of NFs. Moreover, viscosity data are needed to calculate the Reynolds number of the flow of NFs.

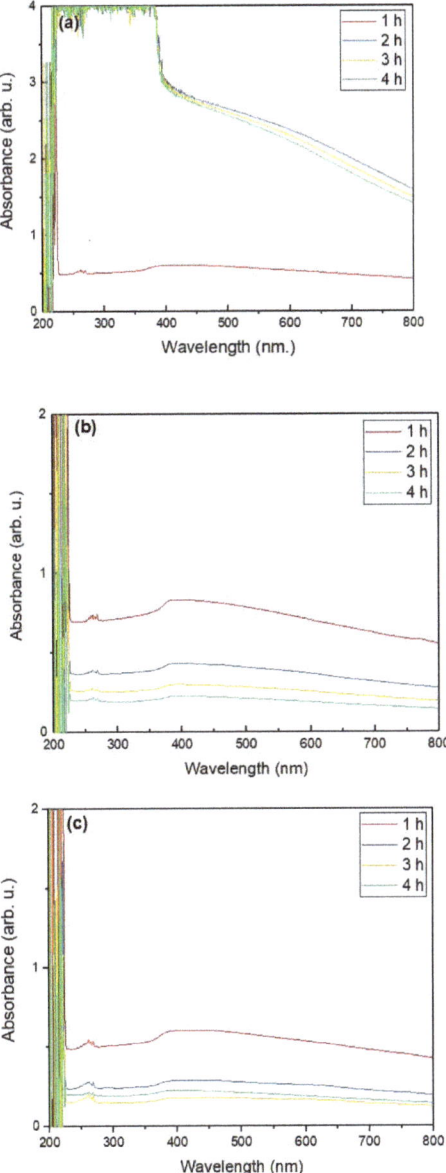

Figure 5. UV–vis spectra of ZnO–EG NFs according to the different time span between 1 and 4 h at different volume percent: (**a**) 0.1 vol%, (**b**) 0.2 vol%, and (**c**) 0.3 vol% of ZnO.

Table 2. The viscosity of BF (EG) and 0.1, 0.2, and 0.3 vol% ZnO–EG NFs at ambient temperature and 100 rpm.

Concentration (vol%)	Viscosity, μ (cP)	Relative Viscosity, μ_r
0 (EG)	7.40	-
0.1	7.60	1.02
0.2	7.85	1.06
0.3	8.46	1.14

3.4. Density of ZnO–EG NFs

The density of ZnO–EG NFs was calculated using the procedure described in the experimental section. Density is another significant parameter of NFs that is needed to know the Reynolds number of flowing NFs. Table 3 shows the calculation of the density of NFs. Data shows that the minimal density occurs due to adding a small amount of ZnO NRs to EG.

Table 3. Density of BF (EG) and 0.1, 0.2, and 0.3 vol% ZnO–EG NFs at ambient temperature, 40, 50, 60 °C.

Sample Conc. (vol%)	Ave. Weight of BF and NFs at Different Temperatures, W_2				Ave. wt. of Blank Pycnometer (50 mL), W_1	Density, ρ (g/m^3) = $(W_2 - W_1)/50$
	Ambient Temperature	40 °C	50 °C	60 °C		
0 (EG)	73.30	72.99	72.77	72.76	26.67	0.932
0.1	73.57	73.31	73.03	72.51	26.67	0.933
0.2	73.25	73.24	73.25	73.28	26.67	0.938
0.3	73.85	73.52	73.27	72.83	26.67	0.944

3.5. HTC of ZnO–EG NFs

Convective heat transfer between a moving fluid and a solid surface can be defined by the following relationship [58]:

$$Q = hA\left(T_s - T_f\right) = hA\Delta T \qquad (1)$$

where Q is the rate of forced convection heat transfer (W), T_s is the solid surface temperature (K), T_f is the fluid temperature (K), A is the area of the surface that is in contact with the fluid (m^2), h is the convective HTC (W/m^2·K).

$$Q = mS\Delta\theta \qquad (2)$$

where m is the mass of hot water, kg, S is the specific heat of water, J/kg·K, $\Delta\theta$ is the difference in water temperature before and after releasing temperature, K.

These relationships can be summarized as follows:

$$h = (Q/A)/(\Delta T) = mS\Delta T/(A \cdot \Delta T) \qquad (3)$$

Let the HTC of the 0 (BF), 0.1, 0.2, 0.3 vol% ZnO–EG NFs be h_0, h_1, h_2, h_3, respectively. The temperature difference before and after release temperature was set as $\Delta\theta_0$, $\Delta\theta_1$, $\Delta\theta_2$, and $\Delta\theta_3$ for the 0 (BF), 0.1, 0.2, and 0.3 vol% ZnO–EG NFs, respectively. After absorbing heat, the temperature difference between H$_2$O and each NFs is ΔT_0, ΔT_1, ΔT_2, and ΔT_3, corresponding to 0 (BF), 0.1, 0.2, and 0.3 vol% ZnO–EG NFs, respectively. All the parameters for the calculation are summarized in Table 4.

Forced convective HTC of BF and ZnO–EG NFs at 0 (BF), 0.1, 0.2, and 0.3 vol% were measured three times under laminar flow conditions. Obtained data is summarized and presented in Table 5.

Average values of forced convective HTC of 0 (BF), 0.1, 0.2, and 0.3 vol% of ZnO–EG NFs were calculated as 219, 1284, 2156, and 2536 Wm^2/K, respectively. Results were plotted in Figure 6 with a standard error bar from standard deviations. As seen from the column graph, the HCT values were increased as the concentration of NRs was increased. The HTC of 0.1, 0.2, and 0.3 vol% NFs were 6, 10, and 12 times higher than that of BF, respectively, indicating the non-linearity of HTC increment. The results indicated that small amounts of ZnO NRs could greatly enhance their HTC values. Notably, 0.1 vol% showed that the best HTC and 0.3 vol% have the lowest HTC compared to BF.

Table 4. Summary of the parameter for HTC of ZnO–EG NFs.

Parameter	Value
Mass of water	0.35 kg
Specific heat of water, S	4178 J/kg·K
The outer diameter of the Cu-tube, d	4.4×10^{-4} m
Heat exchange length of Cu-tube, l	68.58×10^{-2} m
$\Delta\theta_0$	0.01 K
$\Delta\theta_1$	1.3 K
$\Delta\theta_2$	1.3 K
$\Delta\theta_3$	1.2 K
ΔT_0	13.49 K
ΔT_1	4.5 K
ΔT_2	5.9 K
ΔT_3	4.1 K

Table 5. Forced convective HTC of BF and ZnO–EG NFs at 0.1, 0.2, and 0.3 vol% at $Re = 400$.

Concentration (vol%)	Forced Convective HTC (Wm²/K)		
	1st Measurement	2nd Measurement	3rd Measurement
0 (BF)	144	292	220
0.1	1341	1212	1299
0.2	2195	2449	1823
0.3	2656	2452	2500

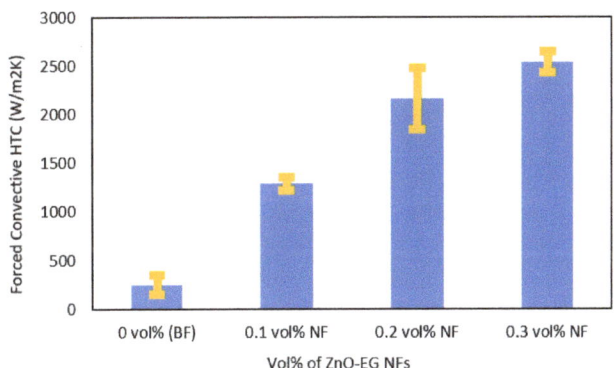

Figure 6. Forced convective HTC of BF and ZnO–EG NFs at different vol% of NRs at room temperature.

The sharp increase in HTC due to the input of the nanorods is apparent, and it can be seen that it is related to the concentration of the NRs in BFs. However, HTC did not increase linearly to the increments of the NRs' concentration. This phenomenon strongly depends on the stability of NFs. The previous section shows that relatively high concentrated NFs (e.g., 0.3%) showed unstable conditions. Thus, local aggregation or agglomeration of ZnO NRs may occur during the HTC measurements. The tendency of the HTC results is consistent with the results of UV–vis shown in Figure 5.

Again, forced convective HTC was measured by varying the temperature of hot water. Figure 7 shows the HTC of 0.1, 0.2, and 0.3 vol% ZnO–EG NFs at different temperatures with standard deviations. If the source of heat (hot water) is kept at higher temperatures, the heat transfer is higher as seen in Figure 7. Average HTC of 2563, 3390, and 3686 were calculated for the hot water temperature of 40 °C, 50 °C, and 60 °C. Therefore, it seems significant to maintain the temperature of heat source fluid at an optimum higher degree.

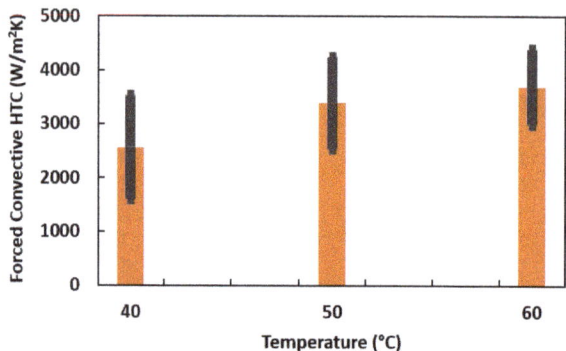

Figure 7. Forced convective HTC of 0.1, 0.2, and 0.3 vol% ZnO–EG NFs at different temperatures.

Another important factor for heat transfer fluid flow is Reynolds numbers (Re), which need to be considered during NF flow. Figure 8 presents the forced convective HTCs of 0.1, 0.2, and 0.3 vol% ZnO–EG NFs at different Re. Experiments were performed Re at 400, 500, and 600. Data indicate that the HTC of the same concentration of NFs increases with the increase in Reynolds numbers.

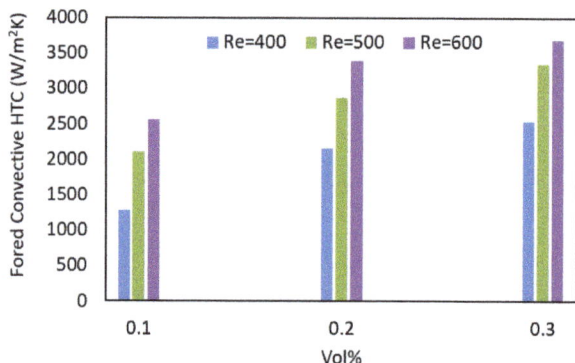

Figure 8. Forced convective HTCs of 0.1, 0.2, and 0.3 vol% ZnO–EG NFs at different Reynolds numbers.

4. Conclusions

The forced convective HTCs of BF (EG) and 0.1, 0.2, and 0.3 vol% ZnO–EG NFs were investigated in this study. For this purpose, ZnO NRS was successfully synthesized by the hydrothermal method. The prepared ZnO was hexagonal wurtzite and had a nano-sized rod shape with an average aspect ratio (length/diameter) of 2.25. ZnO–EG NFs at 0.1, 0.2, and 0.3 vol% were prepared by dispersing ZnO NRs to EG with the aid of 1 h ultrasonication. Time-lapse sedimentation observation, zeta potential measurement, and UV–vis analysis indicated that all the prepared NFs were stable at least for 2 h. Among the NFs, 0.1 vol% ZnO–EG NFs showed the best stability within 3 h. The HTC of ZnO–EG NFs was significantly increased with low loading concentration of ZnO NRs, and those HTCs of 0.1, 0.2, and 0.3 vol% NFs increased for 6, 10, and 12 times compared to BF (EG). At a high ZnO concentration, the HTC was not increased linearly with the increments of NRs' concentration, probably due to the local agglomeration during the measurements. Again, a higher Reynolds number leads to the higher HTC of an NF compared to that of a lower Reynolds number. Therefore, we expect that the well-dispersed ZnO concentration in NFs will significantly increase the HTC in NFs; this will facilitate ZnO–EG NFs in many industrial and other thermal management applications. By selecting a stabilizer under optimal ultrasonication time for enhanced stability, the current study scheme can be explored further to manufacture similar types of aqueous/non-aqueous metal oxide NFs. Moreover, at higher Reynolds, the HTC of current NF systems may need to be investigated further.

Author Contributions: Conceptualization: M.Z.H., Investigation, M.S.A., B.N. and M.A.G.; writing—original draft preparation, M.S.A., review, and editing, M.Z.H. and G.S., visualization, M.Z.H., supervision, M.Z.H. All authors have read and agreed to the published version of the manuscript.

Funding: "This research was funded by the Ministry of Science and Technology, Bangladesh, Grant number Physical Science-14/2018" and "The APC was funded by a JSPS KAKENHI Grant Number JP21K04763".

Institutional Review Board Statement: Not applicable.

Informed Consent Statement: Not applicable.

Data Availability Statement: Not applicable.

Acknowledgments: We acknowledge the Atomic Energy Center, Dhaka, and the Bangladesh Council for Scientific and Industrial Research, Dhaka for instrumental support.

Conflicts of Interest: The authors declare no conflict of interest.

Nomenclature

List of symbols

WmK^{-1}	Thermal conductivity
m	Length of the heat exchanger, l
mV	Zeta potential
J/kg·K	Specific heat of water, S
m	Diameter of Cu-tube
m	Length
K	$\Delta\theta$
K	ΔT
cP	Viscosity, μ
Re	Reynolds number
ζ	zeta potential
μ	Viscosity
T	Temperature
Q	the rate of forced convection heat transfer (W)
T_s	the solid surface temperature (K)
T_f	the fluid temperature (K)
A	the area of the surface that is in contact with the fluid (m^2)
h	the convective HTC (W/m^2·K)
m	the mass of hot water, kg,
S	the specific heat of water, J/kg·K
$\Delta\theta$	the difference in water temperature before and after releasing temperature, K.

Abbreviations

ZnO	Zinc oxide
NR	Nanorod
BF	Base fluid
NF	Nanofluid
XRD	X-ray diffraction
TEM	Transmission electron microscopy
UV–vis	Ultraviolet-visible
HTC	Heat transfer coefficient
EG	Ethylene glycol
NM	Nanomaterial
NaOH	Sodium hydroxide
PVP	Polyvinylpyrrolidone
JCPDS	Joint committee for powder diffraction studies

References

1. Ali, A.R.I.; Salam, B. A Review on Nanofluid: Preparation, Stability, Thermophysical Properties, Heat Transfer Characteristics and Application. *SN Appl. Sci.* **2020**, *2*, 1636. [CrossRef]
2. Okonkwo, E.C.; Wole-Osho, I.; Almanassra, I.W.; Abdullatif, Y.M.; Al-Ansari, T. An Updated Review of Nanofluids in Various Heat Transfer Devices. *J. Therm. Anal. Calorim.* **2021**, *145*, 2817–2872. [CrossRef]
3. Choi, S.U.S. Enhancing Thermal Conductivity of Fluids with Nanoparticles. *Am. Soc. Mech. Eng. Fluids Eng. Div. FED* **1995**, *231*, 99–105.
4. Das, S.K.; Choi, S.U.S.; Patel, H.E. Heat Transfer in Nanofluids—A Review. *Heat Transf. Eng.* **2006**, *27*, 3–19. [CrossRef]
5. Choi, S.U.S.; Eastman, J.A. Enhancing Thermal Conductivity of Fluids with Nanoparticles. In *Developments Applications of Non-Newtonian Flows*; Signer, D.A., Wang, H.P., Eds.; Asme: New York, NY, USA, 1995; pp. 99–105.
6. Hwang, Y.; Lee, J.K.; Lee, J.K.; Jeong, Y.M.; Cheong, S.I.; Ahn, Y.C.; Kim, S.H. Production and Dispersion Stability of Nanoparticles in Nanofluids. *Powder Technol.* **2008**, *186*, 145–153. [CrossRef]
7. Babita; Sharma, S.K.; Gupta, S.M. Preparation and Evaluation of Stable Nanofluids for Heat Transfer Application: A Review. *Exp. Therm. Fluid Sci.* **2016**, *79*, 202–212. [CrossRef]
8. Ponmani, S.; Gupta, P.; Jadhawar, P.; Nagarajan, R.; Sangwai, J. Investigations on the Thermal and Electrical Conductivity of Polyethylene Glycol-Based CuO and ZnO Nanofluids. *Indian Chem. Eng.* **2020**, *62*, 402–412. [CrossRef]
9. Radkar, R.N.; Bhanvase, B.A.; Barai, D.P.; Sonawane, S.H. Intensified Convective Heat Transfer Using ZnO Nanofluids in Heat Exchanger with Helical Coiled Geometry at Constant Wall Temperature. *Mater. Sci. Energy Technol.* **2019**, *2*, 161–170. [CrossRef]
10. Ahmadi, M.H.; Mirlohi, A.; Alhuyi Nazari, M.; Ghasempour, R. A Review of Thermal Conductivity of Various Nanofluids. *J. Mol. Liq.* **2018**, *265*, 181–188. [CrossRef]
11. Rubbi, F.; Das, L.; Habib, K.; Aslfattahi, N.; Saidur, R.; Rahman, M.T. State-of-the-Art Review on Water-Based Nanofluids for Low Temperature Solar Thermal Collector Application. *Sol. Energy Mater. Sol. Cells* **2021**, *230*, 111220. [CrossRef]
12. Asadi, A.; Aberoumand, S.; Moradikazerouni, A.; Pourfattah, F.; Żyła, G.; Estellé, P.; Mahian, O.; Wongwises, S.; Nguyen, H.M.; Arabkoohsar, A. Recent Advances in Preparation Methods and Thermophysical Properties of Oil-Based Nanofluids: A State-of-the-Art Review. *Powder Technol.* **2019**, *352*, 209–226. [CrossRef]
13. Ali, M.; Zeitoun, O.; Almotairi, S.; Al-Ansary, H. The Effect of Alumina-Water Nanofluid on Natural Convection Heat Transfer inside Vertical Circular Enclosures Heated from Above. *Heat Transf. Eng.* **2013**, *34*, 1289–1299. [CrossRef]
14. Ali, H.M.; Ali, H.; Liaquat, H.; Bin Maqsood, H.T.; Nadir, M.A. Experimental Investigation of Convective Heat Transfer Augmentation for Car Radiator Using ZnO-Water Nanofluids. *Energy* **2015**, *84*, 317–324. [CrossRef]
15. Xie, H.; Li, Y.; Yu, W. Intriguingly High Convective Heat Transfer Enhancement of Nanofluid Coolants in Laminar Flows. *Phys. Lett. Sect. A Gen. At. Solid State Phys.* **2010**, *374*, 2566–2568. [CrossRef]
16. Olabi, A.G.; Wilberforce, T.; Sayed, E.T.; Elsaid, K.; Rahman, S.M.A.; Abdelkareem, M.A. Geometrical Effect Coupled with Nanofluid on Heat Transfer Enhancement in Heat Exchangers. *Int. J. Thermofluids* **2021**, *10*, 100072. [CrossRef]
17. Ghozatloo, A.; Rashidi, A.; Shariaty-Niassar, M. Convective Heat Transfer Enhancement of Graphene Nanofluids in Shell and Tube Heat Exchanger. *Exp. Therm. Fluid Sci.* **2014**, *53*, 136–141. [CrossRef]
18. Haddad, Z.; Oztop, H.F.; Abu-Nada, E.; Mataoui, A. A Review on Natural Convective Heat Transfer of Nanofluids. *Renew. Sustain. Energy Rev.* **2012**, *16*, 5363–5378. [CrossRef]
19. Suganthi, K.S.; Leela Vinodhan, V.; Rajan, K.S. Heat Transfer Performance and Transport Properties of ZnO-Ethylene Glycol and ZnO-Ethylene Glycol-Water Nanofluid Coolants. *Appl. Energy* **2014**, *135*, 548–559. [CrossRef]
20. Angayarkanni, S.A.; Sunny, V.; Philip, J. Effect of Nanoparticle Size, Morphology and Concentration on Specific Heat Capacity and Thermal Conductivity of Nanofluids. *J. Nanofluids* **2015**, *4*, 302–309. [CrossRef]
21. Sadripour, S.; Chamkha, A.J. The Effect of Nanoparticle Morphology on Heat Transfer and Entropy Generation of Supported Nanofluids in a Heat Sink Solar Collector. *Therm. Sci. Eng. Prog.* **2019**, *9*, 266–280. [CrossRef]
22. Ferrouillat, S.; Bontemps, A.; Poncelet, O.; Soriano, O.; Gruss, J.A. Influence of Nanoparticle Shape Factor on Convective Heat Transfer and Energetic Performance of Water-Based SiO2 and ZnO Nanofluids. *Appl. Therm. Eng.* **2013**, *51*, 839–851. [CrossRef]
23. Kim, S.H.; Choi, S.R.; Kim, D. Thermal Conductivity of Metal-Oxide Nanofluids: Particle Size Dependence and Effect of Laser Irradiation. *J. Heat Transfer* **2007**, *129*, 298–307. [CrossRef]
24. Ramya, M.; Nideep, T.K.; Nampoori, V.P.N.; Kailasnath, M. Shape Dependent Heat Transfer and Nonlinear Optical Limiting Characteristics of Water Stable ZnO Nanofluid. *Surf. Interfaces* **2021**, *26*, 101345. [CrossRef]
25. Arora, N.; Gupta, M. Stability Evaluation and Enhancement Methods in Nanofluids: A Review. In *AIP Conference Proceedings*; AIP Publishing LLC: New York, NY, USA, 2021; Volume 2341, p. 040022. [CrossRef]
26. Kim, H.J.; Lee, S.H.; Lee, J.H.; Jang, S.P. Effect of Particle Shape on Suspension Stability and Thermal Conductivities of Water-Based Bohemite Alumina Nanofluids. *Energy* **2015**, *90*, 1290–1297. [CrossRef]
27. Ghosh, M.M.; Ghosh, S.; Pabi, S.K. Effects of Particle Shape and Fluid Temperature on Heat-Transfer Characteristics of Nanofluids. *J. Mater. Eng. Perform.* **2013**, *22*, 1525–1529. [CrossRef]
28. Jeong, J.; Li, C.; Kwon, Y.; Lee, J.; Kim, S.H.; Yun, R. Particle Shape Effect on the Viscosity and Thermal Conductivity of ZnO Nanofluids. *Int. J. Refrig.* **2013**, *36*, 2233–2241. [CrossRef]
29. Wang, R.; Qian, S.; Zhang, Z. Investigation of the Aggregation Morphology of Nanoparticle on the Thermal Conductivity of Nanofluid by Molecular Dynamics Simulations. *Int. J. Heat Mass Transf.* **2018**, *127*, 1138–1146. [CrossRef]

30. Lin, J.Z.; Xia, Y.; Ku, X.K. Flow and Heat Transfer Characteristics of Nanofluids Containing Rod-like Particles in a Turbulent Pipe Flow. *Int. J. Heat Mass Transf.* **2016**, *93*, 57–66. [CrossRef]
31. Lin, J.; Xia, Y.; Ku, X. Friction Factor and Heat Transfer of Nanofluids Containing Cylindrical Nanoparticles in Laminar Pipe Flow. *J. Appl. Phys.* **2014**, *116*, 133513. [CrossRef]
32. Milanese, M.; Iacobazzi, F.; Colangelo, G.; de Risi, A. An Investigation of Layering Phenomenon at the Liquid–Solid Interface in Cu and CuO Based Nanofluids. *Int. J. Heat Mass Transf.* **2016**, *103*, 564–571. [CrossRef]
33. Iacobazzi, F.; Milanese, M.; Colangelo, G.; Lomascolo, M.; de Risi, A. An Explanation of the Al_2O_3 Nanofluid Thermal Conductivity Based on the Phonon Theory of Liquid. *Energy* **2016**, *116*, 786–794. [CrossRef]
34. Iacobazzi, F.; Milanese, M.; Colangelo, G.; de Risi, A. A Critical Analysis of Clustering Phenomenon in Al_2O_3 Nanofluids. *J. Therm. Anal. Calorim.* **2019**, *135*, 371–377. [CrossRef]
35. Almuzaiqer, R.; Ali, M.E.; Al-Salem, K. Effect of the Aspect Ratio and Tilt Angle on the Free Convection Heat Transfer Coefficient Inside Al_2O_3—Water-Filled Square Cuboid Enclosures. *Nanomaterials* **2022**, *12*, 500. [CrossRef]
36. Kole, M.; Dey, T.K. Effect of Prolonged Ultrasonication on the Thermal Conductivity of ZnO-Ethylene Glycol Nanofluids. *Thermochim. Acta* **2012**, *535*, 58–65. [CrossRef]
37. Elias, M.M.; Miqdad, M.; Mahbubul, I.M.; Saidur, R.; Kamalisarvestani, M.; Sohel, M.R.; Hepbasli, A.; Rahim, N.A.; Amalina, M.A. Effect of Nanoparticle Shape on the Heat Transfer and Thermodynamic Performance of a Shell and Tube Heat Exchanger. *Int. Commun. Heat Mass Transf.* **2013**, *44*, 93–99. [CrossRef]
38. Deb Majumder, S.; Das, A. A Short Review of Organic Nanofluids: Preparation, Surfactants, and Applications. *Front. Mater.* **2021**, *8*, 630182. [CrossRef]
39. Ma, M.; Zhai, Y.; Yao, P.; Li, Y.; Wang, H. Effect of Surfactant on the Rheological Behavior and Thermophysical Properties of Hybrid Nanofluids. *Powder Technol.* **2021**, *379*, 373–383. [CrossRef]
40. Qamar, A.; Anwar, Z.; Ali, H.; Imran, S.; Shaukat, R.; Mujtaba Abbas, M. Experimental Investigation of Dispersion Stability and Thermophysical Properties of ZnO/DIW Nanofluids for Heat Transfer Applications. *Alexandria Eng. J.* **2022**, *61*, 4011–4026. [CrossRef]
41. Hossain, M.Z.; Hojo, D.; Yoko, A.; Seong, G.; Aoki, N.; Tomai, T.; Takami, S.; Adschiri, T. Dispersion and Rheology of Nanofluids with Various Concentrations of Organic Modified Nanoparticles: Modifier and Solvent Effects. *Colloids Surfaces A Physicochem. Eng. Asp.* **2019**, *583*, 123876. [CrossRef]
42. Qi, J.; Chang, J.; Han, X.; Zhong, R.; Jiang, M.; Chen, Z.; Liu, B. Direct Synthesis of ZnO Nanorods from Solution under Electric Field. *Mater. Chem. Phys.* **2018**, *211*, 168–171. [CrossRef]
43. Ghorbani, H.R.; Mehr, F.P.; Pazoki, H.; Rahmani, B.M. Synthesis of ZnO Nanoparticles by Precipitation Method. *Orient. J. Chem.* **2015**, *31*, 1219–1221. [CrossRef]
44. Singh, R.P.; Shukla, V.K.; Yadav, R.S.; Sharma, P.K.; Singh, P.K.; Pandey, A.C. Biological Approach of Zinc Oxide Nanoparticles Formation and Its Characterization. *Adv. Mater. Lett.* **2011**, *2*, 313–317. [CrossRef]
45. Agarwal, H.; Venkat Kumar, S.; Rajeshkumar, S. A Review on Green Synthesis of Zinc Oxide Nanoparticles—An Eco-Friendly Approach. *Resour. Technol.* **2017**, *3*, 406–413. [CrossRef]
46. Rl, M.; Kv, U.; Naik, D. Synthesis and Characterization of ZnO Nanoparticles: A Review. *J. Pharmacogn. Phytochem.* **2019**, *8*, 1095–1101.
47. Moulahi, A.; Sediri, F. ZnO Nanoswords and Nanopills: Hydrothermal Synthesis, Characterization and Optical Properties. *Ceram. Int.* **2014**, *40*, 943–950. [CrossRef]
48. Mohan, S.; Vellakkat, M.; Aravind, A.; Reka, U. Hydrothermal Synthesis and Characterization of Zinc Oxide Nanoparticles of Various Shapes under Different Reaction Conditions. *Nano Express* **2020**, *1*, 030028. [CrossRef]
49. Elen, K.; Van Den Rul, H.; Hardy, A.; Van Bael, M.K.; D'Haen, J.; Peeters, R.; Franco, D.; Mullens, J. Hydrothermal Synthesis of ZnO Nanorods: A Statistical Determination of the Significant Parameters in View of Reducingthe Diameter. *Nanotechnology* **2009**, *20*, 055608. [CrossRef]
50. Akhoon, S.A.; Rubab, S.; Shah, M.A. A Benign Hydrothermal Synthesis of Nanopencils-like Zinc Oxide Nanoflowers. *Int. Nano Lett.* **2015**, *5*, 9–13. [CrossRef]
51. Alam, S.; Hossain, M.Z. A Simple Hydrothermal Protocol for the Synthesis of Zinc Oxide Nanorods. *Jagannath Univ. J. Sci.* **2021**, *7*, 75–80.
52. Luna, I.Z.; Chowdhury, A.M.S.; Gafur, M.A.; Khan, R.A. Measurement of Forced Convective Heat Transfer Coefficient of Low Volume Fraction CuO-PVA Nanofluids under Laminar Flow Condition. *Am. J. Nanomater.* **2015**, *3*, 64–67. [CrossRef]
53. Liu, F.; Cao, P.J.; Zhang, H.R.; Shen, C.M.; Wang, Z.; Li, J.Q.; Gao, H.J. Well-Aligned Zinc Oxide Nanorods and Nanowires Prepared without Catalyst. *J. Cryst. Growth* **2005**, *274*, 126–131. [CrossRef]
54. Ilyas, S.U.; Pendyala, R.; Marneni, N. Preparation, Sedimentation, and Agglomeration of Nanofluids. *Chem. Eng. Technol.* **2014**, *37*, 2011–2021. [CrossRef]
55. Sati, P.; Shende, R.C.; Ramaprabhu, S. An Experimental Study on Thermal Conductivity Enhancement of DI Water-EG Based ZnO(CuO)/Graphene Wrapped Carbon Nanotubes Nanofluids. *Thermochim. Acta* **2018**, *666*, 75–81. [CrossRef]
56. Choudhary, R.; Khurana, D.; Kumar, A.; Subudhi, S. Stability Analysis of Al_2O_3/Water Nanofluids. *J. Exp. Nanosci.* **2017**, *12*, 140–151. [CrossRef]

57. Mostafizur, R.M.; Saidur, R.; Abdul Aziz, A.R.; Bhuiyan, M.H.U. Thermophysical Properties of Methanol Based Al_2O_3 Nanofluids. *Int. J. Heat Mass Transf.* **2015**, *85*, 414–419. [CrossRef]
58. Asirvatham, L.G.; Vishal, N.; Gangatharan, S.K.; Lal, D.M. Experimental Study on Forced Convective Heat Transfer with Low Volume Fraction of CuO/Water Nanofluid. *Energies* **2009**, *2*, 97–119. [CrossRef]

Preparation of Silicon Hydroxyapatite Nanopowders under Microwave-Assisted Hydrothermal Method

Zully Matamoros-Veloza [1,*], Juan Carlos Rendon-Angeles [2], Kazumichi Yanagisawa [3], Tadaharu Ueda [4,5], Kongjun Zhu [6] and Benjamin Moreno-Perez [1]

1. Graduate Division, Technological Institute of Saltillo, Tecnológico Nacional de México/(I.T. Saltillo), Saltillo 25280, Mexico; bnmoreno24@gmail.com
2. Centre for Research and Advanced Studies of the NPI, Cinvestav-Campus Saltillo, Saltillo 25900, Mexico; jcarlos.rendon@cinvestav.edu.mx
3. Research Laboratory of Hydrothermal Chemistry, Faculty of Science, Kochi University, Kochi 780-8520, Japan; yanagi@kochi-u.ac.jp
4. Department of Marine Resources Science, Faculty of Agricultural and Marine Science, Kochi University, Nankoku 783-8502, Japan; chuji@kochi-u.ac.jp
5. Center for Advanced Marine Core Research, Kochi University, Nankoku 783-8502, Japan
6. State Key Laboratory of Mechanics and Control Mmechanics Structures, Nanjing University of Aeronautics and Astronautics, Nanjing 210016, China; kjzhu@nuaa.edu.cn
* Correspondence: zully.mv@saltillo.tecnm.mx

Abstract: The synthesis of partially substituted silicon hydroxyapatite (Si-HAp) nanopowders was systematically investigated via the microwave-assisted hydrothermal process. The experiments were conducted at 150 °C for 1 h using TMAS ($C_4H_{13}NO_5Si_2$) as a Si^{4+} precursor. To improve the Si^{4+} uptake in the hexagonal structure, the Si precursor was supplied above the stoichiometric molar ratio (0.2 M). The concentration of the TMAS aqueous solutions used varied between 0.3 and 1.8 M, corresponding to saturation levels of 1.5–9.0-fold. Rietveld refinement analyses indicated that Si incorporation occurred in the HAp lattice by replacing phosphate groups (PO_4^{3-}) with the silicate (SiO_4^-) group. FT-IR and XPS analyses also confirmed the gradual uptake of SiO_4^- units in the HAp, as the saturation of Si^{4+} reached 1.8 M. TEM observations confirmed that Si-HAp agglomerates had a high crystallinity and are constituted by tiny rod-shaped particles with single-crystal habit. Furthermore, a reduction in the particle growth process took place by increasing the Si^{4+} excess content up to 1.8 M, and the excess of Si^{4+} triggered the fine rod-shaped particles self-assembly to form agglomerates. The agglomerate size that occurred with intermediate (0.99 mol%) and large (12.16 mol%) Si contents varied between 233.1 and 315.1 nm, respectively. The excess of Si in the hydrothermal medium might trigger the formation of the Si-HAp agglomerates prepared under fast kinetic reaction conditions assisted by the microwave heating. Consequently, the use of microwave heating-assisted hydrothermal conditions has delivered high processing efficiency to crystallize Si-HAp with a broad content of Si^{4+}.

Keywords: hydrothermal microwave assisted synthesis; silicon-hydroxyapatite; nano powders

1. Introduction

The preparation of biomaterials with similar chemical and physical properties to biological hydroxyapatite (HAp), in terms of their chemical and physical properties, involves the uptake of cations and anions in the hexagonal HAp structure. The incorporation of Si^{4+} ions into the PO_4^{3-} unit network of the HAp stimulates both bone formation and resorption processes, which are relevant to both tissue restauration and bone growth [1]. Furthermore, the incorporation of Si^{4+} has been incorporated simultaneously with calcium during the early stage of calcination [2]. Additionally, silicon is also essential for other biological soft tissue functionality, such as cartilage growth, and synthetic calcium-phosphate bioceramics containing low Si^{4+} contents in their structures, which had marked biological properties for bone restauration [3,4]. Hitherto, Si-HAp bioceramics were prepared with a conventional solid-state reaction at 1000 °C for 6 h, employing β-tricalcium phosphate (β-$Ca_3(PO_4)_2$), silicon dioxide (SiO_2), and calcium carbonate ($CaCO_3$). These partially silicon-substituted powders exhibit a good biological dissolution capability in comparison with pure HAp [5].

Moreover, chemical solution methods, such as coprecipitation, neutralization, and sol-gel, are alternative synthetic procedures for producing both HAp and Si-HAp nanoparticles. The increase of silicon incorporation in the HAp structure provokes a marked decrease in the crystallite size [6–11]. Likewise, silicon affects the particle morphology during the embryo precipitation and particle growth processes. Recently, extensive attention has been paid to the appropriated process with a Si^{4+} precursor reagent to overcome the difficulties in handling associated with its reactivity under wet chemical processing [4–15].

Hitherto, tetraethyl orthosilicate (Si $(OCH_2CH_3)_4$, TEOS) in polyethylene glycol/water and silicon tetra-acetate in water (Si $(CH_3CO_2)_4$) have been better reagents for incorporating Si^{4+} in the hexagonal structure [13]. The maximum efficiency of the Si^{4+} incorporation in the apatite structure was 90% according to the nominal stoichiometric content of 8.0 mol%. Wet chemical quantitative analyses revealed the presence of silicon ions hydrolyzed in the remaining mother liquor after precipitation of SiHAp. In contrast, the challenge of producing synthetic Si-HAp has been carried out by various methods, including soft chemistry processes [5–13].

The slow reaction kinetic of the ions species to produce pure HAP and other solid solutions required prolonged processing time in specific systems [14–18].

The hydrothermal process has brought further advantages in terms of chemical reactivity: Higher yield for crystalline products with nanometric size and reaction kinetics enhancement even at relatively low temperatures (100–250 °C) [18,19].

Likewise, a few pioneering research works have reported the synthesis of partially substituted Si-HAp under hydrothermal conditions [12,14]. The synthesis was conducted by two pathways. The first experiments were conducted using the chemical reagents $(NH_4)_3PO_4$ and TEOS as precursor of PO_4^{3-} and SiO_4^{4-} ions. However, the synthesis conducted at 200 °C for 8 h limited the incorporation to only 8.0 mol% Si^{4+}, regardless of the nominal stoichiometric amount intended (9.0 mol%) [13,14]. In other experiments, the uptake of Si^{4+} was further limited by using $(NH_4)_2HPO_4$ to 7.65 mol%. The partially substituted Si-HAp particles also incorporate CO_3^{2-} ions, and the presence of these ions is reported to hinder the incorporation of SiO_4^{4-} during the crystallization and particle coarsening steps [13,20,21].

Similar experiments were recently conducted to attempted the synthesis of Si-HAp under hydrothermal conditions at 150 °C for 10 h by employing tetramethyl ammonium silicate (($C_4H_{13}NO_5Si_2$), TMAS) [1–3,13–16,18–20]. The low silicon reactivity in the hydrothermal alkaline medium at a pH of 10 caused a limited Si^{4+} content in the HAp structure of 30 mol% regarding the stoichiometric amount selected (1–20 mol% Si). In this case, the incorporation of CO_3^{2-} ions was not the cause of the significant Si uptake. The high solubility of the TMAS in the alkaline solution is likely to produce Si complex ions that are highly stable in the hydrothermal medium, giving rise to the decrease in the Si concentration in the embryo and growth steps [22,23].

A similar trend was found in the preparation of Zn-substituted HAp, where the isomorphous incorporation of Zn at the Ca site in the HAp structure was affected by the formation of Zn $(OH)_x^{n+}$ species, which were also stable in alkaline hydrothermal fluids at the standard pH conditions required to crystallize the HAp [20].

Although the detailed effect of the complex ion formation associated with the dopant ions in HAp has not been evaluated yet, it is important from the chemical processing point of view to enhance the control of the stoichiometry of $Ca_{10}(PO_4)_{6-x}(SiO_4)_x(OH)_{2-x}$ solid solutions and the particle growth at nanometer order [21,22]. In the present research work, different approaches for the synthesis of the $Ca_{10}(PO_4)_{6-x}(SiO_4)_x(OH)_{2-x}$ particle were investigated, devoted to investigating the chemical reaction pathway in Si^{4+}-saturated solutions under hydrothermal conditions assisted by microwave heating.

The fast reaction kinetics triggered by the microwave heating in conjunction with the saturation level of Si^{4+} would achieve a broad compositional control to produce $Ca_{10}(PO_4)_{6-x}(SiO_4)_x(OH)_{2-x}$ compounds. The feasibility of controlling the particle size at nanometric order is likely to proceed due to Si complex ions in the hydrothermal medium, which would operate as templates in the particle crystallization process.

2. Materials and Methods

2.1. Materials

Preparation of the reagents for the synthesis of the stoichiometric pure hydroxyapatite (HAp) and silicon-substituted hydroxyapatite (Si-HAp) powders was carried out as follows; all the chemicals of reagent grade (Sigma Aldrich, St. Louis, MO, USA, 99.99% purity) were used without further purification. The 1 M Ca^{2+} and 0.2M P^{5+} stock solutions were prepared by dissolving calcium nitrate tetrahydrate ($Ca(NO_3)_2 4H_2O$) and sodium tripolyphosphate ($Na_5P_3O_{10}$) in distilled water, respectively. The Si^{4+} stock solutions of three different concentrations of 0.3, 0.9, and 1.8 M were prepared by dissolving was tetramethylammonium silicate solution [$(C_4H_{13}NO_5Si_2)$ (TMAS)]. Furthermore, all aqueous TMAS solutions were adjusted to pH = 10 with 7M NH_3 solution before making up the final volume of the TMAS stocks. The 7M of NH_3 stock solution was prepared by mixing 82.6 mL of conc. NH_3 solution with 17.35 mL of water. 2-Propanol was added as a buffer to prevent the formation of another phosphorous species during the reaction [23–25].

2.2. Microwave-Assisted Hydrothermal Synthesis

A mother solution constituted by 17.5 mL of the 1 M Ca^{2+} solution and 15 mL of 2-propanol was magnetically stirred for 5 min. The added 2-propanol is used as a pH buffer to prevent the hydrolysis of calcium tripolyphosphate gel to orthophosphate ions. In parallel, a solution mixture (17.5 mL) containing P^{5+} and Si^{4+} ions was prepared according to the molar mixing Ca/(P + Si) ratio of 1.67. Therefore, the mixture volumes calculated by the molar ratio Ca:P:Si were 17.5:0, 16.45:1.05, 15.75:1.75, and 14.0:3.5, where the molar volumes correspond to the pure HAp and the selected silicon compositions of 6, 10, and 20 mol%, respectively. To investigate the effect of the Si saturation in the mother liquor, the molar volumes calculated were provided with the silicon solutions of 0.3, 0.9, and 1.8 M to investigate the effect of the Si saturation in the mother liquor, respectively. On mixing both solutions instantaneously, a white milky colloid formed, and the colloidal suspension was stirred constantly for 15 min. Them, pH of the colloidal suspension was adjusted to a value of 10.00 ± 0.1 by adding a 7.0 M NH_3 aqueous solution dropwisely [1,14,20]. The suspension (50 mL) was then transferred to a double-walled, Teflon, high-pressure vessel, hermetically closed, and placed in the rotatory device of the microwave oven (MARS-5X, CEM Corp., Manasquan, NJ, USA), and was heated at 150 °C for 1 h. After the reaction, the powders were washed several times with deionized water until a neutral pH was achieved. The powder was dried using a freeze-drier (-47 °C, 3 MPa) to avoid aggregation of the particles. The chemical reaction occurs as Equation (1) to crystallize $Ca_{10}(PO_4^{3-})_{6-x}(SiO_4^{4-})_x(OH)_{2-x}$ nanoparticle crystallization under the proposed hydrothermal reaction assisted by microwave heating. In Equation (1), the ultimate content of $(Si-O-Si)_3O^-$ in the reaction products is equivalent to the subtraction of the Si^{4+} incorporated in the HAp and the nominal content supplied. Furthermore, OH^- deficiency in $Ca_{10}(PO_4^{3-})_{6-x}(SiO_4^{4-})_x(OH)_{2-x}$ results from the charge balance required to compensate the total negative valence of SiO_4^{4-} groups incorporated in the HAp structure.

$$10Ca(NO_3)_{2(aq)} + 2(Na_5P_3O_{10})_{(aq)} + y(C_4H_{13}NO_5Si_2)_{(aq)} + xNH_4OH_{(aq)} + x(CH_3)_2CHOH \rightarrow$$
$$Ca_{10}(PO_4^{3-})_{6-x}(SiO_4^{4-})_y(OH)_{2-x(s)} + 10Na^+ + xOH^-_{(aq)} + xNH_4^{\delta+}NO_3^{\delta-}{}_{(aq)} + xH_2O + yCH_4^{\delta+} + (y-x)[(Si-O-Si)_3O^-]_{(aq)} \quad (1)$$

2.3. Characterization

The crystalline phases of the obtained powders were determined by X-ray powder diffraction (XRD) analyses. Diffraction patterns were collected in a range 2θ from 10 to 80° at a scanning speed of 4°/min and a step size of 0.02° in a 2θ/θ scanning mode using an X-ray diffractometer Rigaku Ultima IV equipped with Cu Kα radiation (α = 1.54056 Å) operated at 40 kV and 20 mA. Furthermore, Rietveld refinement analyses of selected samples were carried out to determine the crystallite size and lattice parameters using the TOPAS 4.2 (Bruker AXS: Karlsruhe, Germany) software [19–21].

Fourier transform infrared (FT-IR) spectra were obtained at a wavelength range or 400–4000 cm^{-1} in the transmittance mode by FT-IR JASCO 4000 Hachioji (Tokyo, Japan) spectrometer, using palletized samples prepared with 5 mg of powder sample and 200 mg of KBr. In addition, Raman spectra analyses were observed in the range 200–4000 cm^{-1} by laze excitation at 514 nm using a Jobin Yvon Labram HR800 Raman Spectrometer (Horiba, Japan).

The content of Ca, Si, and P in the residual powders was quantitatively calculated from the inductively coupled spectrometry analyses data (ICP, ICPE-9000; Shimadzu Co., Kyoto, Japan). XPS spectra were recorded on a Kratos spectrometer (Manchester, UK) operated using an Al Ka (1486.6 eV) monochromatic X-ray source. The XPS analysis was carried out in ESCA Lab 220i-XL equipment (Shimadzu, Kyoto, Japan), at a vacuum of 2×10^{-8} mTorr, and a monochromatic X-ray source with aluminum anode operated at 1486.6 eV was used. The general spectra (survey) were obtained with a step energy of 117.4eV, and the analysis region was 0–1400 (eV) in link energy. Subsequently, high-resolution spectra of the C 1s, Ca 2p, P 2p, and O 1s signals were obtained for each sample. The high-resolution spectra were acquired with a step energy of 11.75 eV. Deconvolution of these spectra was performed by adjusting Gaussian curves, leaving their position and area without restriction. The FWHM value, however, remained fixed in each curve adjusted.

Morphology and particle size distribution were analyzed from the micrographs obtained by field emission scanning electron microscopy (FE-SEM JEOL 6500F JSM-7100F, Akishima, Tokyo, Japan) at 15 kV and 69 µA filament operating conditions. The image analysis was carried out using an Image-Pro® Plus software (Rockville, MD, USA). In addition, crystalline features were investigated by using the high-resolution observations conducted in the transmission electron microscopy (HR-TEM, FEI-TITAN 300, Phillips, Eugene, OR, USA) operated at 300 kV.

3. Results and Discussion

3.1. Effect of Si^{4+} Saturation on the Hydrothermal Synthesis of Si-HAp

Typical XRD patterns of the residual products prepared under microwave-assisted hydrothermal conditions at 150 °C for 1 h are shown in Figure 1. This experimental set was aimed to prepare $Ca_{10}(PO_4)_{6-x}(SiO_4)_x(OH)_{2-x}$ solid solutions with nominal Si^{4+} content above the stoichiometric concentration of 0.2 M. Three levels of Si^{4+} saturation of 1.5, 4.5, and 9-fold were added to the hydrothermal medium employing the TMAS solutions of 0.3, 0.9, and 1.8 M, respectively. The Si^{4+} excess aimed to improve the efficiency in the incorporation of Si^{4+} ions substituting P^{5+} in the apatite hexagonal structure. In general, the diffraction patterns of the HAp and Si-HAp powders were indexed with that of the hexagonal apatite structure with space group $P6_3/m$ (176) (card JCPD 09-0432). The crystallization of the HAp and Si-HAp proceeded via a single-step reaction; this assumption is inferred due to the formation of secondary phases of calcium phosphate or calcium silicate; it did not take place during the hydrothermal treatment. Generally, when the lowest saturation level of Si^{4+} (1.5) was used, the Si-HAp samples intended to incorporate 6 and 10 mol% of Si did not have marked differences of peak intensity and sharpness from those of the pure HAp powders. The sample prepared with the 0.3 M TMAS solution with the volume to supply 0.33 mol% Si exhibited a slight shifting of the peak to a lower diffraction angle (Figure 1a). In contrast, at 4.5 and 9-fold saturation levels, a progressive displacement of the XRD pattern proceeded at small 2θ angles, and also a remarkable peak broadening occurred on the diffraction patterns, as shown in Figure 1b,c. This behavior was markedly evident in the Si-HAp samples prepared with amounts of 5.0 and 12.16 mol% Si with TMAS solutions of 0.9 and 1.8 M. Hence, the Si^{4+} excess in the aqueous phase plays an important role in promoting the uptake of Si during the crystallization of the single-phase Si-HAp powders. These crystalline aspects were not clearly analyzed for the Si-HAp prepared even in hydrothermal conditions elsewhere [14,22,26], which do not bear crystalline structural evidence, indicating that the Si content reported is bulkily incorporated inside the particle rather than near the particle surface.

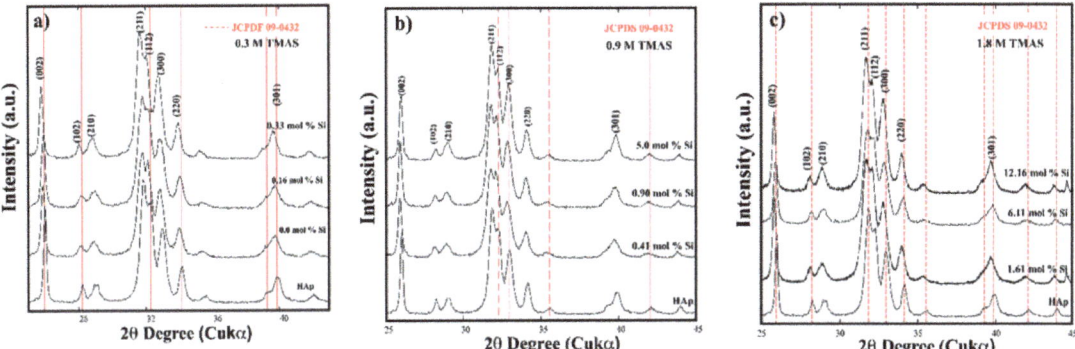

Figure 1. Si-HAp XRD patterns of samples synthesized under microwave-assisted hydrothermal conditions at 150 °C for 1 h, pH = 10, using 0.2 M $Na_5P_3O_{10}$ with different Si^{4+} mol% using (TMAS), which was mixed with a molar excess of (**a**) 1.5 (0.3 M), (**b**) 4.5 (0.9 M), and (**c**) 9 (1.8 M) folds, above the stoichiometric concentration of 0.2 M as shown.

3.2. Crystalline Structural and Chemical Compositional Analyses of Si-HAp Powders

The Rietveld refinement of Si-HA powders conducted with the refinement algorithm includes the Si^{4+} and P^{5+} molar contents determined by wet chemical analyses (Table 1, ICP results), together with the atom occupation, as suggested elsewhere (Figure 2) [1,2,5,8]. Various crystalline structural features were included as the refinement parameters, such as background, lattice parameters, scale factor, profile half width, crystallite size, local strain, thermal isotropic vectors, and spatial coordinates. The parameters selected provided high accuracy for calculating the structural features of the Si-HAp powders. The Rietveld refinement approach provided the lowest values of the goodness-of-fit factor (GOF/χ^2), averaging 1.23 ± 0.6, and low R_{wp} 7.32 ± 1.0 values were obtained (Table 1), which confirm the fine structure of all Si-HAp with excellent accuracy. The calculated diffraction profiles are in good agreement with the experimental ones due to the small residual difference between the observed and calculated patterns depicted by the residual straight line, and the vertical lines correspond to the calculated Bragg peaks positions (Figure 2) [8]. The calculated lattice parameters for pure HAp and Si-HAp are given in Table 1. Both the a_o and c_o axes increase slightly with the Si^{4+} incorporation into the pure HAp powders; whilst the length of both the a_o axis (9.4450 to 9.4365 Å) and the c_o axis (6.8882 to 6.8824 Å) decreased as the Si^{4+} incorporation ratio increased in Si-HAp powders, leading to the decrease in cell volume, specially shown in the Si-HAp powders prepared at a high Si^{4+} saturation level. It should be mentioned that in the case of Si-HAp powders at the maximum Si^{4+} incorporation of 12.16 mol%, the least "a_0" and "c_0" lattice parameters were obtained in comparison with those powders containing below 10 mol% Si contents. These were caused by proportional release of OH^- ions, which compensate the negative charges provided by SiO_4^{4-} ions substituting PO_4^{3-} in tetrahedral positions, located parallel to the c-axis along the tunnels located at the HAp structure honeycomb edges. This phenomenon provokes a bulk shrinkage of the hexagonal unit cell volume (530.75 Å^3) in the HAp structure, which would support the inference that Si is bulkily incorporated in the Si-HAp during the crystallization process, achieved by fast reaction kinetics triggered by the microwave heating under hydrothermal conditions. The maximum content of Si^{4+} (12.16 mol%) incorporated in HAp in the current experiment is above that reached under conventional hydrothermal [22] and coprecipitation [4] conditions.

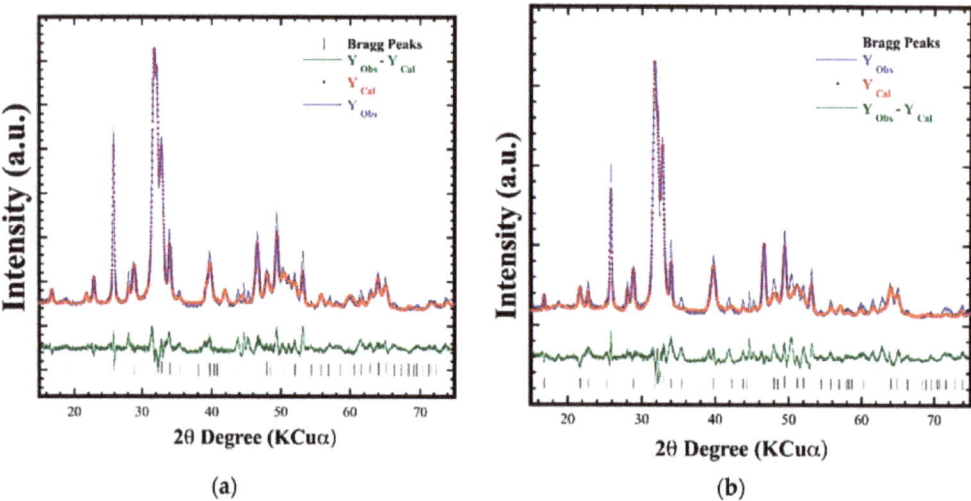

Figure 2. Rietveld refinement plot of Si-HAp powders prepared with different content of Si^{4+}: (a) 6 mol% and (b) 20 mol% using (TMAS) [$(CH3)_4N(OH).2SiO_2$] with a molar ratio of 1:3.0 (1.8 M), precursor, under microwave-assisted hydrothermal conditions at 150 °C for 1h, pH = 10.

FT-IR spectra of the HAp and Si-HAp powders revealed hydroxyl ($OH^−$) groups stretching (3571 $cm^{−1}$) and vibration (631 $cm^{−1}$) bands (Figure 3). Furthermore, the strong bands at 1086, 1014, and 960 $cm^{−1}$ wavenumbers are associated with the stretching vibration modes of $PO_4^{3−}$ tetrahedral group. The doublet band between 593 and 572 $cm^{−1}$ corresponds to O-P-O bond bending mode and these results agree with the data reported elsewhere [14]. The shoulder peaked at 897 $cm^{−1}$, and this band is assigned to the Si-O-Si vibration mode for tetrahedral $SiO_4^{4−}$ groups. A marked distortion of the shoulder peak together with the signals v_1 and v_3 of the P-O-P and $PO_4^{3−}$ major bands was found with progressive increases in the $SiO_4^{4−}$ molar content in the Si-HAp powders, as reported previously [24,26]. Furthermore, v_1 and v_3 bending modes of the P-O-P band gradually decreased in their absorbance by increasing the uptake of Si in the HAp structure, and a slight displacement to lower wavenumbers was revealed on the Si-HAp samples (Figure 3). In comparison with the FT-IR spectrum of pure HAP, the peak at 1086 $cm^{−1}$ ($PO_4^{3−}$) in Si-HAp samples decreased as the silicon content uptake increased, due to the structural change in the HAp lattice [24–27]. Additionally, the intensity of the $OH^−$ group band on the 0.16 and 0.90 mol% Si samples was similar irrespective of the saturation contents of Si^{4+}. On the contrary, the $OH^−$ stretching symmetrical band at 3571 $cm^{−1}$ and bending $OH^−$ 631$cm^{−1}$ markedly deceased in its absorbance in Si-HAp powders prepared with the highest concentration 1.8 M of TMAS (Figure 3c). XRD signals and FT-IT spectra confirm that the bulk Si should be incorporated in the apatite structure rather than partially existing at the particles' surface [26].

Table 1. Chemical and physical features of Si-HAp powders prepared under hydrothermal-assisted microwave treatments at 150 °C for 1 h using different concentrations of TMAS solutions and tripolyphosphate as a P precursor.

Sample * Id Reference	Molar Concentration (±0.001) of Si^{4+}	Nominal Mol % PO_4	Nominal Mol % SiO_4	Chemical Compositon [a]	Molar Ratio Ca/P	GOF $\chi^{2\,b}$	Lattice Parameter [b] a_0 (Å)	Lattice Parameter [b] c_0 (Å)	Cell Volume (Å^3)	Strain [b]	R_{wp}	R_{Bragg}	Crystallite Size (nm)
HAp		100	0	$Ca_{10}(PO_4)_6(OH)_2$	1.667	1.36	9.4248(7)	6.8764(5)	528.97(0.09)	0.5 (0.02)	9.43	1.44	51.69 (2.3)
MM62	0.3	94	6	$Ca_{10}(PO_4)_6(SiO_4)_{0.0}(OH)_2$	1.667	1.28	9.4433(10)	6.8886 (7)	532.00(0.12)	0.68 (0.03)	8.58	1.63	46.14 (3.4)
MM66	0.3	90	10	$Ca_{10}(PO_4)_{5.99}(SiO_4)_{0.01}(OH)_{1.99}$	1.667	1.66	9.4447(18)	6.8869 (8)	532.10(0.22)	0.53 (0.08)	10.89	3.08	46.64 (4.9)
MM63	0.3	80	20	$Ca_{10}(PO_4)_{5.98}(SiO_4)_{0.02}(OH)_{1.99}$	1.667	1.74	9.4439(16)	6.8872 (7)	531.89(0.12)	0.38 (0.04)	11.15	3.93	46.18(2.9)
MM50	0.9	94	6	$Ca_{10}(PO_4)_{5.975}(SiO_4)_{0.025}(OH)_{1.975}$	1.674	1.64	9.4443(13)	6.8866(11)	531.97(0.19)	0.56(0.07)	10.86	3.18	45.03(3.7)
MM51	0.9	90	10	$Ca_{10}(PO_4)_{5.946}(SiO_4)_{0.054}(OH)_{1.946}$	1.682	1.77	9.44413	6.8866(10)	532.00(0.17)	031 (0.05)	11.54	4.11	36.96(2.1)
MM53	0.9	80	20	$Ca_{10}(PO_4)_{5.699}(SiO_4)_{0.301}(OH)_{1.699}$	1.755	1.27	9.4424(9)	6.8863(6)	531.73(0.11)	0.30(0.03)	8.24	1.71	35.94(1.2)
MM64	1.8	94	6	$Ca_{10}(PO_4)_{5.908}(SiO_4)_{0.097}(OH)_{1.908}$	1.692	1.64	9.4475(16)	6.8884(11)	532.46(0.20)	0.61(0.06)	10.56	2.95	48.82(5.1)
MM68	1.8	90	10	$Ca_{10}(PO_4)_{5.633}(SiO_4)_{0.367}(OH)_{1.633}$	1.775	1.30	9.4447(11)	6.8872(8)	532.09(0.14)	0.68(0.04)	8.44	1.81	48.64(3.4)
MM65	1.8	80	20	$Ca_{10}(PO_4)_{5.271}(SiO_4)_{0.729}(OH)_{1.271}$	1.897	1.11	9.4365(9)	6.8824(7)	530.75(0.11)	1.68(0.09)	2.63	1.15	39.87(3.6)

[a] Chemical formula of Si-HAp powders was calculated from Ca, P, and Si contents determined via ICP-AES analysis, and OH^- was calculated by charge balance. [b] Values obtained from the refinement of XRD patterns carried out using the Rietveld method. (* Identification number samples: MM62, MM66, MM63; MM50, MM51, MM53; MM64, MM68 and MM65 correspond to the Si-HAp with 5, 10 and 20 mol% with concentration 0.3, 0.9 and 1.8 M respectively).

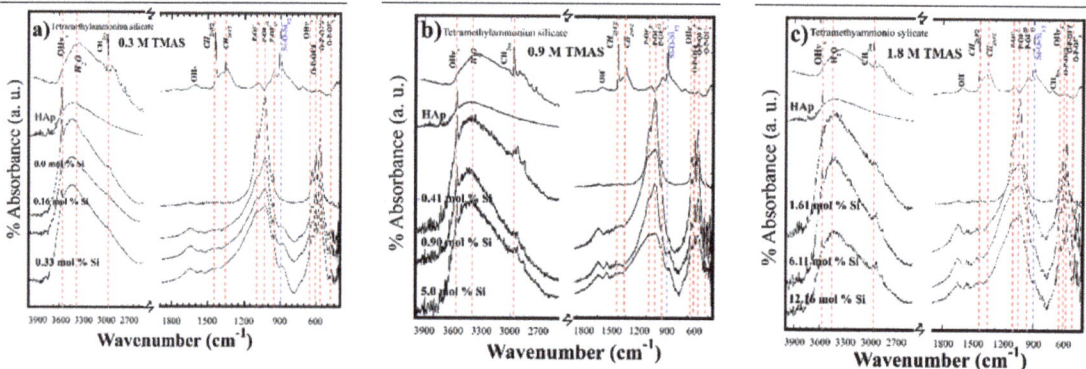

Figure 3. Fourier transform infrared (FT-IR) spectrum analysis of the Si-HAp powders obtained by microwave-assisted hydrothermal process at 150 °C for 1 h using different concentrations of [($C_4H_{13}NO_5Si_2$) (TMAS)]: (a) 0.3 M; (b) 0.9 M, and (c) 1.8 M.

These structural results are consistent with the substitution mechanism proposed, where PO_4^{3-} ions are replaced by SiO_4^{4-} ions, causing a stoichiometric release of OH^- ions, which maintains the total charge balance in the HAp structure (Equation (1)). Under hydrothermal conditions, the saturation of Si^{4+} probably induced a reduction in the amount of hydroxyl groups to compensate for an extra negative electric charge produced by the incorporation of the silicate groups, and the formation of OH^- vacancies (V) might have taken place to maintain the charge balance neutrality, as is described by the following equation: $PO_4^{3-} + OH^- \rightarrow SiO_4^{4-} + V_{(OH)-}$. Indeed, this behavior is consistent with previous research work [13].

Raman analyses were carried out to determine detailed crystalline differences in the bonds. Raman spectra of the SiHAp constituents in Figure 4a show that the v_1 symmetric stretching PO_4 mode at around 996 cm^{-1} corresponds to the HAp structure of the prepared Si-HAp powders samples with both concentrations (0.9 and 1.8 M). Other typical PO_4 peak modes of the bending v_2, asymmetric stretching v_3, and bending v_4 (PO_4) were also observed at around 400, 1100, and 600 cm^{-1}, respectively. The progressive decrease in the peak intensity and their broadening for all PO_4 v_2–v_4 bands confirmed the SiO_4^{4-} substitution by PO_4^{3-} group. The OH^- peak intensity decreased with broadening of the band [20]; these results are in line with those of FT-IR and XRD. Under hydrothermal conditions assisted by microwave heating, the molar percentage of Si^{4+} substitution was larger than 10 mol% by providing a 9-fold saturated Si^{4+} precursor in comparison with the stoichiometric limit of acceptance to keep the stability of the HAp structure. This explains the low variation of the lattice parameter and very small variation in the crystallite size, because only 0.33 mol% of Si^{4+} was incorporated in the HAp structure for samples prepared with a low concentration of TMAS (0.3 M). However, Si-HAp samples synthesized with a greater content of Si^{4+} (12.16 mol%) did not show the presence of the characteristic Si signal, and only small changes were detected in the vibration OH^- signal at 3570 cm^{-1}, which were slightly decreased and broadened under the high Si^{4+} saturation of TMAS (1.8 M) during the hydrothermal reaction.

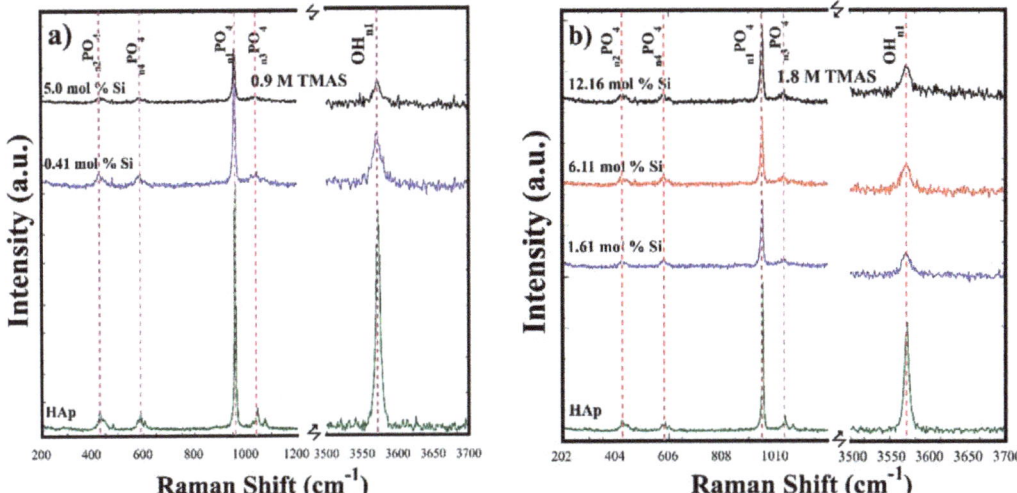

Figure 4. Raman spectra of the Si-HAp powders obtained by microwave assisted hydrothermal process at 150 °C for 1 h using different molar concentrations of the TMAS solutions: (**a**) 0.9 and (**b**) 1.8 M.

Figure 5a–d shows the XPS spectra for HAp and Si-HAp powders prepared with a 0.9 M solution for samples with ultimate contents of 5 and 12.16 mol% Si. The XPS spectra corresponding to the photoelectron core levels of Ca 2p, P 2p, Si 2p, and O 1s without any additional core level of other elements were detected by the XPS analyses in the survey spectrum. Generally, the Ca 2p spectrum recorded in the binding energy (BE) range of 344.0–355.0 eV revealed the doublet associated with Ca-O bonds in the HAp and Si-HAp samples, which are constituted by the core level Ca $2p_{3/2}$ at 347.21 eV and Ca $2p_{1/3}$ at 350.55 eV BE. Furthermore, the P 2p peak is symmetric and its average BE is at 132.9 eV both for HAp and Si-HAp (Figure 5b). Moreover, the O 1s core level peak was deconvoluted into two peaks, and a small peak, which fits the shoulder between 532 and 534 eV, was detected. The deconvolution indicates that the peak average BE energy was of 532.45 eV, and this peak is associated with the SiO_4 units in the prepared powder samples, where the gradual increase in peak intensity in the samples synthesized with different Si^{4+} contents supports our inference (Figure 5c). The second large peak, deconvoluted at an average BE of 530.8 eV, corresponded to the O 1s core level of PO_4^{3-} tetrahedral units. However, the presence of silicon was very clear in powders prepared with 0.41 mol% Si, as shown in Figure 5d. In contrast, a symmetric peak corresponding to the core level Si 2p at BE of 103.3 eV was revealed in the Si-HAp samples obtained with the molar volume corresponding to 5.0 mol% Si using the TMAS solution of 0.9 M.

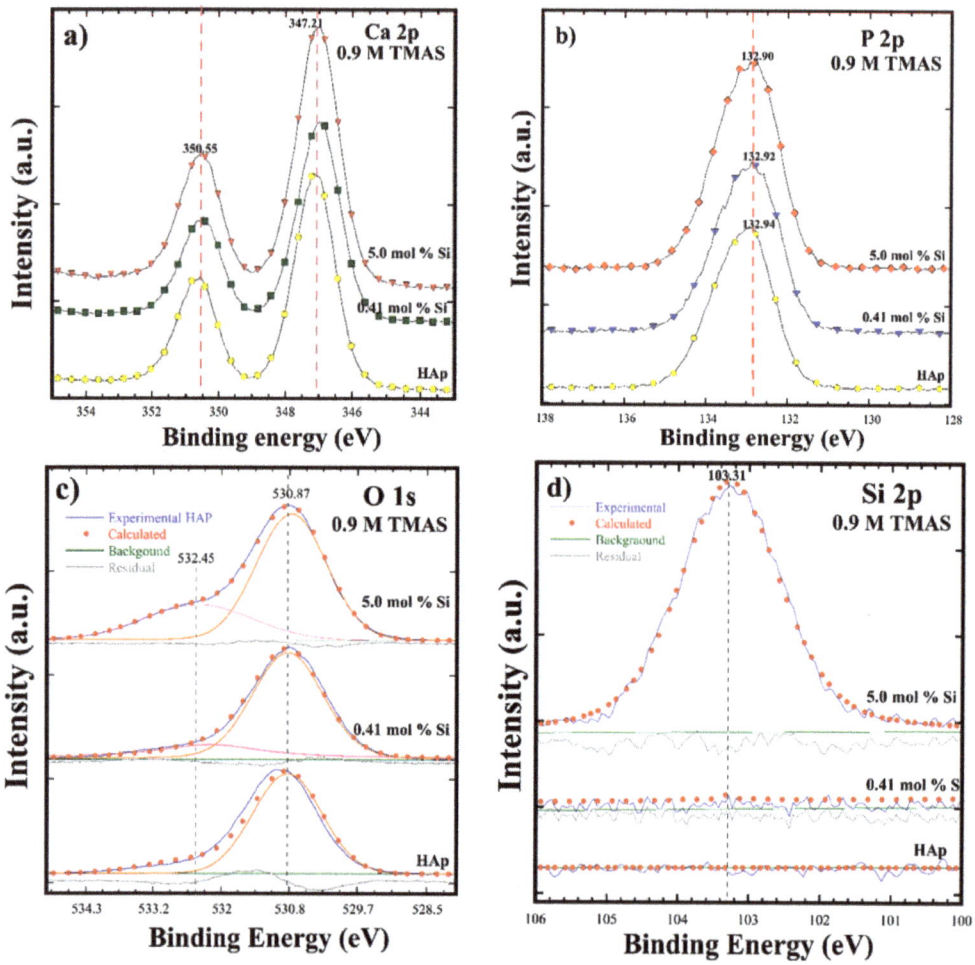

Figure 5. The XPS analysis for HAp and Si-HAp powders obtained by microwave assisted hydrothermal process at 150 °C for 1 h using 0.9 M of TMAS and different mol% of Si: (**a**) Ca 2p; (**b**) P 2p; (**c**) O 1s; (**d**) Si 2p.

The XPS spectra in Figure 6 indicated that HAp and Si-HAp powders prepared in the presence of the highest Si^{4+} saturated TMAS solution (1.8 M) exhibited a gradual uptake of Si. All the samples are constituted by the chemical elements that form the HAp structure. Figure 6a shows the typical doublet peak associated with the core level Ca $2p_{3/2}$ XPS BE at 347.1 eV and Ca $2p_{1/3}$ at 350.07 eV. The doublet peaks increased slightly as the silicon incorporation content increased in the synthesized samples obtained with 1.61 mol% of Si in the Si-HAp powders. Likewise, the symmetric signals of P 2p increased slightly with the increase of Si^{4+} content, achieving a binding energy of 132.92 eV. Furthermore, the binding energy signals for O 2p peaks were nearly symmetric (Figure 6c). The main component at 530.9 eV corresponding to the O^{2-} is linked only to a phosphorus atom as in PO_4^{3-} ions of the HAp structure. With the increase of the mol% of SiO_4^{4-} in the prepared powders, the BE signal shifted slightly into 530.79 eV and a shoulder peak at an average BE of 532.53 eV increased [11]. Therefore, this shoulder peak is attributed to the O^{2-} ions associated with the SiO_4^{4-} tetrahedral units.

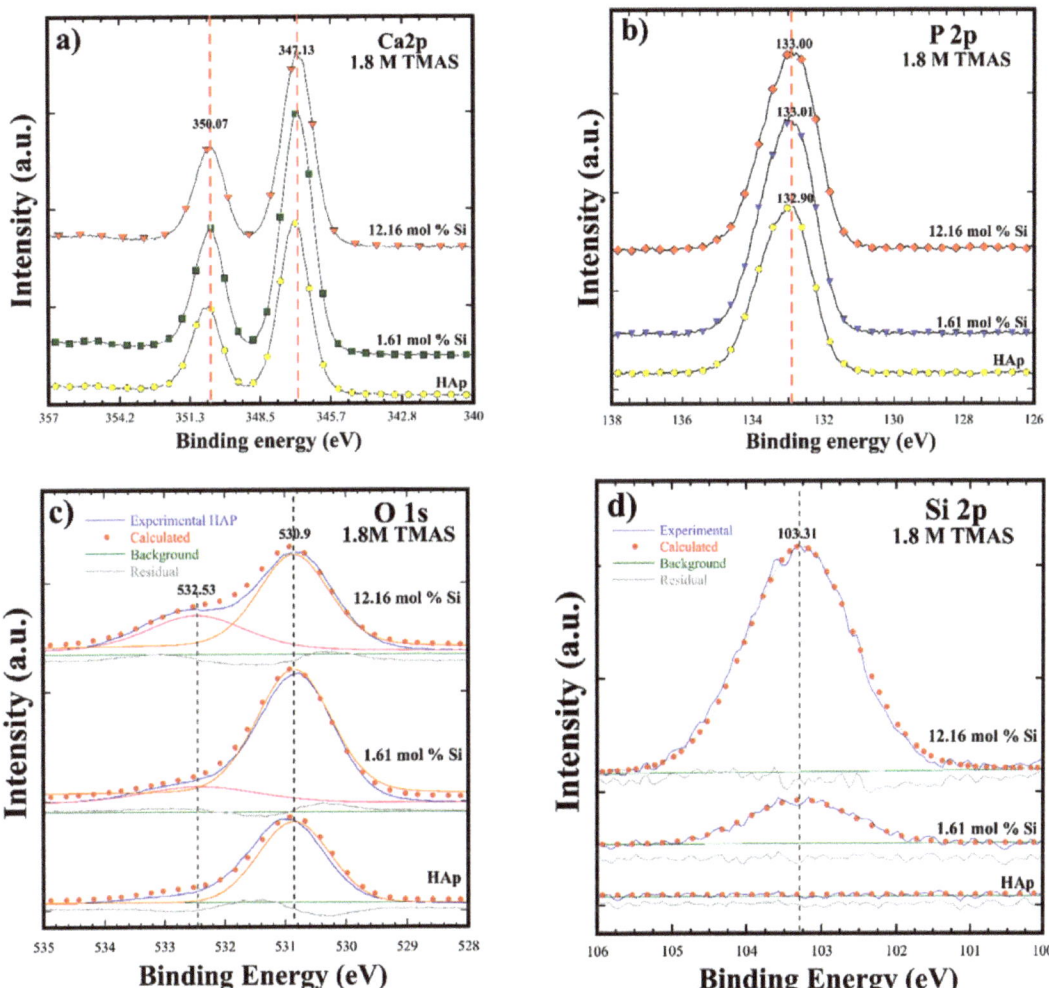

Figure 6. The XPS analysis for HAp and Si-HAp powders obtained by microwave assisted hydrothermal process at 150 °C for 1 h using 1.8 M of the TMAS solutions with different mol% Si: (**a**) Ca 2p; (**b**) P 2p; (**c**) O 1s; (**d**) Si 2p.

In addition, the presence of Si was confirmed in the HAp structure of the powders prepared using the highest saturation of Si (9-fold, Figure 6d). The XPS core level Si 2p corresponded to the symmetric peak at BE of 103.18 eV for the 1.61 mol% Si. The Si 2p peak intensity increased as a result of the improved SiO_4^{4-} incorporation in the HAp structure. The largest peak at a BE of 103.31 eV occurred in the sample containing 12.16 mol% Si, and this SiHAp powder exhibited similar BE behavior as was reported elsewhere [20,21]. The lack of Si uptake in the HAp structure would be caused by the highly soluble species produced by the TMAS precursor, such as $(Si-O-Si)_3O^-$. This anionic specie is preferentially formed in methanolic solutions due to the high solubility property of TMAS [22–27]. Therefore, these species should reduce the solute concentration of SiO_4^{4-} ions at the supersaturation stage reached under microwave-assisted hydrothermal conditions. This behavior is attributed to the fact that the TMAS might dissolve in 2-propanol under the hydrothermal fluid forming silicate ions in the medium. The highly soluble silicate ions themselves promote polymerization of silicate ions, limiting the amount of OH^- ions in the SiHAp powders, because the related vibration OH^- signal in FT-IR spectra decreased slightly under the high Si^{4+} saturation of TMAS (1.8 M) during the hydrothermal reaction.

3.3. Morphological Aspects of the Partially Substituted Si-HAp Particles Prepared Hydrothermally

The morphology of HAp and Si-HAp powders was analyzed by FE-SEM micrographs (Figure 7). Pure HAp was determined to be monodispersed particles with a regular rod-like morphology with an average size of 62 nm. In contrast, the Si-HAP powders were mostly monodispersed Si-HAp agglomerates with a quasi-oval shape. The agglomerates average size of the Si-HAp samples incorporating 0.9 mol% of 5.0 mol% Si contents was between 235.5 ± 29.7and 297.4 ± 19.4 nm. Whilst, when the 9-fold Si^{4+} sutured TMAS solution (1.8 M) was used, the agglomerated average size of the samples with 6.11 mol% and 12.16 mol% Si contents was between 243.9 ± 22.2 and 315.1 ± 22.5 nm. The excess of Si in the hydrothermal medium should trigger the formation of Si-HAp agglomerates prepared under fast kinetic reaction conditions assisted by the microwave heating. These results are supported by variation in the crystallite size, as calculated in the Rietveld refinement results (Table 1).

Figure 7. FE-SEM microphotographs of HAp and Si-HAp powder prepared by microwave-assisted hydrothermal process at 150 °C for 1 h, (**a**) 0 mol%; and using Si saturated [$(C_4H_{13}NO_5Si_2)$ (TMAS)] with different nominal concentrations of TMAS 0.9 M and (**b,c**) and 1.8 M (**d,e**) and different mol% of Si: (**b,c**)10 mol% and (**d,e**) 20 mol%, respectively.

In addition, the detailed crystalline structure features of the quasi-oval agglomerates of Si-HAp powders containing 1.62 mol% Si and 12.16 mol% Si, prepared in the presence of the 1.8 M TMAS solution, were investigated by HR-TEM observations (Figure 8a,d). These images revealed that the bulk morphology of the quasi-oval shaped agglomerates is irrespective of the Si^{4+} saturation in the reaction fluid medium. Generally, the agglomerates are formed by nanosized euhedral rod-shaped crystals with varied sizes. The rod-shaped crystals containing 1.62 mol% Si exhibited a broad monomodal length distribution, and the average length was 32 ± 8.0 nm (Figure 8a,b). On the contrary, a slight reduction in the crystal length occurred in the Si-HAp sample incorporating 12.15 mol% Si, which was shorter than in the sample incorporating 1.62 mol Si, and the length size distribution curve revealed it to be in a vast proportion (27.0 ± 8.0 nm) (Figure 8d,e). The Si-HAp particle size gradually decreased with the increase of Si content. In addition, HR-TEM and SAED provide a high crystallinity of the euhedral rod-shaped SiHAp particles incorporating both 1.62 mol% and 12.15 mol% Si (Figure 8c,f). The SAED pattern (inset) of the squared area in Figure 8c indicates that the preferential stacking of the 1.62 mol% Si-HAp crystals proceeds along the hexagonal structure basal plane with a Miller index of (300), although an irregular atomic stacking occurred in some areas of the agglomerate 12.15 mol% Si. Meanwhile, the SAED pattern of the crystals indicated that the agglomerates containing less silicon were associated with the family plane <112> Miller index. The interplanar spacing calculated for the (112) and (211) planes was 0.32 nm and 0.27 nm, respectively. These values are very close to the interplanar spacing positions in the single-phase hydroxyapatite structure. The SAED patterns confirmed that the fine, rod-like Si-HAp crystals are single crystals.

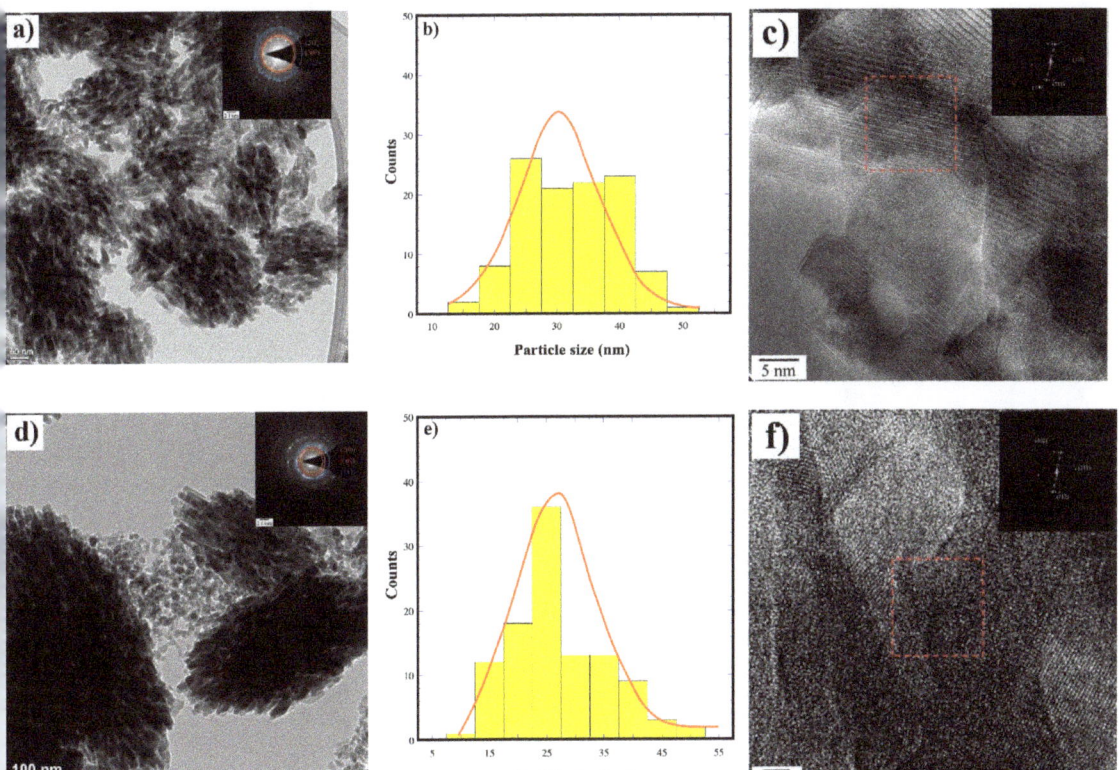

Figure 8. HRTEM micrographs of the Si-HAp powders prepared under an assisted hydrothermal process at 150 °C for 1 h using [(C$_4$H$_{13}$NO$_5$Si$_2$) (TMAS)] 1.8 M: (a–c) 1.62 mol% of Si and (d–f) 12.15 mol% of Si.

SEM and HR-TEM observations indicated that the Si^{4+} excess in the hydrothermal medium led to a marked variation in the growth and the spontaneous assembly process of the euhedral rod-like particles. These differences should be generated from a different dissolution-crystallization mechanism, which achieved rapid reaction kinetics by the microwave heating of the hydrothermal medium at 150 °C. However, the formation of highly soluble anionic silicate tri-branching units $(Si-O-Si)_3O^-$ (Q^3) could form to become dominant under the current hydrothermal conditions due to the addition of 2-propanol [28] The Q^3 units, which can act as fine micelles that trap the Si-HAp nutrients, led to a certain supersaturation state in the hydrothermal reaction conditions, which should be essential for the crystallization of the irregular oval-shaped Si-HAp agglomerates. When a further excess of Si^{4+} was supplied, the molar volume of the Q^3 units increased, leading to an increase in the agglomerate size. The above inference, associated with the reaction pathway, is supported by the fact that no crystalline SiO_2 species were formed as a secondary phase as a result of the crystallization of the Q^3 units. Hence, we surmise that the hydrothermal microwave-assisted method is efficient to produce Si-HAp powders with larger contents of silicon rather than those techniques reported recently [12] including the conventional hydrothermal process [22]. This method coupled with the use of TMAS has the potential for processing Si-HAp to prepare biomaterials with implant applications in medicine.

4. Conclusions

Si-HAp powders were successfully crystallized under fast microwave-assisted hydrothermal synthesis conditions at a low temperature (150 °C) for 1 h using saturated Si^{4+} precursor TMAS solutions.

The maximum amount of Si^{4+} incorporated in the HAp structure was 12.16 mol%, using an excess of 1.8 M of TMAS. The addition of highly concentrated Si^{4+} solutions (0.3–1.8 M) caused differences on the crystalline unit cell of the apatite and produced agglomerates constituted by fine euhedral rod-like crystal with an average length of 27.0 ± 8.0 nm and a single crystal habit. The Si^{4+} excess in the reaction media led to the rod-like crystal self-assembly to produce irregular oval-shaped Si-HAp agglomerates, which were prepared under fast kinetic reaction conditions assisted by the microwave heating and exhibit sizes between 233.5 and 315.1 nm. These agglomerates exhibited a marked size coarsening, which was triggered by the Si^{4+} saturation level supplied in the hydrothermal media. Despite the Si^{4+} ion high level of saturation used in the hydrothermal reaction medium, the lack of Si incorporation in the HAp structure is promoted by the Q_3 species, namely $(Si-O-Si)_3O^-$, which are likely formed in hydrothermal media containing 2-propanol. These highly soluble ions reduced the supersaturation SiO_4^{4-} molar volume in the medium, almost 50% below the ultimate stoichiometric content selected. Furthermore, a remarkable decrease in the O^- ions content was confirmed by FTIR and XPS analyses, and the gradual OH^- lost was caused to compensate for the partial incorporation of SiO_4^{4-} at tetrahedral PO_4^{3-} sites in the HAp structure. The present hydrothermal microwave-assisted method has delivered high processing efficiency to crystallize Si-HAp particles with a control on the Si^{4+} content. This method has potential for processing Si-HAp bioceramic implants in medicine.

Author Contributions: Z.M.-V. conceptualized, designed, and organized the research work; J.C.R.-A. provided the infrastructure and microwave autoclaves; B.M.-P. conducted the hydrothermal experiments; J.C.R.-A. and Z.M.-V. conducted the chemical and crystalline structural characterization and data analysis; K.Z. contributed to the XPS and Raman characterization; K.Y. and T.U. provided the infrastructure to conduct the characterization; Z.M.-V. wrote the manuscript. All authors have read and agreed to the published version of the manuscript.

Funding: This research was financially supported by "Research Projects grants TecNM: 6273.17-P" and partially supported by project grant TecNM 5931.19-P, to support PNPC, SNI-CONACYT of the "Instituto Tecnológico de Saltillo", Mexico. B.M.-P. is indebted to CONACYT, Mexico for financial support in a PhD Scholarship to conduct a part of this research.

Institutional Review Board Statement: Not applicable.

Informed Consent Statement: Not applicable.

Data Availability Statement: The data presented in this study are available in the Hydrothermal Synthesis of Nanoparticles knowledgebase https://www.mdpi.com/journal/nanomaterials/special_issues/hydrotherm_nano, accessed on 10 May 2021.

Acknowledgments: Z.M.-V. and J.C.R.-A. are indebted to the CONACYT-SNI. This study was kindly supported by Center for Advanced Marine Core Research, Kochi University for the use of the FE-SEM and XRD equipment.

Conflicts of Interest: The authors declare no conflict of interest. The funders had no role in the study, or analysis and interpretation of data; in writing the manuscript; or the decision to publish the results.

References

1. Kongjun, Z.; Yanagisawa, K.; Shimanouchi, R.; Onda, A.; Kajiyoshi, K.; Qiu, J. Synthesis and crystallographic study of Pb-Sr hydroxyapatite solid solutions by high temperature mixing method under hydrothermal conditions. *J. Ceram. Soc. Jpn.* **2007**, *115*, 873–876. [CrossRef]
2. Carlisle, E.M. Silicon: A possible factor in bone calcification. *Science* **1970**, *167*, 279–280. [CrossRef] [PubMed]
3. Lee, J.H.; Lee, K.S.; Chang, J.S.; Cho, W.S.; Kim, Y.; Kim, S.R.; Kim, Y.T. Biocompatibility of Si-substituted hydroxyapatite. *J. Key Eng. Mater.* **2004**, *254–256*, 135–138. [CrossRef]
4. Gibson, I.R.; Huang, J.; Best, S.M.; Bonfield, W. Enhanced in vitro cell activity and surface apatite layer formation on novel silicon-substituted hydroxyapatites. *World Sci.* **2009**, *12*, 191–194. [CrossRef]
5. Jamil, M.; Elouatli, B.; Khallok, H.; Jamil, M.; Elouahli, A.; Gourri, E.; Ezzahmouly, M.; Abida, F.; Hatim, Z. Silicon substituted hydroxyapatite: Preparation with solid-state reaction, characterization and dissolution properties. *J. Mater. Environ. Sci.* **2018**, *9*, 2322–2327.
6. Harden, F.J.; Gibson, I.R.; Skakle, J.M.S. Simplification of the synthesis method for silicon-substituted hydroxyapatite: A Raman spectroscopy study. *Key Eng. Mater.* **2012**, *529*, 94–99. [CrossRef]
7. Marchat, D.; Zymelka, M.; Coelho, C.; Gremillard, L.; Jolypottuz, L.; Babonneau, F.; Esnouf, C.; Chevalier, J.; Nernacheasollant, D. Accurate characterization of pure silicon-substituted hydroxyapatite powders synthesized by a new precipitation route. *Acta Biomater.* **2013**, *9*, 6992–7004. [CrossRef]
8. Zou, S.; Huang, J.; Best, S.; Bonfield, W. Rietveld studies on silicon substituted hydroxyapatite. *Key Eng. Mater.* **2006**, *309*, 113–116. [CrossRef]
9. Nakahira, A.; Nakata, K.; Numako, C.; Murata, H.; Matsunaga, K. Synthesis and evaluation of calcium-deficient hydroxyapatite with SiO_2. *Mater. Sci. Appl.* **2011**, *2*, 1194–1198. [CrossRef]
10. Outali, B.E.; Jamil, M.; Elouahli, A.; Ezzahmouly, M.; Abida, F.; Ilou, M.; Hatim, Z. Silicon substitution in biphasic calcium phosphate bioceramics: Crystal structure study. *Int. J. Sci. Eng. Res.* **2016**, *7*, 829–833.
11. Yacoubi, A.E.; Massit, A.; Fathi, M.; Chafik, B.; Idrissi, E.; Yammi, K. Characterization of silicon-substituted hydroxyapatite powders synthesized by a wet precipitation method. *IOSR J. Appl. Chem.* **2014**, *7*, 24–29. [CrossRef]
12. Aminian, A.; Solati, M.; Bakhshi, F.; Fazardi, A. Silicon substitution hydroxyapatite by hydrothermal method. *Adv. Bioceram. Biotechnol.* **2010**, *11*, 59–65.
13. Gibson, I.R.; Best, S.M.; Bonfield, W. Chemical characterization of silicon-substituted hydroxyapatite. *J. Biomed. Mater. Res.* **1999**, *44*, 422–428. [CrossRef]
14. Aminian, A.; Salti-Hashjin, M.; Samadikuchaksaraei, A.; Bakhshi, F.; Gorjipour, F.; Farzadi, A.; Motztarzadath, F.; Schmucker, M. Synthesis of silicon-substitute hydroxyapatite by a hydrothermal method with two different phosphorous sources. *Ceram. Int.* **2011**, *37*, 1219–1229. [CrossRef]
15. Bulina, N.; Chaikina, M.V.; Ishchenko, A.V.; Prosanov, I.Y. Mechanochemical synthesis of SiO_4^{4-}-substituted hydroxyapatite, part II—Reaction mechanism, structure, and substitution limit. *Eur. J. Inorg. Chem.* **2014**, *28*, 4803–4809. [CrossRef]
16. Neira, I.S.; Kolen'ko, V.Y.; Lebedev, O.I.; Tendeloo, G.V.; Gupta, H.S.; Guitian, F.; Yoshimura, M. An effective morphology control of hydroxyapatite crystals via hydrothermal synthesis. *Cryst. Growth Des.* **2009**, *9*, 466–467. [CrossRef]
17. Byrappa, K.; Yoshimura, M. *Handbook of Hydrothermal Technology a Technology for Crystal Growth and Materials Processing*; William Andrew Publishing, LLC: Norwich, NY, USA, 2010; pp. 755–780. ISBN 9781437778366.
18. Yanagisawa, K.; Toya, H.; Feng, Q.; Yamasaki, N. In situ formation of hydroxyapatite crystals under hydrothermal conditions. *Phosphorus Res. Bull.* **1995**, *5*, 43. [CrossRef]
19. Padmanabhan, S.K.; Haq, U.E.; Licciulli, A. Rapid synthesis and characterization of silicon substituted nano hydroxyapatite using microwave irradiation. *Curr. Appl. Phys.* **2014**, *14*, 87–92. [CrossRef]
20. Montoya, K.L.; Rendón-Ángeles, J.C.; Matamoros-Veloza, Z.; Yanagisawa, K. Rapid synthesis and characterization of Zn substituted hydroxyapatite nanoparticles via a microwave-assisted hydrothermal method. *Mater. Lett.* **2017**, *195*, 5–9. [CrossRef]
21. Lamkhao, S.; Phaya, M.; Jansakun, C.; Chandet, N.; Thongkorn, K.; Rujijanagul, G.; Bangrak, P.; Randorn, C. Synthesis of hydroxyapatite with antibacterial properties using a microwave-assisted combustion method. *Sci. Rep. Springer Nat.* **2019**, *9*, 4015. [CrossRef]

22. Moreno Perez, B.; Matamoros Veloza, Z.; Rendon-Angeles, J.C.; Yanagisawa, K.; Onda, A.; Perez Terrazas, J.E.; Mejia Martinez, E.E.; Burciaga Diaz, O.; Rodriguez Reyes, M. Synthesis of silicon-substituted hydroxyapatite using hydrothermal process. *Bol. De La Soc. Española De Cerámica Y Vidr.* **2020**, *59*, 50–64. [CrossRef]
23. Hasegawa, I.; Sakka, S.; Sugahara, Y.; Kuroda, K.; Kato, C. Silicate anions formed in tetramethylammonium silicate methanolic solution as studied by 29 Si nuclear magnetic resonance. *J. Chem. Soc. Chem. Commun.* **1989**, *4*, 208–210. [CrossRef]
24. Yoichiro, M.; Masateru, H.; Masahiko, O.; Toshihiro, K.; Masayuki, N. Large-sized hydroxyapatite whiskers derived from calcium tripolyphosphate gel. *J. Euro. Ceram. Soc.* **2005**, *25*, 3181–3185. [CrossRef]
25. Shing, H.Y.; Jung, J.Y.; Kim, S.W.; Lee, W.K. XPS analysis on chemical properties of calcium phosphate thin films and osteoblastic HOS cell responses. *J. Ind. Eng. Chem.* **2006**, *12*, 476–483.
26. Bianco, A.; Cacciotti, I.; Lombardi, M.; Montanaro, L. Si-substituted hydroxyapatite nanopowders: Synthesis, thermal stability and sinterability. *Mater. Res. Bull.* **2009**, *44*, 345–354. [CrossRef]
27. Younes, B.; Meriame, B.; Ferreira, J.; El Mabrouk, K. Hydrothermal synthesis of Si-doped hydroxyapatite nanopowders: Mechanical and bioactivity evaluation. *Int. J. Appl. Ceram. Technol.* **2015**, *12*, 329–340. [CrossRef]
28. Botelho, C.M.; Lopes, M.A.; Gibson, I.R.; Best, S.M.; Santos, J.D. Structural analysis of Si-substituted hydroxyapatite: Zeta potential and X-ray photoelectron spectroscopy. *J. Mater. Sci. Mater. Med.* **2002**, *13*, 1123–1127. [CrossRef]

Article

Hydrothermal Synthesis of Iridium-Substituted NaTaO$_3$ Perovskites

David L. Burnett [1], Christopher D. Vincent [1], Jasmine A. Clayton [1], Reza J. Kashtiban [2] and Richard I. Walton [1,*]

[1] Department of Chemistry, University of Warwick, Coventry CV4 7AL, UK; d.burnett@bham.ac.uk (D.L.B.); cdv1230@gmail.com (C.D.V.); jasmine.clayton@warwick.ac.uk (J.A.C.)
[2] Department of Physics, University of Warwick, Coventry CV4 7AL, UK; r.jalilikashtiban@warwick.ac.uk
* Correspondence: r.i.walton@warwick.ac.uk

Abstract: Iridium-containing NaTaO$_3$ is produced using a one-step hydrothermal crystallisation from Ta$_2$O$_5$ and IrCl$_3$ in an aqueous solution of 10 M NaOH in 40 vol% H$_2$O$_2$ heated at 240 °C. Although a nominal replacement of 50% of Ta by Ir was attempted, the amount of Ir included in the perovskite oxide was only up to 15 mol%. The materials are formed as crystalline powders comprising cube-shaped crystallites around 100 nm in edge length, as seen by scanning transmission electron microscopy. Energy dispersive X-ray mapping shows an even dispersion of Ir through the crystallites. Profile fitting of powder X-ray diffraction (XRD) shows expanded unit cell volumes (orthorhombic space group *Pbnm*) compared to the parent NaTaO$_3$, while XANES spectroscopy at the Ir L$_{III}$-edge reveals that the highest Ir-content materials contain Ir^{4+}. The inclusion of Ir^{4+} into the perovskite by replacement of Ta^{5+} implies the presence of charge-balancing defects and upon heat treatment the iridium is extruded from the perovskite at around 600 °C in air, with the presence of metallic iridium seen by in situ powder XRD. The highest Ir-content material was loaded with Pt and examined for photocatalytic evolution of H$_2$ from aqueous methanol. Compared to the parent NaTaO$_3$, the Ir-substituted material shows a more than ten-fold enhancement of hydrogen yield with a significant proportion ascribed to visible light absorption.

Keywords: perovskite; tantalate; crystallisation; nanocrystals; photocatalysis; water splitting

1. Introduction

The hydrothermal synthesis of ABO$_3$ perovskite oxides has attracted a large amount of interest in the past decade [1]. This includes families of materials with important properties such as titanates (B = Ti) with dielectric properties [2], piezoelectric zirconate-titanates (B = Zr, Ti) [3], multiferroic chromites (B = Cr) [4], and ferrites (B = Fe) with applications in redox catalysis [5]. The synthesis of this range of compositions work builds on a body of literature on hydrothermal crystallisation of one of the prototypical perovskites BaTiO$_3$ [6]. The attraction of the hydrothermal synthesis method lies in the use of solution chemistry to enable crystallisation from solution directly at mild temperatures typically less than 200 °C: this allows adjustment of the crystal morphology, including size and shape of crystallites on the nanoscale, as well as the possibility of isolating compositions not stable under more extreme conditions [7–12]. This level of control in the synthesis of oxide materials is lacking in traditional high temperature routes, and even in co-precipitation or sol–gel approaches, in which a firing step is needed to induce crystallinity; this means annealing takes place, with control of crystallite size being difficult to achieve.

Many functional oxide materials have been accessed via hydrothermal synthesis routes, some with unique properties arising from their nanostructure. A notable example is the formation of nanowires of Cu$_2$O that have considerably enhanced photoactivity in the visible region of the spectrum [13]. The hydrothermal method also provides a convenient method for the formation of composite materials, where the growth of an oxide on a support in situ provides new functional solids with exceptional properties. Examples include the formation of graphene oxide conjugated Cu_2O nanowires for gas sensing [14], ammonia sensing from Cu_2O nanoparticles decorated with MoS_2 nanosheets [15], and graphene oxide–MnO_2 nanocomposites for supercapacitors [16].

Niobate and tantalate perovskites, $ANbO_3$ and $ATaO_3$ where A = Na or K, have attracted much attention due to their practical applications in two important areas: as potential lead-free electroceramics, as end-members of the piezoelectric material $K_{0.5}Na_{0.5}NbO_3$, and as photocatalysts for applications such as water splitting and carbon dioxide conversion. Hydrothermal synthesis of these materials has been extensively investigated and the pathways during the formation of the perovskite products have been mapped [17–19]. Some control of crystallite morphology has proved possible; for example, it has been shown that a low concentration of Nb_2O_5 as a precursor leads to nanorods or nanoplates of $NaNbO_3$, while a lower concentration of NaOH yields cubes [20]. Crystallite morphology of $NaNbO_3$ can also be influenced by the choice of niobium oxide precursor [21], the pH of the solution [22], and the choice of solvent [23]. For electroceramics, fine-grained ceramics can be produced by annealing the powders from hydrothermal reactions: piezoelectric ceramics formed from hydrothermally prepared alkali niobates and tantalates have shown characteristics comparable to ceramics made by conventional methods, but with the advantage of lower sintering temperatures to achieve densification [24–26]. For photocatalysis, the high surface areas of the nanostructured crystallites offers high reactivity; for example, Shi et al. prepared nanocubes and compared them with nanowires, and found a correlation between crystal shape and photocatalytic activity for hydrogen evolution from water/methanol [27]. High surface areas are also useful for support materials for co-catalysts.

In photocatalysis, $NaNbO_3$ and $NaTaO_3$ are typically doped with substituent metals to tune their band gaps, and then used as supports for precious metals with the particular aim to permit absorption of visible light [28]. Doping of perovskite oxides with precious metal cations is of more general interest, but can present a synthetic challenge as there is a strong tendency for the precious metal substituent to be reduced to the elemental state, even upon heating in air. $LaCr_{1-x}M_xO_3$ (LCMO) ($x = 0.01, 0.05, 0.10$ with M = Pd, Co, Ir) materials were prepared using conventional solid-state synthesis from single metal oxide precursors at 1200 °C, with a reduction in band gap observed for all substituted materials [29]. Rh-doped $BaTiO_3$ nanoparticles, with 2% of the substituent, were prepared by a co-precipitation route using oxalate as solution additive at temperatures between 700 and 900 °C [30]. The Rh was found in the +3 oxidation state and on heating to higher temperatures the rhodium was lost, with phase transformation of the perovskite. 0.5% iridium-doped $SrTiO_3$ was prepared by a solid-state method, and in a second step under reducing conditions, exsolution of the Ir was found that resulted in supported Ir nanoparticles, embedded in the oxide support, showing little agglomeration when used in CO oxidation catalysis [31]. Ir-doped $SrTiO_3$ (1–5% Ir) has also been studied for photocatalysis, and the oxidation state of the Ir was found to dictate the properties towards photocatalytic water splitting, with Ir^{4+} giving the most favourable properties [32]. Kudo et al. studied a number of precious-metal-doped perovskites for photocatalytic water splitting; in the case of $NaNbO_3$ and $NaTaO_3$, prepared by a solid-state method, doping with Ir or Rh creates band gaps suitable for visible light absorption when co-doped with alkaline earth or lanthanide cations, Ba and La, respectively [33,34].

Given the need for convenient synthesis methods for precious-metal-substituted perovskites, which may also allow control of the substituent oxidation state, we have investigated the use of hydrothermal reaction conditions. In this paper we consider the possibility of iridium substitution in $NaTaO_3$ by a direct hydrothermal synthesis and show that a one-step crystallisation is able to form nanocubes with homogeneous distribution of iridium, as proven by various experimental techniques. We chose this composition to study since there is already a body of work on the hydrothermal synthesis of $NaTaO_3$, and the case of iridium substitution has been reported by other synthesis methods, which provide a comparison to the milder solution conditions. To our knowledge, the inclusion of precious metal substituents in an oxide perovskite host by hydrothermal synthesis has not yet been reported. Our aim was to explore the maximum level of inclusion of the precious metal in the perovskite host structure to test the synthetic strategy. We present an assessment of the use of the materials as visible light photocatalysts for hydrogen evolution from water.

2. Materials and Methods

2.1. Materials Synthesis

Chemicals used were sourced from chemical companies: Ta_2O_5 (Alfar Aesar, Heysham, UK, 99%), hydrated $IrCl_3$ (Johnson Matthey, London, UK, 52.29% Ir), NaOH (Fisher Scientific, Loughborough, UK, Laboratory Grade) and H_2O_2 (Sigma Aldrich, Gillingham, UK, 30% in water by volume). The substituted tantalates were synthesised by the reaction of 1.1 mmol tantalum (V) oxide in 10 mL of 10 M NaOH in 40 vol% H_2O_2, with a portion of the oxide replaced by a chosen amount of iridium (III) chloride. The use of H_2O_2 as an oxidant was based on our previous work on hydrothermal synthesis of iridium oxides to prevent the formation of metallic iridium [35]. After stirring the reagents for one hour in a 20 mL polytetrafluoroethylene container, the reaction mixture was sealed in a steel autoclave and heated at 240 °C for three days. After cooling naturally to room temperature, the powders were collected via vacuum filtration, and washed with 20 mL of 3 M HNO_3, followed by 20 mL of acetone, and then dried in air at 70 °C before further study.

2.2. Characterisation

Powder X-ray diffraction (XRD) data were recorded using a Panalytical Empyrean diffractometer (Malvern Panalytical, Malvern, UK) equipped with a Cu target, giving Cu Kα1/2 radiation. Data were recorded in reflection, Bragg–Brentano geometry from samples in silicon plates. The powder diffraction patterns were analysed using the GSAS-II software [36], with Pawley or Rietveld fits performed using published crystal structures of $NaTaO_3$ as a starting point to refine lattice parameters. Powder X-ray thermodiffractometry was performed using a Bruker D8 instrument (Bruker AXS Ltd., Coventry, UK) with Cu Kα1/2 radiation and fitted with an Anton Paar XRK 900 chamber (Anton Paar GmbH, Graz, Austria) and a VÅNTEC solid-state detector (Bruker AXS Ltd., Coventry, UK); this allowed heating of a sample from room temperature to 900 °C and XRD patterns were recorded at intervals of 50 °C. Before each data collection, the temperature was allowed to equilibrate for 5 min.

Scanning electron microscopy was performed using a ZEISS SUPRA 55-VP FEGSEM scanning electron microscope (Oberkochen, Germany) using a field emission gun with an accelerating voltage between 5 and 20 kV and fitted with an Oxford Instruments (Abingdon, UK) energy-dispersive X-ray spectroscopy (EDS) spectrometer that allows elemental composition analysis

Scanning transmission electron microscopy (STEM) was performed using a JEOL ARM200F double aberration corrected instrument (Welwyn Garden City, UK) operating at 200 kV. Specimens were dispersed by ultrasound in ethanol and dropped onto 3 mm lacey carbon grids supplied by Agar Scientific (Stansted, UK). Annular dark field STEM (ADF-STEM) images were obtained using a JEOL annular field detector at a probe current of ~23 pA with a convergence semi-angle of ~25 mrad. Energy-dispersive X-ray spectroscopy (EDS) measurements were carried out with an Oxford Instruments X-MaxN100TLE windowless silicon drift detector (Abingdon, UK) to determine the elemental composition and distribution. The program clTEM [37] was utilised to produce simulations of ADF-STEM images based on the crystal model oriented at (010) zone axis over an area of 1.5 nm by 2.6 nm. Inelastic phonon scattering was applied using the frozen phonon approximation method via an iterative approach to resemble molecular deviation from its equilibrium position under the electron probe at room temperature. The following thermal parameters ($<u^2>$) were used during simulation: 0.0166 Å^2 Na, 0.0048 Å^2 Ta, and 0.0080 Å^2 for O [38].

X-ray absorption near-edge spectroscopy (XANES) spectra at the iridium L_{III}-edge were recorded using Beamline B18 of the Diamond Light Source, Harwell, UK [39]. Data were collected in transmission mode from samples diluted with appropriate amounts of polyethylene powder (~20% sample by mass) and pressed into self-supporting discs around 1 mm thick. Incident energies were selected using a water-cooled, fixed-exit, double-crystal monochromator with Si(111) crystals. The beam was focused horizontally and vertically using a double toroidal mirror, coated with Pt, 25 m from the source, while a pair of smaller plane mirrors were used for harmonic rejection. The raw data were normalised using the software ATHENA (version 0.9.26) [40] to produce XANES spectra.

Diffuse reflectance spectroscopy was performed on powder samples using a Shimadzu UV-2600i UV-Vis spectrophotometer (Milton Keynes, UK). A barium sulfate standard was used as a baseline for the measurements.

The photocatalytic hydrogen production was performed in a Pyrex glass vessel with a top quartz window for vertical illumination in a closed-gas circulation system. In a typical run, 45 mg of catalyst was suspended in 20 vol% methanol solution in water. Then, 1.0 wt.% of Pt was loaded on the photocatalyst particles via photodeposition as a cocatalyst, to provide H_2 evolution sites. The glass reactor vessel was then sealed and repeatedly vacuumed by a rotary pump and purged with argon gas to remove the residual air. Subsequently, the reactor was irradiated with an 800 W Xe-Hg lamp (Newport, RI, USA) from the top, with full-spectrum intensity of 200 mW cm^{-2}. Experiments were also performed with visible light (420 nm cut-off filter) with intensity of 100 mW cm^{-2}. The infrared component in the radiation was removed by a circulating water filter. The temperature in the reactor was maintained at 25 °C by external water circulation. The amount of generated H_2 gas was quantitatively analysed every 2 h by a gas chromatograph (Shimadzu GC-2014; Molecular sieve 5A, TCD detector, Ar carrier gas) (Shimadzu, Duisburg, Germany).

3. Results

The iridium-containing samples are all produced as green powders, in contrast to the white parent $NaTaO_3$. The powder XRD pattern of the unsubstituted $NaTaO_3$ material was initially fitted using the orthorhombic space group, *Pbnm*, but a closer examination of the pattern revealed some mismatched peak intensities, which could be remedied by inclusion of a second *Cmcm* polymorph. A two-phase Rietveld analysis with atom coordinates and temperature factors fixed at the values reported in the literature [41], gave a satisfactory fit to the data, Figure 1a. The refined lattice parameters are in agreement with the literature values, Table 1, and the mixed-phase nature of $NaTaO_3$ has been previously seen, with samples prepared by solid-state synthesis showing ~45% of the *Cmcm* polymorph [38]. The smaller amount of the second polymorph that we observe (24.9%) would be consistent with the different synthesis route that we have used, but it is noteworthy that the hydrothermal method yields a proportion of the *Cmcm* polymorph, that has been defined as a high-temperature phase. The powder XRD patterns of the iridium substituted sodium tantalate materials can all be fitted using the single *Pbnm* polymorph, Figure 1b–f. The variation of lattice parameters with intended Ir content, Table 1 and Figure 2, provides evidence for the inclusion of Ir into the perovskite structure, in particular since all materials with the highest Ir content have larger orthorhombic *c* axes and corresponding larger unit cell volumes. It can be seen that the material with the smallest intended Ir amount has unit cell parameters rather similar to the parent $NaTaO_3$. The ionic radius of both Ir^{4+} (0.625 Å) and Ir^{5+} (0.570 Å) in octahedral coordination are similar to Ta^{5+} (0.640 Å) [42]. Given that it is most likely that the tantalum is replaced in the perovskite, we can propose that a smaller Ir content may be associated with the higher oxidation state, while addition of larger quantities results in inclusion of more Ir^{4+}, with associated charge-balancing oxide-ion vacancies, and hence modification of the lattice parameters. The iridium oxidation state will be discussed further below in the light of spectroscopic evidence. If larger amounts of iridium were added to the synthesis, then no further inclusion of Ir into the perovskite structure was observed and instead poorly crystalline IrO_2 was formed as a byproduct.

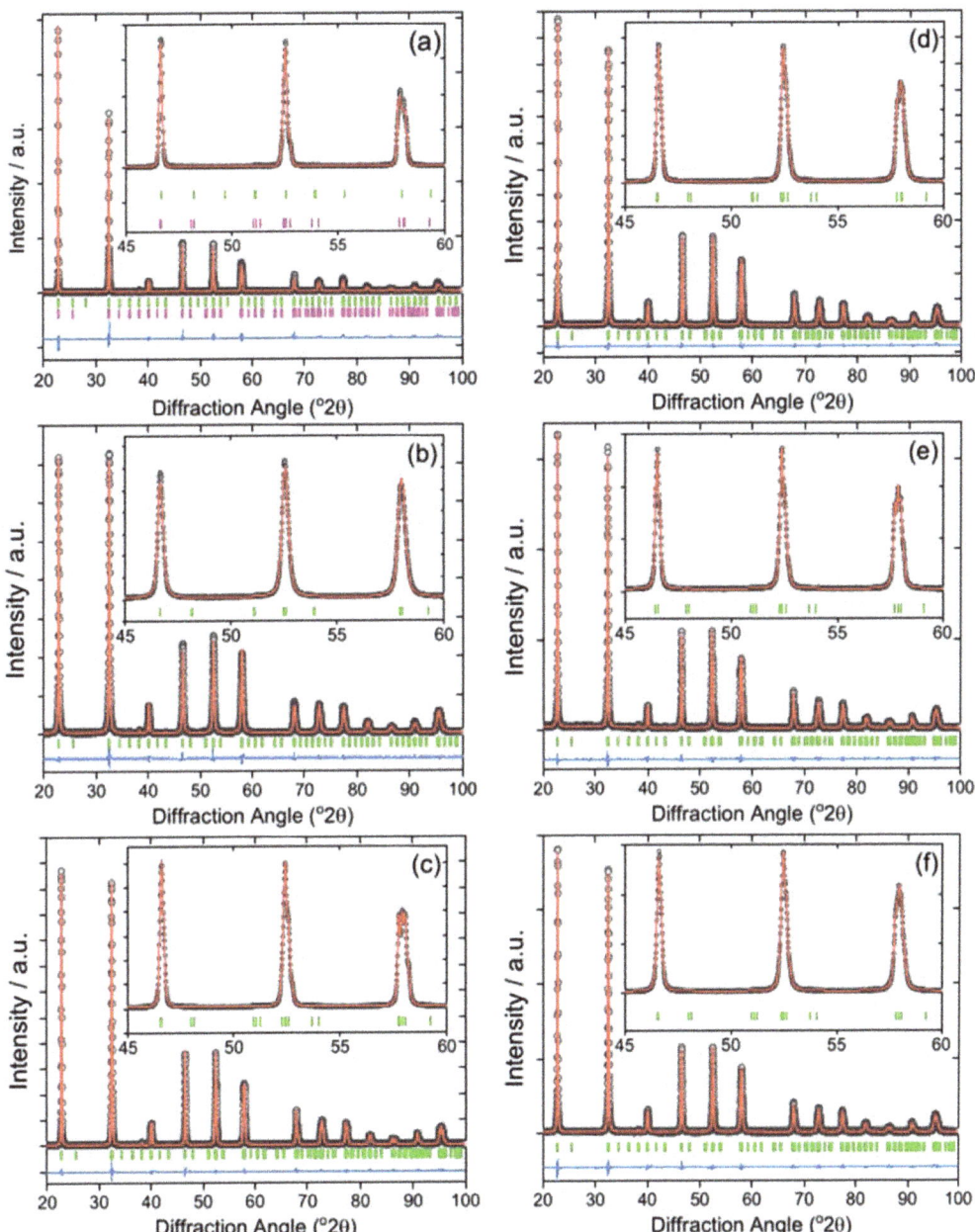

Figure 1. Fitted powder XRD patterns (λ = 1.5418 Å) of Ir-NaTaO$_3$ prepared by hydrothermal synthesis with intended Ir content of (**a**) 0% (R_w = 9.787), (**b**) 10% (R_w = 6.070), (**c**) 20% (R_w = 4.969), (**d**) 30% (R_w = 4.767), (**e**) 40% (R_w = 6.089), (**f**) 50 % (R_w = 6.767). The data are the black circles, the red line is the fitted pattern, the blue line is the difference curve, and the green tick marks are the positions of allowed Bragg reflections (space group, *Pbnm*). In (**a**), the second set of pink tick marks are due to the minor *Cmcm* phase, where a Rietveld fit was used. (**b**–**f**) are Pawley fits.

Table 1. Refined lattice parameters of substituted sodium tantalates from analysis of powder XRD data. The intended level of Ir substitution is indicated.

Material	Lattice Parameters		
	a/Å	b/Å	c/Å
NaTaO$_3$ [41]	5.48109(9)	5.52351(9)	7.79483(12)
NaTaO$_3$ [a]	5.48750(5)	5.52711(6)	7.7993(1)
NaTaO$_3$-10% Ir	5.4934(1)	5.5233(1)	7.7984(2)
NaTaO$_3$-20% Ir	5.49099(4)	5.53154(5)	7.8056(6)
NaTaO$_3$-30% Ir	5.49335(6)	5.53067(9)	7.8053(1)
NaTaO$_3$-40% Ir	5.49212(8)	5.53101(9)	7.8057(1)
NaTaO$_3$-50% Ir	5.4936(1)	5.5295(1)	7.8043(1)

[a]: The major (75.1%) *Pbnm* phase is shown, the second *Cmcm* phase has lattice parameters a = 7.795(2) Å, b = 7.789(3) Å, c = 7.791(5) Å (cf. literature values [38], a = 7.77927(8) Å, b = 7.7815(2) Å, c = 7.7899(1) Å).

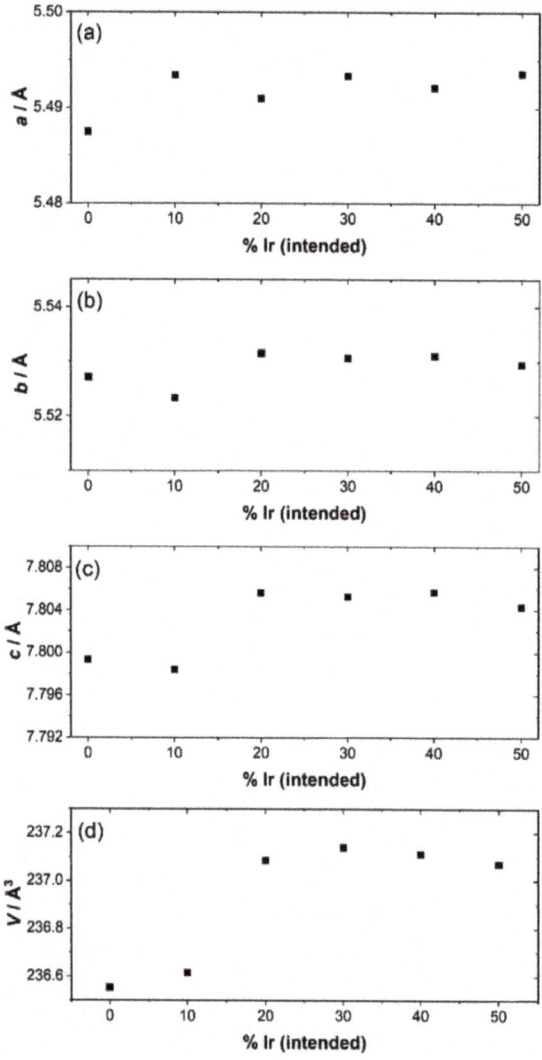

Figure 2. Variation of lattice parameters of Ir-NaTaO$_3$ materials with intended composition. (a) a, (b) b, (c) c, and (d) unit cell volume. The error bars are smaller than the data points.

The iridium-substituted sodium tantalates are formed as nanocubes with edges in the range of up to a few hundreds of nanometres, and typically ~100 nm, as seen by STEM, Figure 3. Although the cubes are not monodisperse in size, they have high crystallinity, as evidenced by high-resolution atomic-scale imaging, Figure 3g. EDS maps measured in STEM show an even dispersion of iridium in the substituted materials, with no evidence of clustering of iridium, nor any distinct particles of separate iridium-rich material, such as iridium oxide, or iridium metal, Figure 4. The cube-shaped particles are similar to hydrothermally synthesised sodium tantalate reported in literature [43–45], including those substituted with Bi^{3+} [46] and Cu^{2+} [47].

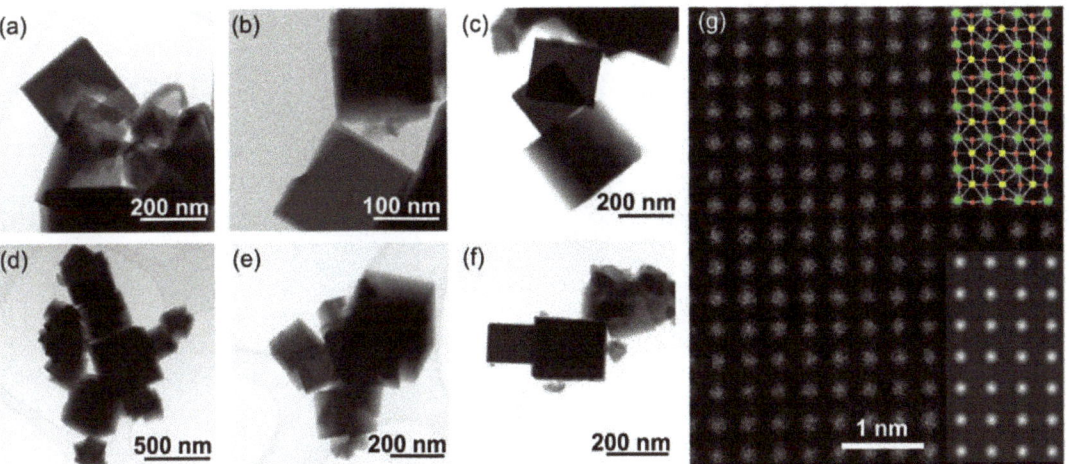

Figure 3. TEM images of Ir-substituted $NaTaO_3$ materials with intended Ir content of (**a**) 0%, (**b**) 10%, (**c**) 20%, (**d**) 30%, (**e**) 40% and (**f**) 50%. See Supporting Information for further images of the specimens. (**g**) shows a high-resolution TEM image of the 40% Ir material, with the simulated image (lower inset) and the corresponding crystal structure (upper inset, where green atoms are Ta(Ir), yellow are sodium, and red are oxygen.

Quantification of the EDS performed using scanning electron microscopy reveals that the iridium content of the materials is somewhat lower than the amounts used in the reactions used to prepare them, Table 2. This is consistent with the strong colour of the filtrate during washing of the samples, which suggests not all the precious metal was incorporated into the final material. Nevertheless, the amount of iridium does increase proportionally with increasing amount used in synthesis.

Three representative samples were studied in more detail to understand the chemical state of iridium in the samples. XANES spectra recorded at the Ir L_{III}-edge, Figure 5, were used to determine the average oxidation state of iridium by comparison to reference materials. Here, the position of the white line was used as a measure of edge shift, as in our previous work [48], and the materials $IrCl_3$, IrO_2, and $BaNa_{0.5}Ir_{0.5}O_{3-x}$ ($x = 0.525$) were used as calibrants for oxidation states +3, +4, and +4.9, respectively. This shows that the iridium in the materials with higher levels of Ir has an oxidation state of close to +4, as in IrO_2, Figure 6. The sample with the lowest iridium content appears to have a higher average Ir oxidation state, but still lower than +5. The chemistry of Ir^{4+} and Ir^{5+} in oxides is almost exclusively associated with octahedral coordination [49] and so it is anticipated that iridium occupies the B-site of the perovskite structure.

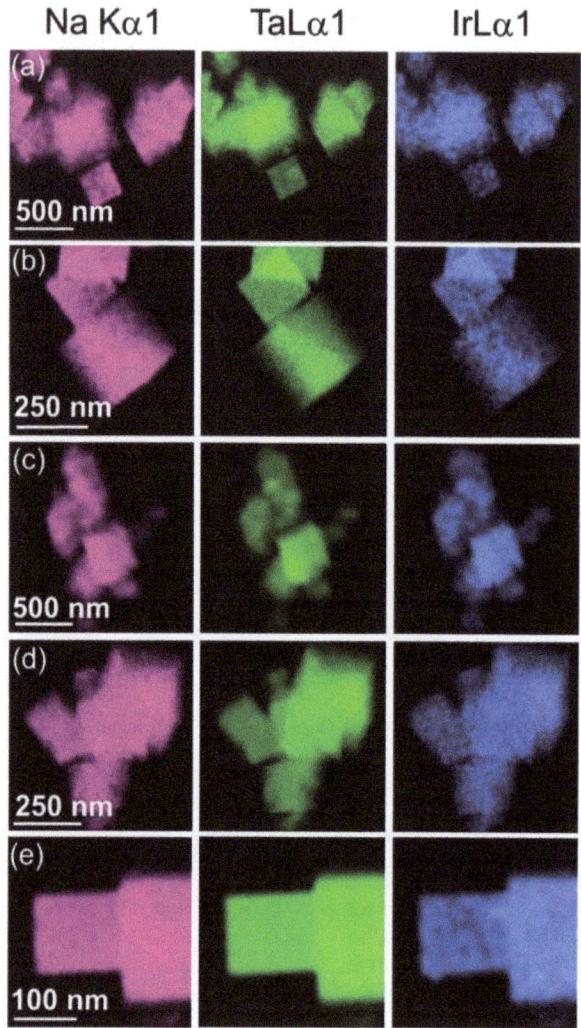

Figure 4. EDS maps performed using TEM of Ir-substituted $NaTaO_3$ materials with intended Ir content of (**a**) 10%, (**b**) 20%, (**c**) 30%, (**d**) 40%, and (**e**) 50%.

Table 2. EDXA of Ir-substituted $NaTaO_3$ materials performed using scanning electron microscopy.

Intended Ir Substitution	EDS Results Tantalum/%	EDS Results Iridium/%	Determined Formula
10%	97.6	2.4	$NaTa_{0.98}Ir_{0.02}O_3$
20%	95.3	4.7	$NaTa_{0.95}Ir_{0.05}O_3$
30%	91.6	8.4	$NaTa_{0.92}Ir_{0.08}O_3$
40%	87.8	12.2	$NaTa_{0.88}Ir_{0.12}O_3$
50%	85.1	14.9	$NaTa_{0.85}Ir_{0.15}O_3$

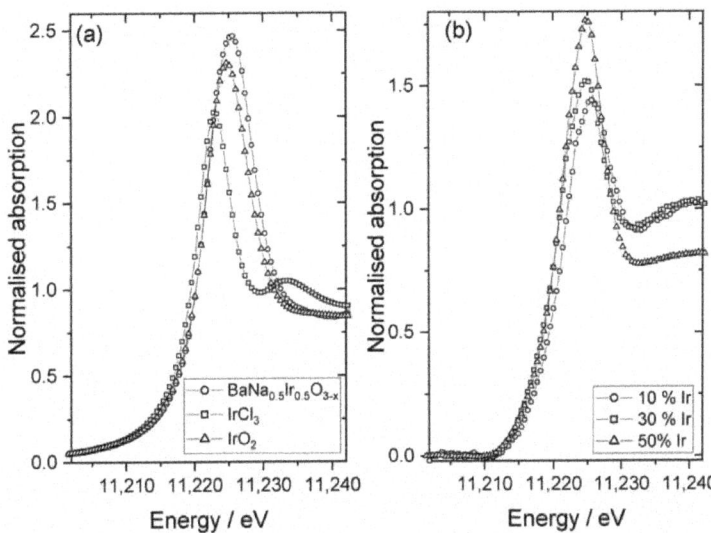

Figure 5. Ir L_{III}-edge XANES spectra: (**a**) reference materials for oxidation state calibration and (**b**) Ir-substituted NaTaO$_3$ materials labelled according to the proportion of Ir added in synthesis.

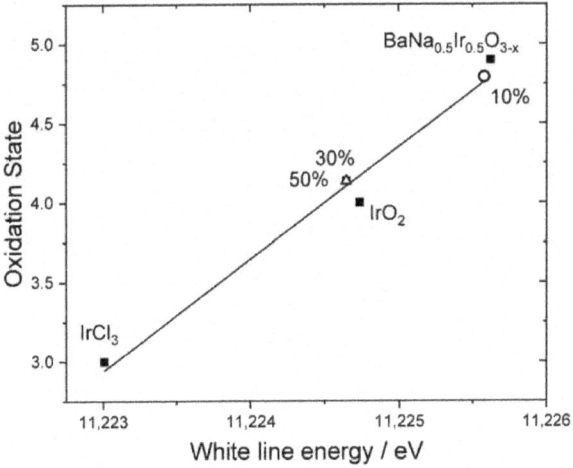

Figure 6. Oxidation states of Ir-substituted NaTaO$_3$ materials labelled according to the proportion of Ir added in synthesis, plotted with reference materials. The line is the linear regression fit to the points from the reference materials.

The results of the XANES analysis are consistent with the results from powder XRD, as described above. The sample with the smallest amount of iridium contains a significant proportion of Ir^{5+}, which would be consistent with the small changes in lattice parameter, but the materials with greater Ir content contain Ir^{4+}, which is larger in ionic radius and would require charge-balancing oxide-ion vacancies, and hence rather different lattice parameters. The reason for this difference is not apparent from the data we have measured, and it may be the case that a greater surface accumulation of Ir at the higher amounts used leads to a change in average oxidation state of iridium, or that the redox chemistry in solution leads to only small amounts of Ir^{5+} that are available for inclusion in the perovskite structure.

Figure 7 shows diffuse-reflectance UV-vis spectra measured from the same three samples, along with the spectrum of NaTaO$_3$ for comparison. The parent perovskite NaTaO$_3$ shows a spectrum very similar to that reported from the literature [50], and a Tauc plot analysis gives a direct band gap of 4.0 eV, as expected. The iridium-containing materials show significant absorption in the visible part of the spectrum, which illustrates the green colour of the powdered samples, with strong absorptions in the 450 nm and 700 nm regions. Interestingly, the sample with smallest iridium content shows different absorption maxima to the samples with the higher iridium content, suggesting a different electronic state for the iridium cations in the solid. This is entirely consistent with the XANES and powder XRD analysis presented above.

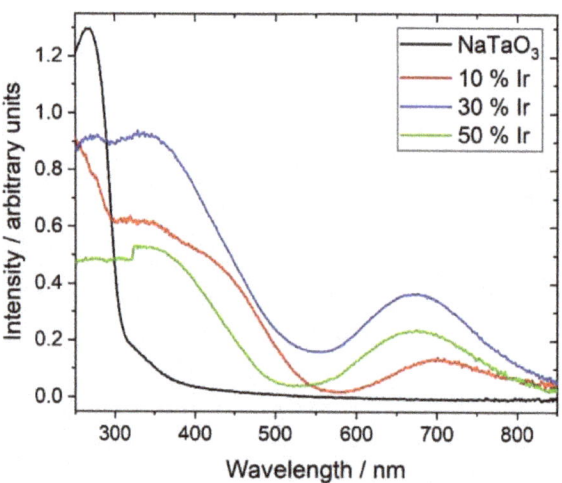

Figure 7. Diffuse-reflectance UV-vis spectra of Ir-substituted NaTaO$_3$ materials labelled according to the proportion of Ir added in synthesis.

We examined the materials use in photocatalysis by studying hydrogen generation from water using the material with highest iridium content. To optimise the activity, the powders were heated in air at 500 °C to remove any surface-bound water and minimise any possible hydroxyl defects that may be anticipated in hydrothermally produced oxide perovskites [51–53]. To verify that this did not result in phase separation, X-ray powder thermodiffractometry was carried out with heating from room temperature to 900 °C, Figure 8. This shows that the perovskite structure remains unchanged until above 600 °C and only then are the strongest Bragg peaks of face-centred cubic iridium metal [54] observed.

Photocatalysis results reveal that the parent NaTaO$_3$ when loaded with 1 wt.% Pt shows a low activity in full-spectrum irradiation, but no detectable activity under visible light towards hydrogen evolution in aqueous methanol, Figure 9. In contrast, the full-spectrum irradiation of Na(Ta,Ir)O$_3$ yields approximately 15 times the yield of hydrogen, and, notably, under visible light shows yields of hydrogen comparable to the un-substituted material in UV + visible light. These preliminary results demonstrate the effectiveness of iridium substitution in tuning optical properties of NaTaO$_3$ to provide visible light activity. We note that the catalytic production of hydrogen is not linear, but this has been reported previously for Bi-containing NaTaO$_3$ studied under the same conditions [55]. The cause of this is not known at present, but may be related to a change of the catalyst surface with time, such as restructuring of surface defects.

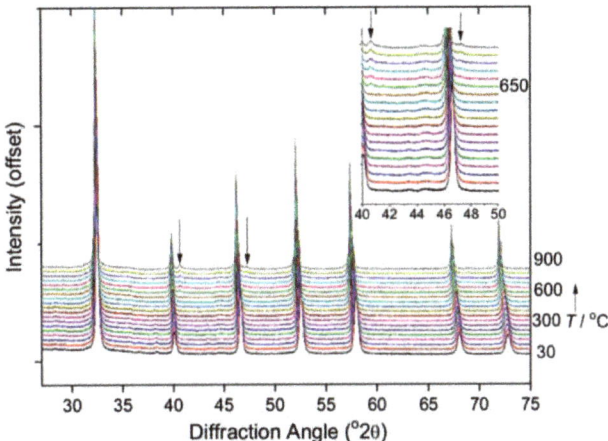

Figure 8. Variable temperature XRD on heating in air showing the stability of Na(Ta,Ir)O$_3$ until above 600 °C when iridium metal is seen, with its two strongest Bragg peaks, (111) and (200), indicated by the arrows.

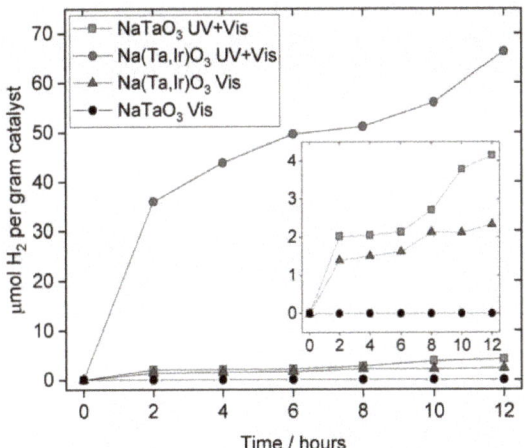

Figure 9. Photocatalytic hydrogen evolution observed from NaTaO$_3$ and Ir-substituted sample.

4. Discussion

The observation of iridium in oxidation states between +4 and +5 in the perovskite structure is consistent with other iridium oxides that contain octahedrally coordinated Ir and that have been prepared under similar hydrothermal conditions in alkali solutions in the presence of peroxide as oxidant [35,48]. It is noteworthy that for the material with the lowest iridium concentration a higher oxidation state is observed spectroscopically, while increasing Ir concentration lowers the average Ir oxidation state (corroborated indirectly by powder XRD). Interestingly, for other iridium-substituted perovskites reported in the literature via other synthesis methods, different oxidation states of Ir may be found. For example, Calì et al. observed Ir^{3+} using X-ray photoelectron spectroscopy in 5 mol% Ir-substituted SrTiO$_3$ that had been prepared by solid-state synthesis at 1340 °C [31]. On the other hand, Kawasaki et al. found Ir^{4+} in samples of Ir-substituted SrTiO$_3$ prepared by solid-state synthesis at 1100 °C [32]. While the valence states of the host perovskite material might influence the Ir oxidation state, the synthesis method is likely to play an important role, and hence the hydrothermal method offers an alternative approach to solid-state synthesis that might allow a more tunable synthesis of Ir-containing perovskites.

The presence of Ir^{4+} in $NaTaO_3$ means that charge-balancing defects must be necessary. In other precious-metal-doped perovskites, surface oxide defects have been inferred, for example, in $Ir-SrTiO_3$ [32]. In the case of $NaTaO_3$ prepared by hydrothermal synthesis, inclusion of Bi^{3+} was accompanied by oxide defects, based on evidence from photoluminescence spectra [46]. It is also a possibility that the hydrothermal route allows charge-balance by inclusion of hydroxide ions in place of oxide: these have been detected in other perovskites prepared by this method, such as $Na_{0.5}Bi_{0.5}TiO_3$ [51,53] and $KNbO_3$ [52]. Defects such as anion vacancies may be in part responsible for the enhanced light absorption and photocatalytic activity of the substituted materials [56]. Further in-depth experimental work is needed to explore the role of defects and how these are modified in the presence of the co-catalyst, and with heat treatment. Methods such X-ray photoelectron spectroscopy would be useful to example the nature of oxide defects, such as lattice vacancies [57].

The materials we have prepared show stability on heating in air to only 600 °C, which demonstrates how it would be impossible to prepare the same samples by conventional solid-state synthesis, or even coprecipitation methods that require an annealing step. It is interesting to note that Rh^{3+}-containing $BaTiO_3$ prepared by an oxalate-aided coprecipitation route showed extrusion of rhodium metal when heated above its synthesis temperature [30]. This illustrates the tendency for the precious metals to be reduced, but also how this may be aided by embedding in a host lattice by replacement of a cation of higher charge.

Our photocatalysis results show the beneficial effect of Ir-inclusion in $NaTaO_3$ with greater than ten-fold enhancement of activity towards hydrogen evolution in UV+visible radiation, a significant part of which can be ascribed to visible light absorption. The results obtained are a similar order of magnitude to those produced from Bi^{3+}-$NaTaO_3$, studied using the same experimental protocol [55], although they do not reach such a high hydrogen yield. However, there is scope for further optimisation, as the previous work showed how Pt loading and substitution level of the perovskite should be adjusted to improve yield, and that the surface area of the materials should be controlled for maximum catalytic efficiency. In optimising photocatalytic properties for hydrogen evolution of $Ir-SrTiO_3$ the oxidation state of Ir is important [32], but also loading with surface Ir metal can optimise properties [58]. Hence, our work provides a convenient synthesis method that may allow further tuning of properties of materials for visible-light photocatalysis.

5. Conclusions

We have presented a hydrothermal synthesis route to introduce iridium into a prototypical perovskite structure $NaTaO_3$ that uses mild reaction conditions in a single-step process. A comprehensive set of characterisation data shows that the iridium replaces tantalum to give small adjustments in lattice parameters. For the samples that contain the most iridium, the substituent is present in the +4 oxidation state, which implies the presence of defects for charge balance, but this is in line with other perovskites that contain precious-metal substituents, and for other oxides that have been prepared by hydrothermal chemistry. Our preliminary photocatalysis results show promising properties for visible-light generation of hydrogen from water, but beyond this the materials may prove useful for other applications in heterogeneous catalysis, either as prepared, or upon reduction to extrude the iridium as supported nanocrystals. Finally, we note that the redox chemistry taking place during synthesis as the Ir^{3+} precursor is oxidised is likely to be complex, and this is where the origin of the substitutional chemistry occurs. Further work is needed to understand the mechanisms of hydrothermal crystallisation of oxides, and the evolution of the solution chemistry as a solid is formed is largely unexplored at present.

Author Contributions: Conceptualization, R.I.W.; materials synthesis, D.L.B. and C.D.V.; data collection and analysis; D.L.B., J.A.C. and R.J.K.; writing—original draft preparation, R.I.W.; writing—reviewing and editing, D.L.B., J.A.C., R.I.W.; funding acquisition, R.I.W. All authors have read and agreed to the published version of the manuscript.

Funding: The research work was supported by EPSRC (University of Warwick Doctoral Training account EP/K503204/1).

Data Availability Statement: The research data underpinning this article can be accessed at: http://wrap.warwick.ac.uk/153846/ (accessed on 9 June 2021).

Acknowledgments: We thank the EPSRC for funding a PhD studentship to DLB in collaboration with Johnson Matthey, and the University of Warwick Undergraduate Research Support Scheme for provision of a scholarship to CDV. We are grateful to Chen Zhong and Qiuling Tay at NTU Singapore for measurement of photocatalysis, Diamond Light Source for provision of beamtime (SP8708), Tom Chamberlain for measuring UV-Vis spectra, and Giannantonio Cibin for assistance with measuring the XANES spectra. Some of the equipment used in this work was provided by the University of Warwick's Research Technology Platforms.

Conflicts of Interest: The authors declare no conflict of interest.

References

1. Walton, R.I. Perovskite Oxides Prepared by Hydrothermal and Solvothermal Synthesis: A Review of Crystallisation, Chemistry, and Compositions. *Chem. Eur. J.* **2020**, *26*, 9041–9069. [CrossRef] [PubMed]
2. Sato, Y.; Aoki, M.; Teranishi, R.; Kaneko, K.; Takesada, M.; Moriwake, H.; Takashima, H.; Hakuta, Y. Atomic-Scale Observation of Titanium-Ion Shifts in Barium Titanate Nanoparticles: Implications for Ferroelectric Applications. *ACS Appl. Nano Mater.* **2019**, *2*, 5761–5768. [CrossRef]
3. Morita, T. Piezoelectric Materials Synthesized by the Hydrothermal Method and Their Applications. *Materials* **2010**, *3*, 5236–5245. [CrossRef] [PubMed]
4. Sardar, K.; Lees, M.R.; Kashtiban, R.J.; Sloan, J.; Walton, R.I. Direct Hydrothermal Synthesis and Physical Properties of Rare-Earth and Yttrium Orthochromite Perovskites. *Chem. Mat.* **2011**, *23*, 48–56. [CrossRef]
5. Diodati, S.; Walton, R.I.; Mascotto, S.; Gross, S. Low-Temperature wet chemistry synthetic approaches towards ferrites. *Inorg. Chem. Front.* **2020**, *7*, 3282–3314. [CrossRef]
6. Eckert, J.O.; HungHouston, C.C.; Gersten, B.L.; Lencka, M.M.; Riman, R.E. Kinetics and mechanisms of hydrothermal synthesis of barium titanate. *J. Amer. Ceram. Soc.* **1996**, *79*, 2929–2939. [CrossRef]
7. Sōmiya, S.; Roy, R. Hydrothermal synthesis of fine oxide powders. *Bull. Mater. Sci.* **2000**, *23*, 453–460. [CrossRef]
8. Riman, R.E.; Suchanek, W.L.; Lencka, M.M. Hydrothermal crystallization of ceramics. *Ann. Chim. Sci. Mater.* **2002**, *27*, 15–36. [CrossRef]
9. Walton, R.I. Subcritical solvothermal synthesis of condensed inorganic materials. *Chem. Soc. Rev.* **2002**, *31*, 230–238. [CrossRef] [PubMed]
10. Komarneni, S. Nanophase materials by hydrothermal, microwave-hydrothermal and microwave-solvothermal methods. *Curr. Sci.* **2003**, *85*, 1730–1734.
11. Yoshimura, M.; Byrappa, K. Hydrothermal processing of materials: Past, present and future. *J. Mater. Sci.* **2008**, *43*, 2085–2103. [CrossRef]
12. Kumada, N. Preparation and crystal structure of new inorganic compounds by hydrothermal reaction. *J. Ceram. Soc. Jap.* **2013**, *121*, 135–141. [CrossRef]
13. Shi, J.; Li, J.; Huang, X.; Tan, Y. Synthesis and enhanced photocatalytic activity of regularly shaped Cu_2O nanowire polyhedra. *Nano Res.* **2011**, *4*, 448–459. [CrossRef]
14. Deng, S.; Tjoa, V.; Fan, H.M.; Tan, H.R.; Sayle, D.C.; Olivo, M.; Mhaisalkar, S.; Wei, J.; Sow, C.H. Reduced Graphene Oxide Conjugated Cu_2O Nanowire Mesocrystals for High-Performance NO_2 Gas Sensor. *J. Am. Chem. Soc.* **2012**, *134*, 4905–4917. [CrossRef] [PubMed]
15. Ding, Y.; Guo, X.; Du, B.; Hu, X.; Yang, X.; He, Y.; Zhou, Y.; Zang, Z. Low-operating temperature ammonia sensor based on Cu_2O nanoparticles decorated with p-type MoS_2 nanosheets. *J. Mater. Chem. C* **2021**, *9*, 4838–4846. [CrossRef]
16. Chen, S.; Zhu, J.W.; Wu, X.D.; Han, Q.F.; Wang, X. Graphene Oxide-MnO_2 Nanocomposites for Supercapacitors. *ACS Nano* **2010**, *4*, 2822–2830. [CrossRef]
17. Modeshia, D.R.; Darton, R.J.; Ashbrook, S.E.; Walton, R.I. Control of polymorphism in $NaNbO_3$ by hydrothermal synthesis. *Chem. Commun.* **2009**, 68–70. [CrossRef] [PubMed]
18. Skjaervø, S.L.; Sommer, S.; Norby, P.; Bojesen, E.D.; Grande, T.; Iversen, B.B.; Einarsrud, M.A. Formation mechanism and growth of $MNbO_3$, M = K, Na by insitu X-ray diffraction. *J. Amer. Ceram. Soc.* **2017**, *100*, 3835–3842. [CrossRef]
19. Skjaervø, S.L.; Wells, K.H.; Sommer, S.; Vu, T.D.; Tolchard, J.R.; van Beek, W.; Grande, T.; Iversen, B.B.; Einarsrud, M.A. Rationalization of Hydrothermal Synthesis of $NaNbO_3$ by Rapid in Situ Time-Resolved Synchrotron X-ray Diffraction. *Cryst. Growth Des.* **2018**, *18*, 770–774. [CrossRef]
20. Song, H.W.; Ma, W.H. Hydrothermal synthesis of submicron $NaNbO_3$ powders. *Ceram. Int.* **2011**, *37*, 877–882. [CrossRef]
21. Kumada, N.; Dong, Q.; Yonesaki, Y.; Takei, T.; Kinomura, N. Hydrothermal synthesis of $NaNbO_3$-morphology change by starting compounds. *J. Ceram. Soc. Jap.* **2011**, *119*, 483–485. [CrossRef]

22. Shi, G.D.; Wang, J.H.; Wang, H.L.; Wu, Z.J.; Wu, H.P. Hydrothermal synthesis of morphology-controlled KNbO$_3$, NaNbO$_3$, and (K,Na)NbO$_3$ powders. *Ceram. Int.* **2017**, *43*, 7222–7230. [CrossRef]
23. Nakashima, K.; Toshima, Y.; Kobayashi, Y.; Kakihana, M. Effects of raw materials on NaNbO$_3$ nanocube synthesis via the solvothermal method. *J. As. Ceram. Soc.* **2019**, *7*, 36–41. [CrossRef]
24. Kanie, K.; Numamoto, Y.; Tsukamoto, S.; Takahashi, H.; Mizutani, H.; Terabe, A.; Nakaya, M.; Tani, J.; Muramatsu, A. Hydrothermal Synthesis of Sodium and Potassium Niobates Fine Particles and Their Application to Lead-Free Piezoelectric Material. *Mater. Trans.* **2011**, *52*, 2119–2125. [CrossRef]
25. Pan, H.; Zhu, G.S.; Chao, X.L.; Wei, L.L.; Yang, Z.P. Properties of NaNbO$_3$ powders and ceramics prepared by hydrothermal reaction. *Mater. Chem. Phys.* **2011**, *126*, 183–187. [CrossRef]
26. Fukada, M.; Shibata, K.; Imai, T.; Yamazoe, S.; Hosokawa, S.; Wada, T. Fabrication of lead-free piezoelectric NaNbO$_3$ ceramics at low temperature using NaNbO$_3$ nanoparticles synthesized by solvothermal method. *J. Ceram. Soc. Jap.* **2013**, *121*, 116–119 [CrossRef]
27. Shi, H.F.; Li, X.K.; Wang, D.F.; Yuan, Y.P.; Zou, Z.G.; Ye, J.H. NaNbO$_3$ Nanostructures: Facile Synthesis, Characterization, and Their Photocatalytic Properties. *Catal. Lett.* **2009**, *132*, 205–212. [CrossRef]
28. Kanhere, P.; Chen, Z. A Review on Visible Light Active Perovskite-Based Photocatalysts. *Molecules* **2014**, *19*, 19995–20022 [CrossRef] [PubMed]
29. Polat, O.; Durmus, Z.; Coskun, F.M.; Coskun, M.; Turut, A. Engineering the band gap of LaCrO$_3$ doping with transition metals (Co, Pd, and Ir). *J. Mater. Sci.* **2018**, *53*, 3544–3556. [CrossRef]
30. Lontio Fomekong, R.; You, S.; Enrichi, F.; Vomiero, A.; Saruhan, B. Impact of Oxalate Ligand in Co-Precipitation Route on Morphological Properties and Phase Constitution of Undoped and Rh-Doped BaTiO$_3$ Nanoparticles. *Nanomaterials* **2019**, *9*, 1697 [CrossRef] [PubMed]
31. Calì, E.; Kerherve, G.; Naufal, F.; Kousi, K.; Neagu, D.; Papaioannou, E.I.; Thomas, M.P.; Guiton, B.S.; Metcalfe, I.S.; Irvine, J.T.S.; et al. Exsolution of Catalytically Active Iridium Nanoparticles from Strontium Titanate. *ACS Appl. Mater. Interfaces* **2020**, *12*, 37444–37453. [CrossRef]
32. Kawasaki, S.; Takahashi, R.; Akagi, K.; Yoshinobu, J.; Komori, F.; Horiba, K.; Kumigashira, H.; Iwashina, K.; Kudo, A.; Lippmaa, M. Electronic Structure and Photoelectrochemical Properties of an Ir-Doped SrTiO$_3$ Photocatalyst. *J. Phys. Chem. C* **2014**, *118*, 20222–20228. [CrossRef]
33. Iwase, A.; Saito, K.; Kudo, A. Sensitization of NaMO$_3$ (M: Nb and Ta) Photocatalysts with Wide Band Gaps to Visible Light by Ir Doping. *Bull. Chem. Soc. Jap.* **2009**, *82*, 514–518. [CrossRef]
34. Kudo, A.; Yoshino, S.; Tsuchiya, T.; Udagawa, Y.; Takahashi, Y.; Yamaguchi, M.; Ogasawara, I.; Matsumoto, H.; Iwase, A. Z-scheme photocatalyst systems employing Rh- and Ir-doped metal oxide materials for water splitting under visible light irradiation. *Farad. Disc.* **2019**, *215*, 313–328. [CrossRef] [PubMed]
35. Sardar, K.; Fisher, J.; Thompsett, D.; Lees, M.R.; Clarkson, G.J.; Sloan, J.; Kashtiban, R.J.; Walton, R.I. Structural variety in iridate oxides and hydroxides from hydrothermal synthesis. *Chem. Sci.* **2011**, *2*, 1573–1578. [CrossRef]
36. Toby, B.H.; von Dreele, R.B. GSAS-II: The genesis of a modern open-source all purpose crystallography software package. *J. Appl. Cryst.* **2013**, *46*, 544–549. [CrossRef]
37. Peters, J.J.P. clTEM. Available online: https://jjppeters.github.io/clTEM/ (accessed on 19 March 2021).
38. Knight, K.S.; Kennedy, B.J. Phase coexistence in NaTaO$_3$ at room temperature; a high resolution neutron powder diffraction study *Solid State Sci.* **2015**, *43*, 15–21. [CrossRef]
39. Dent, A.J.; Cibin, G.; Ramos, S.; Smith, A.D.; Scott, S.M.; Varandas, L.; Pearson, M.R.; Krumpa, N.A.; Jones, C.P.; Robbins, P.E. B18: A core XAS spectroscopy beamline for Diamond. *J. Phys. Conf. Ser.* **2009**, *190*, 012039. [CrossRef]
40. Ravel, B.; Newville, M. ATHENA, ARTEMIS, HEPHAESTUS: Data analysis for X-ray absorption spectroscopy using IFEFFIT *J. Synchrotron Rad.* **2005**, *12*, 537–541. [CrossRef] [PubMed]
41. Mitchell, R.H.; Liferovich, R.P. A structural study of the perovskite series Ca$_{1-x}$Na$_x$Ti$_{1-x}$Ta$_x$O$_3$. *J. Solid State Chem.* **2004**, *177*, 4420–4427. [CrossRef]
42. Shannon, R. Revised effective ionic radii and systematic studies of interatomic distances in halides and chalcogenides. *Acta Crystallogr. Sect. A* **1976**, *32*, 751–767. [CrossRef]
43. Gao, Y.; Su, Y.G.; Meng, Y.; Wang, S.W.; Jia, Q.Y.; Wang, X.J. Preparation and Photocatalytic Mechanism of Vanadium Doped NaTaO$_3$ Nanoparticles. *Integr. Ferroelectr.* **2011**, *127*, 106–115. [CrossRef]
44. Gao, R.; Zhou, S.X.; Li, W.; Chen, M.; Wu, L.M. Facile synthesis of uniform and well-defined single-crystal sodium tantalate cubes and their assembly into oriented two-dimensional nanofilm. *Cryst. Eng. Commun.* **2012**, *14*, 7031–7035. [CrossRef]
45. Grewe, T.; Meier, K.; Tuysuz, H. Photocatalytic hydrogen production over various sodium tantalates. *Catal. Today* **2014**, *225*, 142–148. [CrossRef]
46. Wang, X.J.; Bai, H.L.; Meng, Y.; Zhao, Y.H.; Tang, C.H.; Gao, Y. Synthesis and Optical Properties of Bi^{3+} Doped NaTaO$_3$ Nano-Size Photocatalysts. *J. Nanosci. Nanotechnol.* **2010**, *10*, 1788–1793. [CrossRef]
47. Liu, Y.L.; Su, Y.G.; Han, H.; Wang, X.J. Hydrothermal Preparation of Copper Doped NaTaO$_3$ Nanoparticles and Study on the Photocatalytic Mechanism. *J. Nanosci. Nanotechnol.* **2013**, *13*, 853–857. [CrossRef]

28. Sardar, K.; Petrucco, E.; Hiley, C.I.; Sharman, J.D.B.; Wells, P.P.; Russell, A.E.; Kashtiban, R.J.; Sloan, J.; Walton, R.I. Water-Splitting Electrocatalysis in Acid Conditions Using Ruthenate-Iridate Pyrochlores. *Angew. Chem. Int. Edit.* **2014**, *53*, 10960–10964. [CrossRef]
29. Müller-Buschbaum, H. On the crystal chemistry of Oxoiridates. *Z. Anorg. Allg. Chem.* **2005**, *631*, 1005–1028. [CrossRef]
30. Kato, H.; Kudo, A. Highly efficient decomposition of pure water into H_2 and O_2 over $NaTaO_3$ photocatalysts. *Catal. Lett.* **1999**, *58*, 153–155. [CrossRef]
31. Kumada, N.; Morozumi, Y.; Yonesaki, Y.; Takei, T.; Kinomura, N.; Hayashi, T. Preparation of $Na_{0.5}Bi_{0.5}TiO_3$ by hydrothermal reaction. *J. Ceram. Soc. Jap.* **2008**, *116*, 1238–1240. [CrossRef]
32. Handoko, A.D.; Goh, G.K.L.; Chew, R.X. Piezoelectrically active hydrothermal $KNbO_3$ thin films. *Cryst. Eng. Commun.* **2012**, *14*, 421–427. [CrossRef]
33. O'Brien, A.; Woodward, D.I.; Sardar, K.; Walton, R.I.; Thomas, P.A. Inference of oxygen vacancies in hydrothermal $Na_{0.5}Bi_{0.5}TiO_3$. *Appl. Phys. Lett.* **2012**, *101*, 142902. [CrossRef]
34. Singh, H.P. Determination of thermal expansion of germanium, rhodium and iridium by X-rays. *Acta Crystallogr. Sect. A* **1968**, *24*, 469–471. [CrossRef]
35. Kanhere, P.; Zheng, J.; Chen, Z. Visible light driven photocatalytic hydrogen evolution and photophysical properties of Bi^{3+} doped $NaTaO_3$. *Int. J. Hydro. Ener.* **2012**, *37*, 4889–4896. [CrossRef]
36. Onishi, H. Sodium Tantalate Photocatalysts Doped with Metal Cations: Why Are They Active for Water Splitting? *ChemSusChem* **2019**, *12*, 1825–1834. [CrossRef]
37. Alves, G.A.S.; Centurion, H.A.; Sambrano, J.R.; Ferrer, M.M.; Gonçalves, R.V. Band Gap Narrowing of Bi-Doped $NaTaO_3$ for Photocatalytic Hydrogen Evolution under Simulated Sunlight: A Pseudocubic Phase Induced by Doping. *ACS Appl. Energy Mater.* **2021**, *4*, 671–679. [CrossRef]
38. Suzuki, S.; Matsumoto, H.; Iwase, A.; Kudo, A. Enhanced H_2 evolution over an Ir-doped $SrTiO_3$ photocatalyst by loading of an Ir cocatalyst using visible light up to 800 nm. *Chem. Commun.* **2018**, *54*, 10606–10609. [CrossRef]

Article

Fabrication of Liquid Scintillators Loaded with 6-Phenylhexanoic Acid-Modified ZrO$_2$ Nanoparticles for Observation of Neutrinoless Double Beta Decay

Akito Watanabe [1,*], Arisa Magi [1], Akira Yoko [2], Gimyeong Seong [3], Takaaki Tomai [4], Tadafumi Adschiri [2,3,4], Yamato Hayashi [1], Masanori Koshimizu [1], Yutaka Fujimoto [1] and Keisuke Asai [1]

1. Department of Applied Chemistry, Graduate School of Engineering, Tohoku University, 6-6-07 Aoba, Aramaki, Aoba-ku, Sendai 980-8579, Japan; arsm723.tmsj@gmail.com (A.M.); hayashi@aim.che.tohoku.ac.jp (Y.H.); koshi@qpc.che.tohoku.ac.jp (M.K.); fuji-you@qpc.che.tohoku.ac.jp (Y.F.); asai@qpc.che.tohoku.ac.jp (K.A.)
2. WPI-AIMR, Tohoku University, 2-1-1 Katahira, Aoba-ku, Sendai 980-8577, Japan; akira.yoko.c7@tohoku.ac.jp (A.Y.); tadafumi.ajiri.b1@tohoku.ac.jp (T.A.)
3. New Industry Creation Hatchery Center, Tohoku University, 6-6-10 Aoba, Aramaki, Aoba-ku, Sendai 980-8579, Japan; kimei.sei.c6@tohoku.ac.jp
4. Institute of Multidisciplinary Research for Advanced Materials, Tohoku University, 2-1-1 Katahira, Aoba-ku, Sendai 980-8577, Japan; takaaki.tomai.e6@tohoku.ac.jp
* Correspondence: akito.watanabe.p7@dc.tohoku.ac.jp; Tel.: +81-22-795-7219

Abstract: The observation of neutrinoless double beta decay is an important issue in nuclear and particle physics. The development of organic liquid scintillators with high transparency and a high concentration of the target isotope would be very useful for neutrinoless double beta decay experiments. Therefore, we propose a liquid scintillator loaded with metal oxide nanoparticles containing the target isotope. In this work, 6-phenylhexanoic acid-modified ZrO$_2$ nanoparticles, which contain ^{96}Zr as the target isotope, were synthesized under sub/supercritical hydrothermal conditions. The effects of the synthesis temperature on the formation and surface modification of the nanoparticles were investigated. Performing the synthesis at 250 and 300 °C resulted in the formation of nanoparticles with smaller particle sizes and higher surface modification densities than those prepared at 350 and 400 °C. The highest modification density (3.1 ± 0.2 molecules/nm^2) and Zr concentration of (0.33 ± 0.04 wt.%) were obtained at 300 °C. The surface-modified ZrO$_2$ nanoparticles were dispersed in a toluene-based liquid scintillator. The liquid scintillator was transparent to the scintillation wavelength, and a clear scintillation peak was confirmed by X-ray-induced radioluminescence spectroscopy. In conclusion, 6-phenylhexanoic acid-modified ZrO$_2$ nanoparticles synthesized at 300 °C are suitable for loading in liquid scintillators.

Keywords: ZrO$_2$; liquid scintillator; neutrinoless double beta decay; hydrothermal synthesis; nanoparticles; 6-phenylhexanoic acid

1. Introduction

The identification of neutrinoless double beta decay ($0\nu\beta\beta$) events has become important toward understanding neutrino properties ever since neutrino mass was confirmed through the observation of neutrino oscillation [1,2]. $0\nu\beta\beta$ is a radioactive event in which an even–even nucleus transforms into a lighter isobar containing two more protons with the emission of only two electrons. The occurrence of $0\nu\beta\beta$ would confirm that a neutrino has finite mass and is a Majorana particle, which means that a neutrino is its own antiparticle. However, the detection of $0\nu\beta\beta$ is challenging, because the estimated half-lives of the candidate isotopes for the decay mode are extremely long, typically more than 10^{25} years [3].

The performances of $0\nu\beta\beta$ detectors are parameterized with the experimental sensitivity $F_D^{0\nu}$ [4]. The experimental sensitivity is shown below:

$$F_D^{0\nu} = \ln 2 \frac{\eta \epsilon N_A}{A} \sqrt{\frac{T_m M}{B \Delta}} \qquad (1)$$

where η is the isotopic abundance of the candidate isotope, ϵ is the detection efficiency, N_A is the Avogadro's number, A is the mass number, T_m is the measurement time, M is the total detector mass, B is the background level, and Δ is the energy resolution. Therefore, to detect $0\nu\beta\beta$, which has an extremely low number of events, a large and efficient detector containing tens or hundreds of kilograms of candidate isotopes is required. Therefore, the measurement of $0\nu\beta\beta$ requires a large and efficient detector containing tens or hundreds of kilograms of the candidate isotopes for detection. $0\nu\beta\beta$ can be distinguished from ordinary double beta decay with two neutrino emissions based on the difference in the total energy of the two beta rays, which requires detectors with high energy resolution. To achieve these specifications, various types of detector systems, such as bolometers, Ge semiconductor detectors, and liquid scintillators, have been proposed [3].

Candidate isotope-loaded liquid scintillators are one of the detector systems useful for $0\nu\beta\beta$ experiments. A liquid scintillator, which is mainly composed of an organic solvent and phosphor, is a liquid that converts ionizing radiation into visible photons. The advantage of liquid scintillators is that large detectors with high uniformity can be constructed at a low cost; in addition, limited energy resolution can be sufficient. However, to accumulate adequate detection events to provide sufficient statistics in a reasonable time period, the liquid scintillator must contain the candidate isotope at concentrations in the range of 0.1–10 wt.% [5] without severe degradation of the optical properties of the liquid scintillator. Several research groups have developed metal-loaded liquid scintillators by dissolving organometallic molecules that contain the candidate isotope [5–8]; however, the excited state can be quenched by these incorporated molecules, resulting in the reduction of the scintillation light yield. Moreover, the concentration of candidate isotopes is limited by the solubility of the metal compound and some of the incorporated molecules may be degraded via photo-induced oxidation in the presence of oxygen [5].

As an alternative approach, we propose liquid scintillators loaded with metal oxide nanoparticles that contain the candidate isotopes for $0\nu\beta\beta$. In this approach, we expect that a large amount of the candidate isotope can be loaded into the liquid scintillator, because a dispersion of nanoparticles in an organic solvent is reportedly stable for a long period at remarkably high concentrations (up to 77 wt.%) [9,10]. Nanoparticles with diameters less than one-tenth of the wavelength of light can reduce Rayleigh scattering. Therefore, liquid scintillators with optical transparency can be fabricated upon the addition of nanoparticles. Surface modification of the nanoparticles is effective for dispersing inorganic nanoparticles in hydrophobic solutions used for liquid scintillators. It has been reported that the aggregation of nanoparticles can be suppressed and that the nanoparticles can be dispersed in an organic solvent by tuning the affinity of the nanoparticle surface upon surface modification using a hydrophobic organic solvent [10–13].

Sub/supercritical hydrothermal methods are suitable for the synthesis of such nanoparticles whose surface is modified by organic molecules. These methods provide unique reaction conditions for nanoparticle formation, allowing in situ modification of their surface. The dielectric constant of water decreases upon increasing the temperature, slowly approaching that of organic compounds. As the temperature of the aqueous solution increases, the solubility of the metal oxide decreases causing a high degree of supersaturation and the formation of nanoparticles [14–17]. Organic molecules are miscible with water at high temperatures under sub/supercritical conditions, which allows the surface modification of nanoparticles [18–20]. To date, many studies have been conducted toward the sub/supercritical hydrothermal synthesis of surface-modified metal oxide nanoparticles [20–23].

The solvents used in liquid scintillators are generally benzene derivatives, particularly alkyl benzenes, such as toluene, xylene, and cumene. Although many studies have been reported on long-chain carboxylic acid-modified nanoparticles, which are highly dispersible in saturated aliphatic organic solvents such as cyclohexane, their dispersibility in toluene is very low with a stable dispersion concentration of only 0.01 wt.% [9,23]. Therefore, our group proposed a method of modifying the surface of nanoparticles using aromatic carboxylic acids in which a benzene ring is introduced at the tip of a linear carboxylic acid to improve the dispersibility in aromatic organic solvents. Using 3-phenylpropionic acid or 6-phenylcaproic acid as a modifier, we successfully dispersed the nanoparticles in toluene [24,25].

In this study, we focused on the synthesis of organically modified ZrO_2 nanoparticles, which contain ^{96}Zr as a candidate isotope for $0\nu\beta\beta$. Since the double beta decay of ^{96}Zr has a high Q value of 3350 keV [8], it is less susceptible to gamma-ray environmental background and a clear $0\nu\beta\beta$ peak in the energy spectrum is expected. Since we have previously shown that the modification of nanoparticles improves their dispersibility in toluene [24], the sub/supercritical hydrothermal synthesis of ZrO_2 nanoparticles was performed using 6-phenylhexanoic acid. The effect of the synthesis temperature on the formation and surface modification of the nanoparticles is discussed. To develop a scintillator containing a large amount of the candidate isotope, the dispersion was used as the liquid scintillator. It has been reported that adding a large number of nanoparticles to organic scintillators results in increased light scattering and reduces the scintillation light yield [26]. Therefore, X-ray-induced radioluminescence spectroscopy was used to determine whether the scintillation peak of the fabricated scintillator was present.

2. Experiment

ZrO_2 nanoparticles were synthesized using subcritical/supercritical hydrothermal methods [14–17] in batch reactors (inner volume, 5 mL; AKICO Corp., Tokyo, Japan) made of Hastelloy C, which is a nickel-based alloy with high resistance to corrosion at high temperature. The precursor solution consisting of ZrO_2 nanoparticles was prepared by dissolving $ZrOCl_2 \cdot H_2O$ (99.0%, FUJIFILM Wako Pure Chemical Corp., Osaka, Japan) in distilled water at a concentration of 0.1 mol/L. The pH was adjusted to 5.8, which is greater than the pKa of 6-phenylhexanoic acid (=4.78), by adding KOH (FUJIFILM Wako Pure Chemical Corp., Osaka, Japan), following the procedure in previous studies [24,25]. This precursor solution (4.12, 3.75, 3.22, or 1.75 mL) was added to the batch reactors for synthesis at 250, 300, 350, or 400 °C, respectively. These volumes correspond to a final pressure of 30 MPa at the specified reaction temperatures. As an organic modifier, 6-phenylhexanoic acid (PHA; Tokyo Chemical Industry Co., LTD., Tokyo, Japan) was added to the precursor solution such that the molar ratio of PHA to Zr was 6:1. The reactions were performed at each temperature for 10 min while shaking and then cooled to room temperature in water. The reactors were rinsed several times using 3.75 mL of toluene (99.5%; FUJIFILM Wako Pure Chemical Corp., Osaka, Japan) and water to collect the products. The collected organic and aqueous phases were separated upon standing overnight. The organic phase was recovered and used for further experiments.

To obtain the powdered nanoparticles, ethanol (99.5%; FUJIFILM Wako Pure Chemical Corp., Osaka, Japan) was added to the organic phase, the solution was subjected to centrifugation, and the supernatant was removed to separate the unreacted chemicals. This procedure was repeated three times. The remaining particles were dried on a petri dish at room temperature and used for further characterization.

To determine the crystallographic phases of the nanoparticles, the samples were analyzed using X-ray diffraction (XRD; Ultima IV, Rigaku Corp., Tokyo, Japan) [27] with Cu Kα radiation in a 2θ–θ set-up; the scan interval and rate were 0.02° and 4.0°/min, respectively.

The crystallite sizes were estimated using the Halder–Wagner method [28] via Rietveld fitting with the RIETAN-FP [29] code. The Halder–Wagner equation is shown below:

$$\left(\frac{\beta}{\tan \theta}\right)^2 = \frac{K\lambda}{D} \times \frac{\beta}{\tan \theta \sin \theta} + 16\varepsilon^2 \quad (2)$$

where β is the full width at half maximum of the diffraction peak (in radians), θ is the Bragg angle, K is the shape factor for the mean volume-weighted size of spherical crystallites (4/3) [30], λ is the wavelength of the X-ray (0.154184 nm), D is the crystallite size, and ε is the microstrain of the crystal. The shapes and sizes of the nanoparticles were observed using transmission electron microscopy (TEM; HD-2700, Hitachi High-Technologies Corp, Tokyo, Japan) operated at 200 kV [31]. The nanoparticle diameters were obtained as the average diameter of 50 particles with error bars representing the standard deviation. Fourier-transform infrared (FTIR) spectroscopy was carried out on an FTIR spectrometer (Nicolet 6700, Thermo Fisher Scientific K. K., Tokyo, Japan) [32]. Measurements were conducted in the range of 4000–650 cm^{-1}, which is the measurable wavenumber of the device. The thermal properties of the nanoparticle powders were measured by thermogravimetric analysis (TGA; SDT Q600, TA Instruments Japan Inc., Tokyo, Japan) [33] in the temperature range from room temperature to 600 °C at a heating rate of 10 °C/min under a flow of air at 100 mL/min.

Dynamic light scattering (DLS; Nano-ZS, Malvern Panalytical Ltd., Malvern, UK) measurements [34] were performed to investigate the aggregated state of the nanoparticles in the toluene dispersion. An inductively coupled plasma atomic emission spectroscopy (ICP-AES) analysis [35] was performed (iCAP6500, Thermo Fisher Scientific K. K., Tokyo, Japan) at an emission wavelength of 339 nm to determine the Zr concentration in the nanoparticle dispersion. The absorption spectra of the dispersions were measured using a spectrophotometer (U-3500, Hitachi Ltd., Tokyo, Japan). The measurement was performed using a quartz cell with a 1-mm optical length.

Nanoparticle dispersions were prepared by concentrating the organic phase containing dispersed particles. Concentration was carried out using 15 mL of toluene dispersion obtained by collecting the organic phase. The dispersions were placed in screw vials, in which the spouts were covered with a perforated aluminum foil and stored at 50 °C to evaporate the solvent until precipitation occurred, which results in the maximum soluble state of the nanoparticles in toluene [10,36]. After the precipitation occurred, the supernatants were extracted to obtain nanoparticle dispersions of about 5 mL.

The liquid scintillator was fabricated using a concentrated nanoparticle dispersion. 2,5-Diphenyloxazole (DPO; Dojindo Laboratories, Kumamoto, Japan) and 1,4-bis(5-phenyl-2-oxazolyl)benzene (POPOP; Dojindo Laboratories, Kumamoto, Japan) were used as phosphors in the liquid scintillator. DPO and POPOP were first mixed homogeneously in a weight ratio of 80:1, and the mixed powder was then added to the toluene dispersion at a concentration of 100 g/L. The X-ray-induced radioluminescence spectra of the liquid scintillator loaded with ZrO$_2$ nanoparticles was measured with the same measurement system as in the previous study [37]. The liquid scintillator was contained in a quartz cell and irradiated with X-rays from an X-ray generator (D2300-HK, Rigaku Corp., Tokyo, Japan) equipped with a Cu target operated at 40 kV and 40 mA. A charge-coupled device (CCD)-based spectrometer (QE Pro Spectrometer, Ocean Insight Inc., Tokyo, Japan) was used to record the radioluminescence spectra.

3. Results and Discussion

Figure 1 shows the XRD patterns of the ZrO$_2$ nanoparticles synthesized at different temperatures. The XRD patterns of the nanoparticles synthesized at 250 and 300 °C can be attributed to the tetragonal ZrO$_2$ phase, whereas those synthesized at 350 and 400 °C were attributed to both tetragonal and monoclinic phases. The diffraction peaks of the XRD patterns of the nanoparticles synthesized at 250 and 300 °C were broader than those synthesized at 350 and 400 °C. These results indicate that the crystallite size decreased when the synthesis was performed at 250 and 300 °C. The crystallite diameters were calculated by fitting using Equation (2). Figure 2 shows Halder-Wanger plots of ZrO$_2$ nanoparticles synthesized at different temperatures. The crystallite sizes of the nanoparticles synthesized at 250, 300, 350, and 400 °C were 1.5 ± 0.1, 1.8 ± 0.1, 4.4 ± 2.2, and 5.4 ± 2.4 nm, respectively, suggesting that single-nanometer-sized crystallites were synthesized under all the temperature conditions studied.

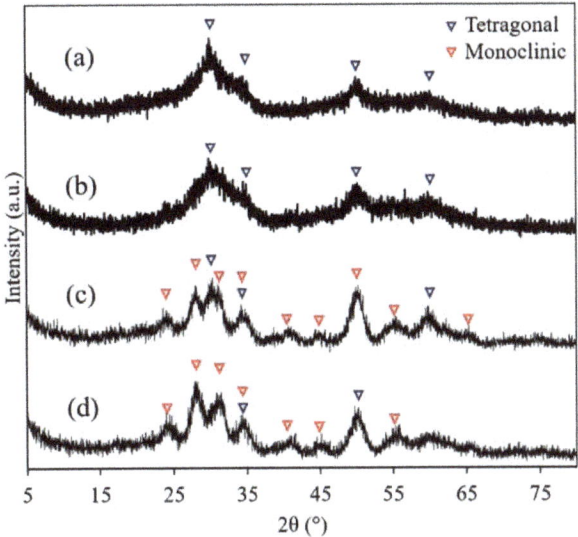

Figure 1. XRD patterns obtained for ZrO$_2$ nanoparticles synthesized at various temperatures: (**a**) 250 °C, (**b**) 300 °C, (**c**) 350 °C, and (**d**) 400 °C.

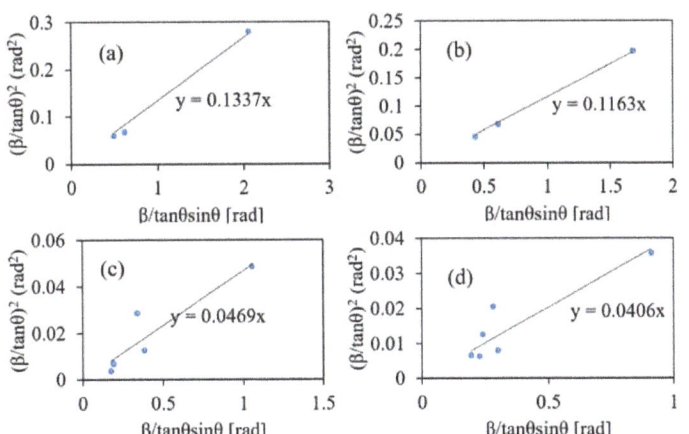

Figure 2. Halder–Wanger plot of the ZrO$_2$ nanoparticles synthesized at (**a**) 250 °C, (**b**) 300 °C, (**c**) 350 °C, and (**d**) 400 °C obtained using Equation (2).

Figure 3 shows the TEM images of the ZrO$_2$ nanoparticles synthesized at different temperatures. Spherical nanoparticles were formed under all the reaction temperatures investigated. The diameters of the nanoparticles synthesized at 250, 300, 350, and 400 °C were 4.0 ± 1.3, 3.7 ± 1.2, 6.7 ± 1.8, and 5.6 ± 1.4 nm, respectively. The particle sizes of the nanoparticles synthesized at 250 and 300 °C were smaller than those synthesized at 350 and 400 °C. The particle sizes formed at different reaction temperatures showed a similar trend as the corresponding crystallite sizes estimated from the XRD patterns. In addition, the aggregation of nanoparticles was observed in the TEM images of all the samples.

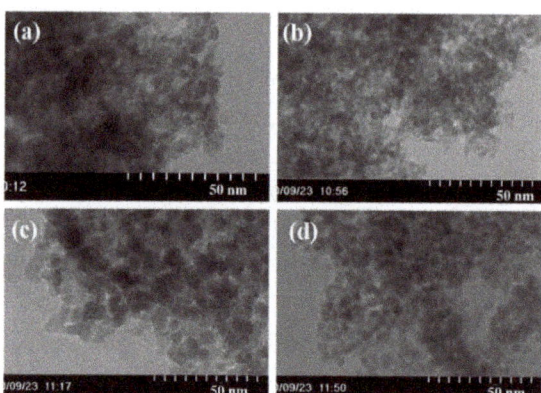

Figure 3. TEM images obtained for ZrO$_2$ nanoparticles synthesized at various temperatures. (**a**) 250 °C, (**b**) 300 °C, (**c**) 350 °C, and (**d**) 400 °C.

The chemical states of the surface-modifying molecules on the nanoparticles were evaluated using FTIR spectroscopy, and the results are shown in Figure 4. All the ZrO$_2$ nanoparticles exhibit bands at 650–850 cm^{-1} assigned to the Zr–O modes of the ZrO$_2$ nanoparticles [38], in addition to weak bands at 2950 and 2820 cm^{-1} assigned to the asymmetric and symmetric stretching modes of $-CH_2-$ in the alkyl chain of PHA, respectively [39,40]. Two strong peaks at 1540 and 1410 cm^{-1} were also observed, which were assigned to the asymmetric and symmetric stretching modes of the carboxylate group ($-COO^-$) in PHA, respectively [38,41]. The presence of these bands indicates that the surface-modifying molecules are attached to the surface of the ZrO$_2$ nanoparticles via coordination bonds to the carboxyl group [39,40,42]. In addition, no peak was detected at ~1708 cm^{-1}, which corresponds to the stretching mode of the $-COOH$ group. These results indicate that the nanoparticles were almost free from unreacted organic surface modifier on their surface. The strongest peak in the spectra was observed for the nanoparticles synthesized at 300 °C, which corresponds to the carboxylate group, while the sample synthesized at 250 °C exhibited the next strongest peak. The spectra obtained for the samples synthesized at 350 and 400 °C showed relatively weaker peaks. These results indicate that the number of organic surface modifier on the surface of the nanoparticles synthesized at 250 and 300 °C is larger than those synthesized at 350 and 400 °C.

TGA was performed to quantitatively analyze the number of surface-modified molecules present on the nanoparticles, and the TGA curves obtained for the nanoparticles are presented in Figure 5. All the TGA curves showed a significant weight loss at >250 °C, which corresponds to the weight of the surface modifiers attached to the surface of the nanoparticles [39,40]. The weight loss assigned to the desorption of the surface modifiers ended near 500 °C for any of the samples. The surface organic modification density of the ZrO$_2$ nanoparticles is estimated from the weight loss observed from 250 to 500 °C and presented in Table 1. These results were consistent with those obtained using FTIR spectroscopy. As the synthesis temperature increased from 250 to 300 °C, the modification density of the nanoparticles increased from 2.8 ± 0.0 to 3.1 ± 0.2 molecules/nm^2. On the other hand, the modification densities of the nanoparticles synthesized at 350 and 400 °C were <1 molecules/nm^2. Consequently, the modification density decreases as the synthetic conditions approach the supercritical state, which was similar to the results obtained for HfO$_2$ nanoparticles in a previous study [43]. The difference in the affinity between the crystalline phase and modifier molecules appears to be related to the decrease in the modification density. Since the surface of tetragonal ZrO$_2$ has more than three times the number of basic sites on the surface of monoclinic ZrO_2 [44], its reactivity with 6-phenylhexanoic acid is considered to be higher. Therefore, the nanoparticles synthesized at 350 and 400 °C, which are rich in monoclinic phases, may reduce the modification density.

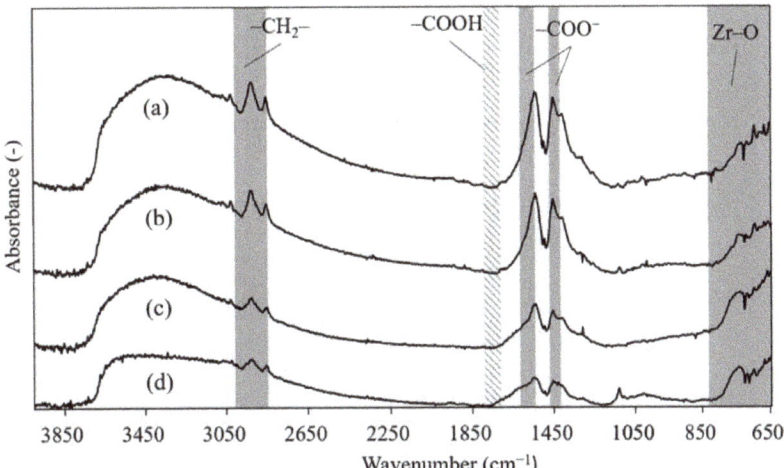

Figure 4. FTIR spectra obtained for PHA-modified ZrO_2 nanoparticles synthesized at various temperatures: (**a**) 250 °C, (**b**) 300 °C, (**c**) 350 °C, and (**d**) 400 °C.

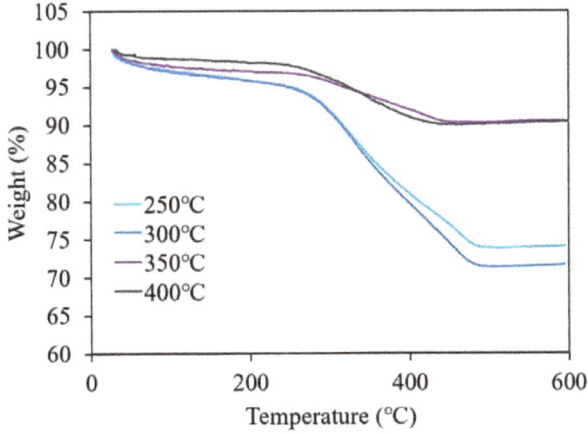

Figure 5. TGA curves obtained for PHA-modified nanoparticles synthesized at 250, 300, 350, and 400 °C under an air atmosphere and heating rate of 10 K/min.

Table 1. The surface PHA modification density of the ZrO_2 nanoparticles and Zr concentrations of the nanoparticle dispersions.

Product	Weight Loss (%) from TGA Measurements	Modification Density (Molecules/nm²)	Zr Concentration (wt.%) from ICP-AES
250 °C	23.0 ± 0.1	2.8 ± 0.0	$1.1 \pm 0.1 \times 10^{-1}$
300 °C	24.3 ± 1.3	3.1 ± 0.2	$3.3 \pm 0.4 \times 10^{-1}$
350 °C	8.63 ± 1.67	0.9 ± 0.2	$5.7 \pm 0.3 \times 10^{-2}$
400 °C	8.42 ± 0.24	0.9 ± 0.3	$9.2 \pm 0.1 \times 10^{-3}$

Figure 6 shows the photographic images obtained for the concentrated nanoparticle dispersions prepared using samples synthesized at different temperatures. The dispersions were stable for at least one month, and no precipitation was observed. The dispersions synthesized at 250–350 °C showed pale turbidity, while those synthesized at 400 °C were transparent. When the dispersions were irradiated with green laser light to observe the Tyndall scattering, significant scattering was observed in the dispersions synthesized at 250–350 °C but not at 400 °C, which was strongly related to the concentration of ZrO_2 nanoparticles.

Figure 6. Photographic images of the PHA-modified ZrO_2 nanoparticle dispersions in toluene synthesized at 250, 300, 350, and 400 °C.

The number size distributions of the ZrO_2 nanoparticles in the toluene dispersions measured using DLS are shown in Figure 7. The particle sizes of the nanoparticles in the dispersions synthesized at 250, 300, 350, and 400 °C were estimated to be 38.7 ± 6.6, 26.5 ± 5.5, 26.9 ± 5.9, and 31.9 ± 7.7 nm, respectively. When compared with the crystallite sizes estimated from the XRD patterns and the particle sizes estimated using TEM, DLS indicated the presence of significantly larger particles, which indicates that the nanoparticles were aggregated in the dispersions. The aggregate size was homogeneous and maintained a dispersive state without further agglomeration or phase separation.

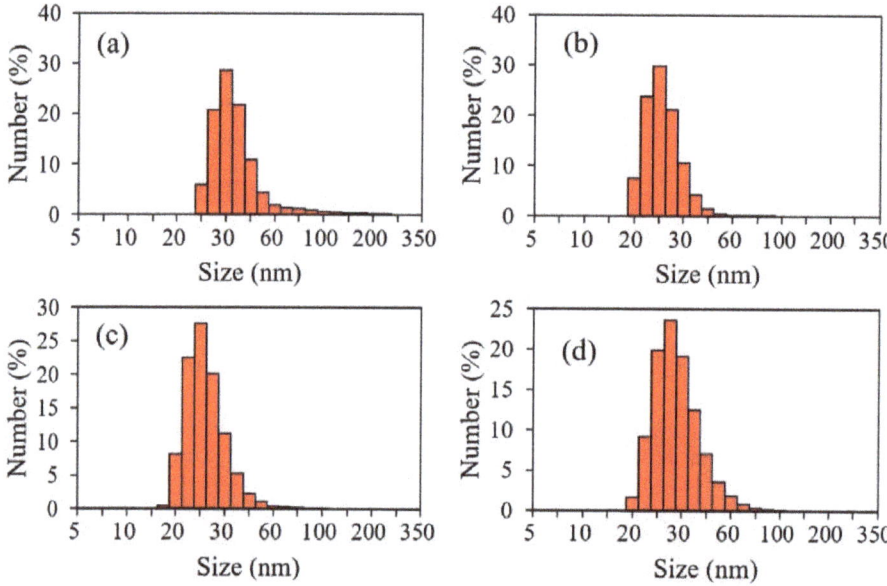

Figure 7. Number size distributions of the ZrO_2 nanoparticles in the toluene dispersions synthesized at (**a**) 250 °C, (**b**) 300 °C, (**c**) 350 °C, and (**d**) 400 °C obtained using dynamic light scattering.

The Zr concentrations in the nanoparticle dispersions are summarized in Table 1. The highest Zr concentration of 0.33 ± 0.04 wt.% was obtained for the nanoparticles synthesized at 300 °C. The Zr concentrations of the dispersions obtained from the particles synthesized at 250 and 300 °C were an order of magnitude higher than those of the particles synthesized at 350 and 400 °C. The dispersibility and solubility of these nanoparticles can be improved by increasing the surface modification density [9,10,36]. Table 1 shows that the surface modification densities of nanoparticles synthesized at 250 and 300 °C were higher than those synthesized at 350 and 400 °C. These results demonstrate the correlation between the dispersed ZrO_2 concentration and surface modification density.

The liquid scintillators were fabricated using the ZrO_2 nanoparticle dispersions. Figure 8 shows the absorption spectra of the toluene dispersion of ZrO_2 nanoparticles. In all dispersions, the absorbance below 500 nm was enhanced upon loading the ZrO_2 nanoparticles in toluene. This increase in the absorbance was attributed to Rayleigh scattering by the aggregated nanoparticles, which was observed using DLS. The absorbance in this range was the highest in the dispersion of particles synthesized at 350 °C, followed by those at 250 and 300 °C. The bandgap of tetragonal and monoclinic ZrO_2 nanoparticles are 3.6 and 4.5 eV, respectively, according to a previous paper [45], and the wavelengths corresponding to the bandgap energies are 344 and 276 nm, respectively. Since the ZrO_2 nanoparticles synthesized at 350 °C contain a monoclinic phase, the absorbance increased significantly in the range of 300 to 350 nm. In addition, it is considered that the dispersion of particles synthesized at 400 °C has a low absorbance, because the Zr concentration was small, which also correlates with the results shown in Figure 6. The absorbance was <0.1 for wavelengths >410 nm, which suggests that the ZrO_2 nanoparticle dispersions have high transparency in the region of POPOP emission. The liquid scintillators were prepared by dissolving the DPO and POPOP phosphors in these dispersions. Figure 9 shows the X-ray-induced radioluminescence spectra obtained for the liquid scintillators. The energy of the X-ray is 8.048 keV. In all samples, a dominant band attributed to the emission from POPOP is observed at 425 nm. No significant decrease in the peak intensity of emission was observed in any of the samples. This result indicates that the energy transfer from toluene to POPOP via DPO occurs successfully even in the liquid scintillator, which contains 0.33 ± 0.04 wt.% of Zr. A similar scintillation spectrum is expected for high-energy β-rays, because the scintillation was caused by high-energy electrons produced via the photoelectric effect. Considering the absorption spectra shown in Figure 8, the liquid scintillator is transparent to the scintillation wavelength.

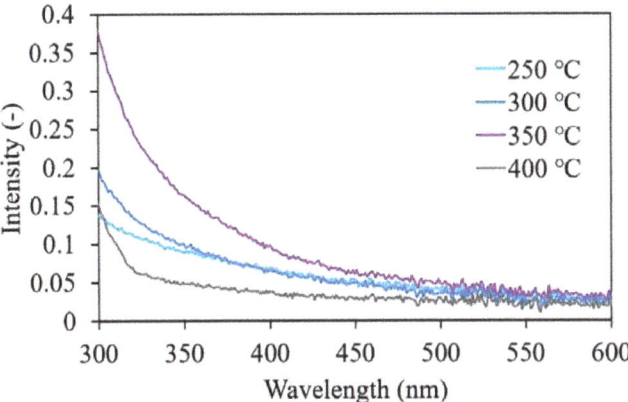

Figure 8. Absorption spectra obtained for the PHA-modified ZrO_2 nanoparticle–toluene dispersion synthesized at 250, 300, 350, and 400 °C.

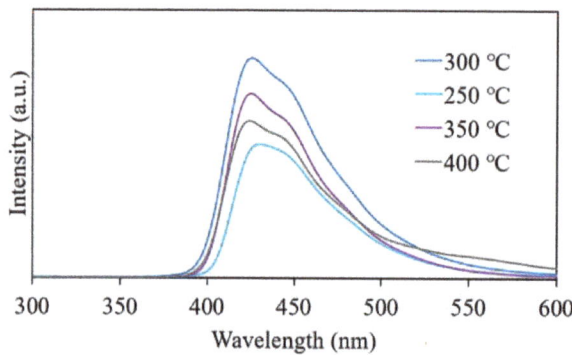

Figure 9. X-ray-induced radioluminescence spectra obtained for the liquid scintillator incorporating PHA-modified ZrO$_2$ nanoparticles synthesized at 250, 300, 350, and 400 °C.

4. Conclusions and Outlook

6-phenylhexanoic acid-modified ZrO$_2$ nanoparticles were synthesized to assess their suitability for use in double beta decay experiments. The ZrO$_2$ nanoparticles synthesized at lower temperatures (250 and 300 °C) have smaller particle sizes and larger surface modification densities than those synthesized at higher temperatures (350 and 400 °C). The highest modification density of 3.1 ± 0.2 molecules/nm^2 was obtained at 300 °C, which resulted in the highest concentration of dispersed nanoparticles in toluene (0.33 ± 0.04 wt.%-Zr). The ZrO$_2$ nanoparticle-loaded liquid scintillator was transparent to the scintillation wavelength, and an apparent scintillation peak from the phosphor emission was observed at 425 nm in the X-ray-induced radioluminescence spectra. As a result, a toluene-based liquid scintillator with a relatively high concentration of ZrO$_2$ nanoparticles was successfully fabricated.

Currently, the maximum dispersed concentration of Zr was 0.33 wt.%. With a natural isotope ratio of Zr, 9.24 × 10^{-3} wt.% of ^{96}Zr is present in the liquid scintillator. Even for a 100-ton liquid scintillator, 9.24 kg of ^{96}Zr can be loaded. The dispersion concentration of nanoparticles needs to be enhanced, because tens or hundreds of kilograms of candidate isotopes are required for the measurement of $0\nu\beta\beta$. In addition, it is necessary to use raw materials enriched with ^{96}Zr. If enriched raw materials of 90% ^{96}Zr is were used, the Zr dispersion concentration of 1 to 2 wt.% is required for loading the target amount of candidate isotopes. Therefore, further enhancement in the ZrO$_2$ loading concentration is necessary.

When observing 0ubb of ^{96}Zr, β-rays generated from the daughter nuclei of ^{238}U and ^{232}Th are expected to be a source of serious background. According to a previous report, the precursor ZrOCl$_2$·8H$_2$O contains U and Th at several ppm [46]. Therefore, high-purity raw materials will be required. From ^{238}U and ^{232}Th, α-rays of 4.27 and 4.80 MeV are generated. Therefore, it is necessary to consider the influence of these α-rays on the background. According to a previous paper, the ratio of the scintillation efficiency of α-rays to β-rays (R(α/β)) of liquid scintillators is about 0.1 [47]. Hence, they would be observed in the energy spectrum at 0.42 and 0.48 MeVee for α-rays from ^{238}U and ^{232}Th, respectively. Hence, it is considered that the background of α rays near the Q value is limited. However, since the influence of β-rays generated from the daughter nuclei of ^{238}U and ^{232}Th cannot be ignored, it is necessary to purify the Zr raw material.

Furthermore, it is necessary to estimate the energy resolution. To estimate the energy resolution of these scintillators, it is necessary to measure with gamma rays with energy close to the Q value of $0\nu\beta\beta$. However, since the liquid scintillator used in this study had a small volume, a gamma-ray photoelectric peak was not observed. Since it is necessary to increase the amount of liquid scintillator to observe the photoelectric peak, the immediate challenge is to establish the technology to construct a large-scale scintillator.

Finally, for double beta decay search, it is necessary to establish a technology for mass synthesis of nanoparticles. Since nanoparticles were synthesized in a batch process in this study, it is necessary to establish a flow-type continuous synthesis process in the future.

Author Contributions: Writing—original draft preparation and investigation, A.W.; investigation, A.M.; methodology, validation, formal analysis, and writing—review and editing, A.Y., G.S., T.T. and T.A.; validation and formal analysis, Y.H.; conceptualization, methodology, validation, formal analysis, writing—review and editing, supervision, project administration, and funding acquisition, M.K.; and writing—review and editing, supervision, and project administration, Y.F. and K.A. All authors have read and agreed to the published version of the manuscript.

Funding: This research was funded by JSPS, grant Nos. JP18H03890 and JP20K20917.

Data Availability Statement: The data presented in this study are available within this article.

Acknowledgments: This research was supported by JSPS KAKENHI (Grants No. JP18H03890 and JP20K20917). Part of this research was based on the Cooperative Research Project of Research Center for Biomedical Engineering, Ministry of Education, Culture, Sports, Science and Technology.

Conflicts of Interest: The authors declare no conflict of interest.

References

1. Dell'Oro, S.; Marcocci, S.; Viel, M.; Vissani, F. Neutrinoless Double Beta Decay: 2015 Review. *Adv. High Energy Phys.* **2016**, *2016*, 2162659. [CrossRef]
2. Vergados, J.D.; Ejiri, H.; Šimkovic, F. Neutrinoless double beta decay and neutrino mass. *Int. J. Mod. Phys. E* **2016**, *25*, 1630007. [CrossRef]
3. Dolinski, M.J.; Poon, A.W.P.; Rodejohann, W. Neutrinoless Double-Beta Decay: Status and Prospects. *Annu. Rev. Nucl. Part. Sci.* **2019**, *69*, 219–251. [CrossRef]
4. Beretta, M.; Pagnanini, L. Development of Cryogenic Detectors for Neutrinoless Double Beta Decay Searches with CUORE and CUPID. *Appl. Sci.* **2021**, *11*, 1606. [CrossRef]
5. Buck, C.; Yeh, M. Metal-loaded organic scintillators for neutrino physics. *J. Phys. G Nucl. Part. Phys.* **2016**, *43*, 093001. [CrossRef]
6. Nemchenok, I.B.; Brudanin, V.B.; Kochetov, O.I.; Timkin, V.V.; Shurenkova, A.A. A Neodymium-Loaded Liquid Scintillator. *Bull. Russ. Acad. Sci. Phys.* **2011**, *75*, 1007–1010. [CrossRef]
7. Hwang, M.J.; Kwon, Y.J.; Kim, H.J.; Kwak, J.W.; Kim, S.C.; Kim, S.K.; Kim, T.Y.; Kim, S.Y.; Lee, H.S.; Myung, S.S.; et al. A search for $0\nu\beta\beta$ decay of ^{124}Sn with tin-loaded liquid scintillator. *Astropart. Phys.* **2009**, *31*, 412–416. [CrossRef]
8. Fukuda, Y.; Moriyama, S.; Ogawa, I. Development of liquid scintillator containing a zirconium complex for neutrinoless double beta decay experiment. *Nucl. Instrum. Methods Phys. Res. Sect. A* **2013**, *732*, 397–402. [CrossRef]
9. Arita, T.; Ueda, Y.; Minami, K.; Naka, T.; Adschiri, T. Dispersion of Fatty Acid Surface Modified Ceria Nanocrystals in Various Organic Solvents. *Ind. Eng. Chem. Res.* **2010**, *49*, 1947–1952. [CrossRef]
10. Arita, T.; Yoo, J.; Ueda, Y.; Adschiri, T. Highly Concentrated Colloidal Dispersion of Decanoic Acid Self-assembled Monolayer-protected CeO$_2$ Nanoparticles Dispersed to a Concentration of up to 77 wt% in an Organic Solvent. *Chem. Lett.* **2012**, *41*, 1235–1237. [CrossRef]
11. Bagwe, R.P.; Hilliard, L.R.; Tan, W. Surface Modification of Silica Nanoparticles to Reduce Aggregation and Nonspecific Binding. *Langmuir* **2006**, *22*, 4357–4362. [CrossRef] [PubMed]
12. Kamiyama, H.; Iijima, M. Surface modification and characterization for dispersion stability of inorganic nanometer-scaled particles in liquid media. *Sci. Technol. Adv. Mater.* **2010**, *11*, 044304. [CrossRef]
13. Arita, T.; Yoo, J.; Adschiri, T. Relation between the Solution-State Behavior of Self-Assembled Monolayers on Nanoparticles and Dispersion of Nanoparticles in Organic Solvents. *J. Phys. Chem. C* **2011**, *115*, 3899–3909. [CrossRef]
14. Adschiri, T.; Kanazawa, K.; Arai, K. Rapid and Continuous Hydrothermal Crystallization of Metal Oxide Particles in Supercritical Water. *J. Am. Ceram. Soc.* **1992**, *75*, 1019–1022. [CrossRef]
15. Byrappa, K.; Adschiri, T. Hydrothermal technology for nanotechnology. *Prog. Cryst. Growth Charact. Mater.* **2007**, *53*, 117–166. [CrossRef]
16. Adschiri, T.; Hakuta, Y.; Arai, K. Hydrothermal Synthesis of Metal Oxide Fine Particles at Supercritical Conditions. *Ind. Eng. Chem. Res.* **2000**, *39*, 4901–4907. [CrossRef]
17. Adschiri, T.; Hakuta, Y.; Sue, K.; Arai, K. Hydrothermal Synthesis of Metal Oxide Nanoparticles at Supercritical Conditions. *J. Nanoparticle Res.* **2001**, *3*, 227–235. [CrossRef]
18. Akiya, N.; Savage, P.E. Roles of Water for Chemical Reactions in High-Temperature Water. *Chem. Rev.* **2002**, *102*, 2725–2750. [CrossRef] [PubMed]
19. Zhang, J.; Ohara, S.; Umetsu, M.; Naka, T.; Hatakeyama, Y.; Adschiri, T. Colloidal Ceria Nanocrystals: A Tailor-Made Crystal Morphology in Supercritical Water. *Adv. Mater.* **2007**, *19*, 203–206. [CrossRef]

20. Adschiri, T. Supercritical Hydrothermal Synthesis of Organic–Inorganic Hybrid Nanoparticles. *Chem. Lett.* **2007**, *36*, 1188–1193. [CrossRef]
21. Yoko, A.; Aida, T.; Aoki, N.; Hojo, D.; Koshimizu, M.; Ohara, S.; Seong, G.; Takami, S.; Togashi, T.; Tomai, T.; et al. Application 53-Supercritical Hydrothermal Synthesis of Nanoparticles. In *Nanoparticle Technology Handbook*, 3rd ed.; Elsevier: Amsterdam, The Netherlands, 2018; pp. 683–689.
22. Yoko, A.; Seong, G.; Tomai, T.; Adschiri, T. Continuous Flow Synthesis of Nanoparticles Using Supercritical Water: Process Design, Surface Control, and Nanohybrid Materials. *KONA Powder Part. J.* **2020**, *37*, 28–41. [CrossRef]
23. Tomai, T.; Tajima, N.; Kimura, M.; Yoko, A.; Seong, G.; Adschiri, T. Solvent accommodation effect on dispersibility of metal oxide nanoparticle with chemisorbed organic shell. *J. Colloid Interface Sci.* **2021**, *587*, 574–580. [CrossRef]
24. Takigawa, S.; Koshimizu, M.; Noguchi, T.; Aida, T.; Takami, S.; Adschiri, T.; Fujimoto, Y.; Yoko, A.; Seong, G.; Tomai, T.; et al. Synthesis of ZrO_2 nanoparticles for liquid scintillators used in the detection of neutrinoless double beta decay. *J. Radioanal. Nucl. Chem.* **2017**, *314*, 611–615. [CrossRef]
25. Arai, S.; Noguchi, T.; Aida, T.; Yoko, A.; Tomai, T.; Adcshiri, T.; Koshimizu, M.; Fujimoto, Y.; Asai, K. Development of liquid scintillators loaded with alkaline earth molybdate nanoparticles for detection of neutrinoless double-beta decay. *J. Ceram. Soc. Jpn.* **2019**, *127*, 28–34. [CrossRef]
26. Hiyama, F.; Noguchi, T.; Koshimizu, M.; Kishimoto, S.; Haruki, R.; Nishikido, F.; Fujimoto, Y.; Aida, T.; Takami, S.; Adschiri, T.; et al. X-ray detection properties of plastic scintillators containing surface-modified Bi_2O_3 nanoparticles. *Jpn. J. Appl. Phys.* **2018**, *57*, 052203. [CrossRef]
27. Warren, B.E. *X-ray Diffraction*; Addison Wesley: New York, NY, USA, 1968.
28. Halder, N.C.; Wanger, C.N.J. Separation of particle size and lattice strain in integral breadth measurements. *Acta Crystallogr.* **1966**, *20*, 312–313. [CrossRef]
29. Izumi, F.; Momma, K. Three-Dimensional Visualization in Powder Diffraction. *Solid State Phenom.* **2007**, *130*, 15–20. [CrossRef]
30. Ida, T.; Shimazaki, S.; Hibino, H.; Toraya, H. Diffraction peak profiles from spherical crystallites with lognormal size distribution. *J. Appl. Crystallogr.* **2003**, *36*, 1107–1115. [CrossRef]
31. Hirsch, P.B.; Howie, A.; Nicholson, R.B.; Pashley, D.W.; Whelan, M.J. *Electron Microscopy of Thin Crystals*; Butterworth: London, UK, 1965.
32. Faix, O. Fourier Transform Infrared Spectroscopy. In *Methods in Lignin Chemistry*; Lin, S.Y., Dence, C.W., Eds.; Springer: Berlin/Heidelberg, Germany, 1992.
33. Coats, A.W.; Redfern, J.P. Thermogravimetric analysis. A review. *Analyst* **1963**, *88*, 906–924. [CrossRef]
34. Berne, B.J.; Pecora, R. *Dynamic Light Scattering: With Application to Chemistry, Biology, and Physics*; Dover Publications: New York, NY, USA, 2000.
35. Houk, R.S.; Fassel, V.A.; Flesch, G.D.; Svec, H.J.; Gray, A.L.; Taylor, C.E. Inductively coupled argon plasma as an ion source for mass spectrometric determination of trace elements. *Anal. Chem.* **1980**, *52*, 2283–2289. [CrossRef]
36. Hossainm, M.Z.; Hojo, D.; Yoko, A.; Seong, G.; Aoki, N.; Tomai, T.; Takami, S.; Adschiri, T. Dispersion and rheology of nanofluids with various concentrations of organic modified nanoparticles: Modifier and solvent effects. *Colloids Surf. A* **2019**, *583*, 123876. [CrossRef]
37. Fujimoto, Y.; Yanagida, T.; Koshimizu, M.; Asai, K. Photoluminescence and scintillation Propaties of SiO_2 Glass Activated with Eu^{2+}. *Sens. Mater.* **2015**, *27*, 263–268.
38. Zhou, S.; Garnweitner, G.; Niederberger, M.; Antonietti, M. Dispersion Behavior of Zirconia Nanocrystals and Their Surface Functionalization with Vinyl Group-Containing Ligands. *Langmuir* **2007**, *23*, 9178–9187. [CrossRef]
39. Taguchi, M.; Takami, S.; Adschiri, T.; Nakane, T.; Sato, K.; Naka, T. Synthesis of surface-modified monoclinic ZrO_2 nanoparticles using supercritical water. *CrystEngComm* **2012**, *14*, 2132–2138. [CrossRef]
40. Taguchi, M.; Takami, S.; Adschiri, T.; Nakane, T.; Sato, K.; Naka, T. Supercritical hydrothermal synthesis of hydrophilic polymer-modified water-dispersible CeO_2 nanoparticles. *CrystEngComm* **2011**, *13*, 2841–2848. [CrossRef]
41. Pawsey, S.; Yach, K.; Halla, J.; Reven, L. Self-Assembled Monolayers of Alkanoic Acids: A Solid-State NMR Study. *Langmuir* **2000**, *16*, 3294–3303. [CrossRef]
42. Zhang, T.; Ge, J.; Hu, Y.; Yin, Y. A General Approach for Transferring Hydrophobic Nanocrystals into Water. *Nano Lett.* **2007**, *7*, 3203–3207. [CrossRef]
43. Sahraneshin, A.; Takami, S.; Hojo, D.; Minami, K.; Arita, T.; Adschiri, T. Synthesis of shape-controlled and organic–hybridized hafnium oxide nanoparticles under sub- and supercritical hydrothermal conditions. *J. Supercrit. Fluids* **2012**, *62*, 190–196. [CrossRef]
44. Ma, Z.-Y.; Yang, C.; Wei, W.; Li, W.-H.; Sun, Y.-H. Surface properties and CO adsorption on zirconia polymorphs. *J. Mol. Catal. A* **2005**, *227*, 119–124. [CrossRef]
45. Ciuparu, D.; Ensuque, A.; Shafeev, G.; Bozon-verduraz, F. Synthesis and apparent bandgap of nanophase zirconia. *J. Mater. Sci. Lett.* **2000**, *19*, 931–933. [CrossRef]
46. Tuyen, N.V.; Quang, V.T.; Huong, T.G.; Anh, V.H. Preparation of High Quality Zirconium Oxychloride from Zircon of Vietnam. In *The VAEC-Annual Report for 2007*; U.S. Department of Energy: Oak Ridge, TN, USA, 2007; Volume VAEC-AR-07, pp. 286–291.
47. Seliger, H.H. Liquid scintillation counting of alpha;-particles and energy resolution of the liquid scintillator for alpha;- and β-particles. *Int. J. Appl. Radiat. Isot.* **1960**, *8*, 29–34. [CrossRef]

Article

Direct Observation Techniques Using Scanning Electron Microscope for Hydrothermally Synthesized Nanocrystals and Nanoclusters

Natsuko Asano [1], Jinfeng Lu [1], Shunsuke Asahina [1,*] and Seiichi Takami [2]

1. EP Business Unit, EP Application Department, SEM Team, JEOL Ltd., 3-1-2 Musashino, Akishima, Tokyo 196-8558, Japan; nasano@jeol.co.jp (N.A.); lujf@jeol.co.jp (J.L.)
2. Graduate School of Engineering, Nagoya University, Furo-cho, Chikusa-ku, Nagoya 464-8603, Japan; takami.seiichi@material.nagoya-u.ac.jp
* Correspondence: sasahina@jeol.co.jp

Abstract: Metal oxide nanocrystals have garnered significant attention owing to their unique properties, including luminescence, ferroelectricity, and catalytic activity. Among the various synthetic methods, hydrothermal synthesis is a promising method for synthesizing metal oxide nanocrystals and nanoclusters. Because the shape and surface structure of the nanocrystals largely affect their properties, their analytical methods should be developed. Further, the arrangement of nanocrystals should be studied because the properties of nanoclusters largely depend on the arrangement of the primary nanocrystals. However, the analysis of nanocrystals and nanoclusters remains difficult because of their sizes. Conventionally, transmission electron microscopy (TEM) is widely used to study materials in nanoscale. However, TEM images are obtained as the projection of three-dimensional structures, and it is difficult to observe the surface structures and the arrangement of nanocrystals using TEM. On the other hand, scanning electron microscopy (SEM) relies on the signals from the surface of the samples. Therefore, SEM can visualize the surface structures of samples. Previously, the spatial resolution of SEM was not enough to observe nanoparticles and nanomaterials with sizes of between 10 and 50 nm. However, recent developments, including the low-landing electron-energy method, improved the spatial resolution of SEM, which allows us to observe fine details of the nanocluster surface directory. Additionally, improved detectors allow us to visualize the elemental mapping of materials even at low voltage with high solid angle. Further, the use of a liquid sample holder even enabled the observation of nanocrystals in water. In this paper, we discuss the development of SEM and related observation technologies through the observation of hydrothermally prepared nanocrystals and nanoclusters.

Keywords: hydrothermal synthesis; scanning electron microscope; nanocluster; in situ observation; characterization

1. Introduction

In the last 30 years, many synthetic methods have been proposed and developed to produce nanomaterials with controlled sizes and morphologies [1–3]. The particle size, surface area, geometry, and chemical properties of the nanocrystals play a significant role in the interaction of these materials with biological systems, which affects their characteristics and applications in various disciplines [4–6]. Among the various synthetic methods, hydrothermal synthesis has gained attention and been subjected to research and development due to its capacity to produce small and shape-controlled nanocrystals in water [7,8]. Moreover, in situ surface modification during synthesis controls the surface of the nanocrystals and makes them easily dispersible in various solvents [8–10]. Generally, transmission electron microscopy (TEM) has been used to observe the size, shape, and heterogeneity of the nanocrystals. However, as the TEM images are obtained by projecting the signals on a screen, it is difficult to confirm the overlapping of nanoparticles. It is also difficult to observe the three-dimensional surface structure of nanocrystals from a two-dimensional projected image. On the other hand, scanning electron microscopy (SEM) uses the electrons that are emitted from the surface of samples. Therefore, SEM is a suitable method to visualize surface structures. One of the drawbacks of SEM is the spatial resolution of images. However, with the development of observation techniques at lower accelerating voltages and high voltage specimen bias, the resolution of SEM has been dramatically improved to directly observe the morphology and surface of nanocrystals, together with improvements in the lens detector [11–16]. In this study, we report the use of low-voltage high-resolution SEM and ultra-high solid angle energy-dispersive X-ray analyzer (EDS) to understand the 10–50 nm sized CeO_2 nanocrystals and nanoclusters that were synthesized by a hydrothermal reaction.

2. Materials and Methods

2.1. Synthesis of CeO_2 Nanoclusters

We have synthesized CeO_2 nanoclusters using a flow-type reactor. The details are described in [17]. In short, the flow of preheated water was mixed with the flow of the precursor solution, that is, aqueous solutions of $Ce(NO_3)_3$ (0.01 M) with or without hexanedioic acid (0.05 M) at a junction in the tubular reactor. The reaction temperature was set to 250 °C, and the pressure was maintained at 25 MPa. The typical reaction time was 1.9 s. The product solution was depressurized with a backpressure regulator and collected at the outlet. The products were washed and dried.

2.2. Synthesis of Organic Ligand-Modified CeO_2 Nanocrystals

Surface-modified CeO_2 nanocrystals were synthesized in supercritical water, according to previous reports [18]. Briefly, cerium hydroxides (2.5 mL of 0.05 M cerium hydroxides) and 0.09 g of decanoic acid were transferred to a pressure-resistant reactor, and the reaction was performed at 400 °C for 10 min. The obtained organic ligand-modified CeO_2 nanoparticles were washed with ethanol and toluene. The products were dispersed in 10 mL cyclohexane and were freeze-dried.

Silicon (100) substrates were modified with organic molecules to assemble the modified CeO_2 nanocrystals on them. First, the silicon substrates were treated with ozone for 30 min to produce hydroxylated silicon oxide. Next, the silicon substrate was immersed in a solution of 1.15 g 3-aminopropyltriethoxysilane (3APTS), 1.26 mL 28% ammonia aqua, 45.6 mL ethanol, and 0.76 mL water to produce amine group termination. To further attach APTS to the substrate, the substrates were heated at 130 °C for 2 h. Then, the substrate was immersed in 5 mL N,N-dimethylformamide (DMF) containing 0.2 M of 3,4-dihydroxyhydrocinnamic acid (DHCA), 0.2 M of 1-ethyl-3-(3-dimethylaminopropyl) carbodiimide hydrochloride (EDC), and 0.02 M of N,N-dimethyl-4-aminopyridine (DMAP) for 15 h to produce catechol termination on the surface of substrates. Finally, the surface-modified silicon substrates were immersed in 0.2 mL cyclohexane containing 1 mg of organic ligand-modified CeO_2 nanocrystals and then sonicated for 1 h to immobilize the CeO_2 nanocrystals on the substrates. Subsequently, the substrates were rinsed with fresh cyclohexane solvent and dried.

2.3. Synthesis of Composite of Mesoporous SBA-15 and CeO_2

SBA-15 and mesoporous CeO_2 were synthesized using the following procedure [19] First, 5.14 g of Pluronic P123 was dissolved in a mixture of 30.96 g of aqueous 37% HCl and water (144 g). Then, 11.12 g of tetraethoxysilane (TEOS) was added to this solution and stirred for 20 h at 40 °C. The mixture was then transferred to a stainless steel reactor with a Teflon lining and heated for 24 h at 100 °C. After cooling, the mixture was filtered, washed, and air-dried. Finally, mesoporous SBA-15 was obtained by calcining the mixture at 550 °C for 5 h.

Next, the obtained SBA-15 was used as a solid template to prepare mesoporous CeO_2. SBA-15 (0.50 g) and $Ce(NH_4)_2(NO_3)_6$ (0.5 g) were dispersed in ethanol (5.0 mL). The mixture was stirred at 50 °C until the ethanol was completely removed. The dried mixture was calcined in air at 450 °C for 5 h.

2.4. Low-Voltage SEM

The development of low-voltage (LV) field emission (FE) SEM realized a spatial resolution of 1 nm or less at an acceleration voltage of 1 kV. Furthermore, by choosing appropriate observation conditions, different information, such as material topology and composition, are also selectively obtained. In this report, we focused on the observation of extremely low impact electron energy to collect information from the surface. The mean free path of an electron with an energy of 100 eV in a solid sample is below 1 nm [14]. Therefore, the observation of the surface morphology at low electron energies is favorable. However, this low electron energy resulted in a larger diameter of the probe due to the chromatic aberration (Cc). The diameter of the electron probe size (dc) is described as $dc = Cc(\Delta V/V_{acc})\alpha$, where ΔV is the energy spread, V_{acc} is the acceleration voltage, and α is the angle of the beam(rad) [20]. Therefore, smaller ΔV and Cc values are necessary to minimize the probe diameter. Currently, we can use combined electrostatic and magnetic lenses to minimize Cc as well as field emission-type emitters with small ΔV. These so-called super hybrid lenses (SHL) are equipped on a FE-SEM (JSM-IT800SHL, JEOL, Tokyo, Japan). In addition, a negative surface potential can be applied to the sample surface to lower the impact electron energy, which reduces the scattering of electrons in the sample. Therefore, the lowest impact electron energy to the surface is reduced to 10 eV, while maintaining a small probe size with high coherency. This low landing voltage technique also suppressed the charging up and damage of the sample, leading to a clear observation of the shape and size of the nanocrystals and nanoclusters. Figure 1 shows an SEM image of CeO_2 nanoclusters observed at a landing voltage of 1 kV (sample bias: −5 kV, probe current: 8 pA, detector: in-column detector). The shape and size of the CeO_2 nanocrystals were clearly observed at both landing voltages. However, the image at 1 kV landing voltage had much finer details of the surface steps because of the smaller penetration depth.

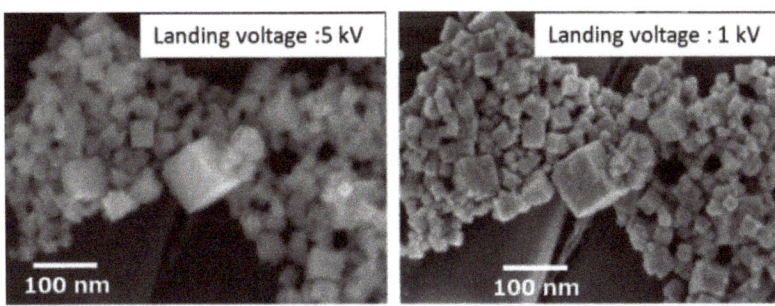

Figure 1. SEM images of CeO_2 nanoclusters with different electron landing voltages.

2.5. Cross Section Polisher

A novel sample preparation technique for SEM observation of these materials was also developed. We used a cross section polisher (IB-19520CCP, JEOL, Tokyo, Japan) to irradiate the sample with an Ar ion beam to display the internal structures without causing serious artificial effects. The design of this technique is shown in Figure 2. A shield plate was placed on top of the specimen to determine its position. Only the portions left uncovered by the shield plate were milled using an Ar ion beam.

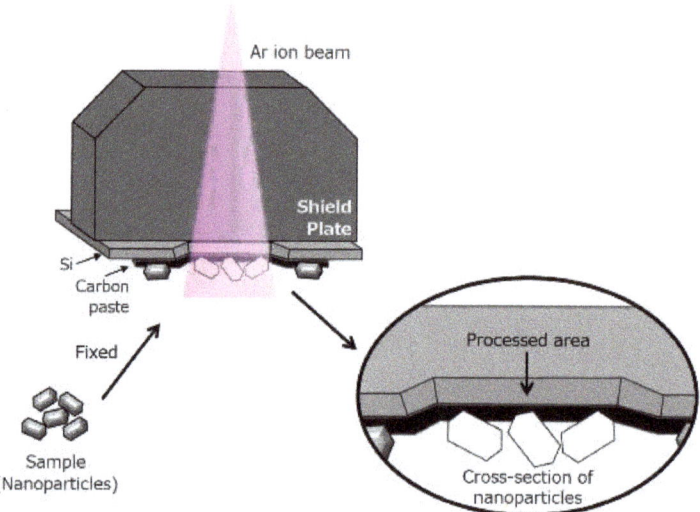

Figure 2. Precise cross sectioning using broad Ar ion beam.

2.6. Multi-Energy Dispersive X-ray Spectroscopy (EDS)

The recent development of a silicon drifted detector (SDD) detector significantly enlarged the detection area of characteristic X-rays, which enabled a large solid angle for X-ray detection. The SDD also improved the electronic noise and dynamic range. Owing to these improvements, more characteristic X-rays can be detected, which allowed the detection of faint characteristic X-ray signals with improved signal to noise (S/N) ratio during SEM measurement. Thus, the necessary acquisition time for EDS mapping is shortened dramatically. In this experiment, we employed a multi-EDS system with a total detection area of 440 mm^2, with a total solid angle of more than 0.15 sr from Oxford Instruments.

2.7. In Situ Holder

In most cases, SEM observation is performed in vacuo because electrons are scattered by molecules in air or in water. Therefore, the observation of nanomaterials in water remains difficult. Recently, we attempted to use an in situ holder (FlowVIEW Tek) with a 30 nm thick Si$_3$N$_4$ window. Figure 3 shows a schematic of the in situ holder. This holder enables SEM observation of nanomaterials in water by detecting electrons penetrated through the window [21]. This window has sufficient strength to hold water during the SEM observation. Si$_3$N$_4$ is composed of lighter elements; thus, electrons are less scattered by this window.

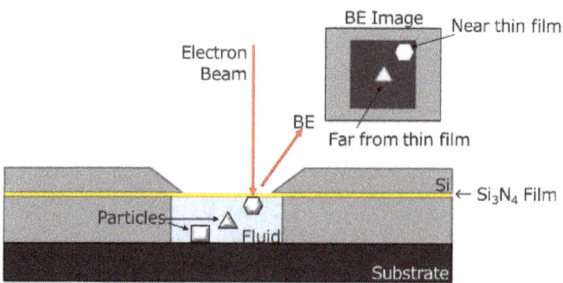

Figure 3. Schematic draw of in situ holder.

3. Results and Discussion

Figure 4 shows a high spatial resolution image of hydrothermally produced CeO_2 nanocrystals on a Si substrate. The cubic shape of the surface-modified CeO_2 nanocrystals was well resolved, even at several nanometers. This image was taken with a landing energy of 1 keV, and the sample bias was set to −5 kV. It is clearly confirmed that the low-landing voltage technique visualizes more details of the samples.

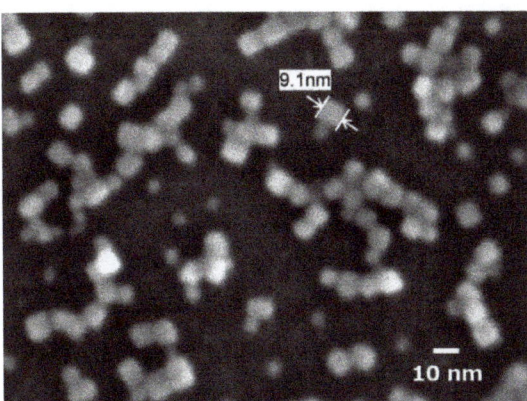

Landing Voltage: 1 kV Sample bias: -5 kV
Sample: CeO_2 Nano particle on Si

Figure 4. Low-voltage high spatial resolution imaging.

This technique is also applicable to mesoporous materials. Figure 5a shows SBA-15 mesoporous silica, where CeO_2 was nano-cast. The SEM image was obtained with a landing voltage of 1 kV and sample bias of −5 kV. Because of the reduced landing voltage, CeO_2 nanoclusters on or just below the surface of SBA-15 were selectively visualized as lighter contracts [16]. Moreover, to observe the CeO_2 nanoclusters located deep inside SBA-15, the sample was processed using an Ar ion beam cross-sectional polisher (Figure 2). Cross-sectional images are shown in Figure 5b,c. During the observation of the cross section, an energy filter (2 kV) was used to selectively collect the back-scattered electron signals. The landing voltage was set to 2.0 kV, and a −0.5 kV bias voltage was added to the sample. The SEM images showed that the hexagonal mesochannels of SBA-15 were clearly confirmed from the cross section. At the same time, many CeO_2 nanoclusters were confirmed in the mesopores of SBA-15. The shape and size of the prepared CeO_2 nanocrystals were in good agreement with those of the pores of mesoporous silica used as the template. From the longitudinal directional cross section image (Figure 5b), the distribution of CeO_2 clusters in SBA-15 and their sizes can be estimated.

Furthermore, the effect of the low-voltage technique was remarkable in the EDS analyses. Generally, high electron probe currents are required to generate characteristic X-rays from samples for EDS analyses. SEM images of the CeO_2 nanocrystals (Figure 6) were obtained with a landing voltage of 8 kV. The elemental maps of Ce and O were clearly visible. However, the elemental map of C was blurred because the incident electrons were not fully scattered by the CeO_2 nanocrystals with sizes less than 50 nm. The penetrated electrons interacted with the carbon substrate under CeO_2 and emitted characteristic X-rays. As a result, X-ray signals were detected from the entire area of the carbon substrate and formed an EDS map of C with a slight view of the nanocrystals. On the other hand, the electrons with a lower landing voltage (1.5 kV) could only interact with the surface of the sample to emit the characteristic X-rays. Therefore, the EDS map of C did not show any signals from the substrate under the nanocrystals.

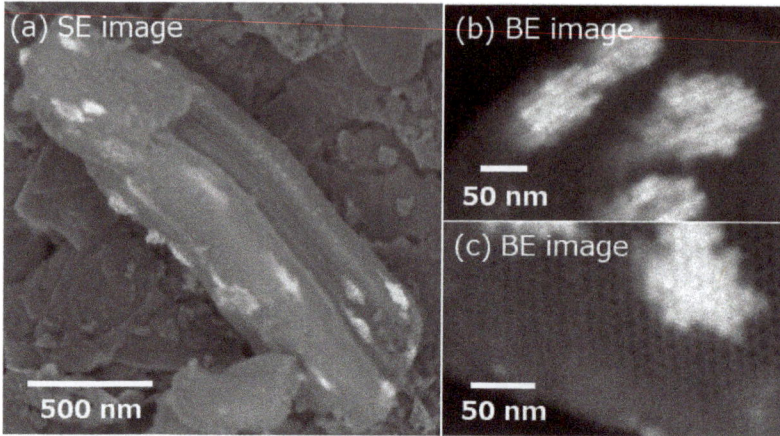

SEM Condition for (a)
Landing Voltage: 1.0 kV
Acc vol: 6.0 kV
Sample bias: -5 kV

SEM condition for (b) and (c)
Landing Voltage: 2.0 kV
Acc vol: 2.5 kV,
Sample bias: -0.5 kV
UED filter: 2.0kV

Figure 5. SEM images of mesoporous silica SBA-15 embedded with nano-casted CeO_2. (**a**) Surface image of mesoporous silica SBA-15 which has been nano-casted with CeO_2. (**b**) Cross section of longitudinal direction for CeO_2 in meso pores. (**c**) Cross section of CeO_2 in meso pores.

The simulation results of the electron penetration depth in 50 nm CeO_2 nanocrystals are also shown in Figure 7. The penetration depth of electrons with a landing voltage of 1.5 kV is less than 50 nm. This indicates that even for the EDS analysis, the low-voltage technique can also promote the selective acquisition of surface information.

Figure 8a shows an image of the CeO_2 nanocrystals in water. Usually, nanomaterials which are smaller than 50 nm in water cannot be observed using SEM. However, as shown in Figure 3, the use of an in situ holder enabled observation. As confirmed in Figure 8a, the size and dispersive of the CeO_2 nanocrystals in water were visualized. By comparing the SEM images obtained in vacuum (Figure 8b), we noticed that the measurement of nanomaterials in water well reproduced the size and dispersive of the CeO_2 nanocrystals.

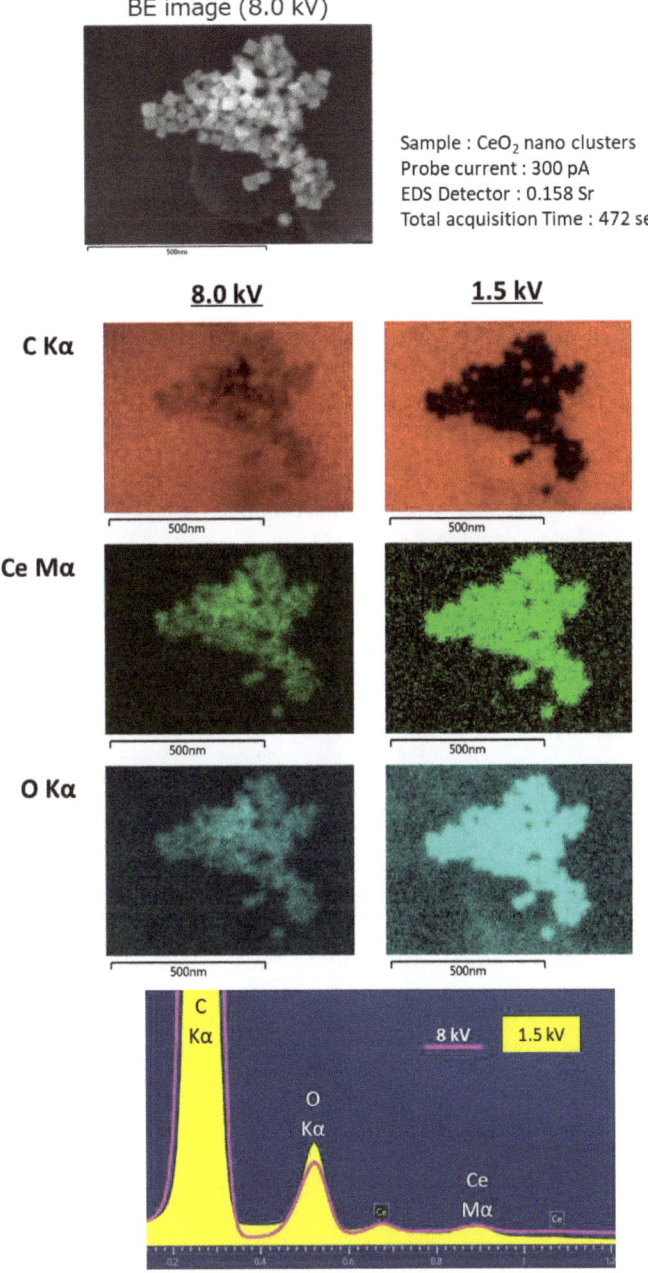

Figure 6. Results of large solid angle Energy Dispersive X-ray Spectroscopy (EDS).

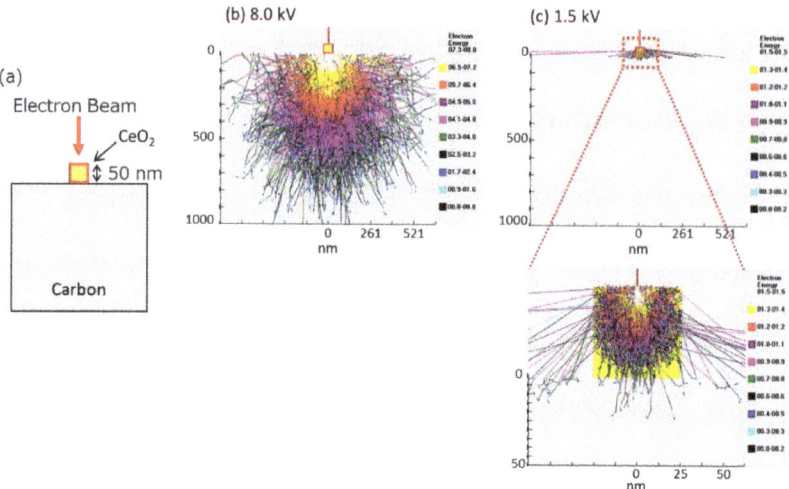

Figure 7. Calculation of electron penetration depth in 50 nm CeO_2. (**a**) A schematic image of this simulation. (**b**,**c**) Results of simulations in different acceleration voltages, 8.0 kV, 1.5 kV.

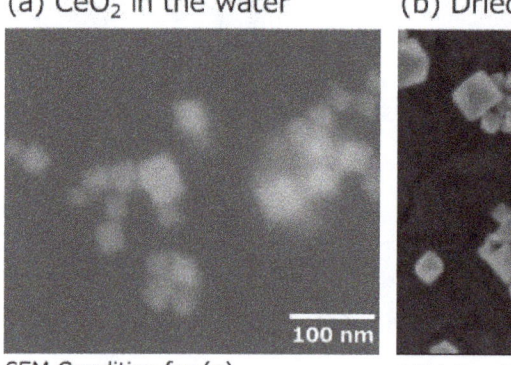

SEM Condition for (a)
Sample : CeO_2 nano crystals in water
Acceleration voltage: 10.0 kV
Vacuum condition: 50 Pa
Detector : BSE detector

SEM Condition for (b)
Sample : CeO_2 nano crystals
Acceleration voltage: 1.0 kV
Vacuum condition: high vacuum
Detector : SE detector

Figure 8. BSE image of in situ observation for CeO_2 in water. (**a**) An image of the CeO_2 nanocrystals in water. (**b**) An image of the dried CeO_2 nanocrystals.

4. Conclusions

Various shapes of hydrothermally synthesized CeO_2 nanocrystals and nanoclusters were successfully observed using recent SEM techniques. SEM images are sensitive to the surface structure of the samples, and thus the detailed morphology that related to crystal growth at the nanometer scale can be visualized. Furthermore, the cross-sectional images of mesoporous silica embedded with CeO_2 nanoparticles clearly show how the nanoparticles were produced in the mesopores. In addition, by using an in situ holder with an ultrathin Si_3N_4 window, CeO_2 nanocrystals in water were directly observed. It is possible to directly observe the structure that is thought to exhibit the function of the nanoclusters of CeO_2 synthesized under hydrothermal conditions by using recent SEM techniques. Most researchers are currently using TEM for the analysis of nanocrystals and nanoclusters. However, the development of SEM with improved performance and usability enabled us to obtain the high-resolution images of hydrothermally synthesized nanoparticles. In addition to the size and morphology, the surface structure can be clearly confirmed for the samples with the size of 10–100 nm. Moreover, it is possible to perform high-quality elemental analysis when the particle size is larger than 50 nm. By providing high-resolution imaging and high-quality analysis, SEM can now deepen the understanding of the essence of nanotechnology.

5. Patents

The cross section method which was used patented as JP 4922632.

Author Contributions: Conceptualization, S.A. and S.T.; investigation, N.A.; writing—original draft preparation, S.A.; writing—review and editing, J.L.; visualization, N.A.; supervision, S.T.; All authors have read and agreed to the published version of the manuscript.

Funding: This research received no external funding.

Data Availability Statement: Not applicable.

Conflicts of Interest: The authors declare no conflict of interest.

References

1. Nunes, D.; Pimentel, A.; Santos, L.; Barquinha, P.; Pereira, L.; Fortunato, E.; Martins, R. *Metal Oxide Nanostructures*, 1st ed.; Elsevier Science: Cambridge, MA, USA, 2018; pp. 59–102.
2. Nikam, A.V.; Prasad, B.L.V.; Kulkarni, A.A. Wet Chemical Synthesis of Metal Oxide Nanoparticles: A Review. *CrystEngComm* **2018**, *20*, 5091–5107. [CrossRef]
3. Chavali, M.S.; Nikolova, M.P. Metal Oxide Nanoparticles and Their Applications in Nanotechnology. *SN Appl. Sci.* **2019**, *1*, 1–30. [CrossRef]
4. Oskam, G. Metal oxide nanoparticles: Synthesis, characterization and application. *J. Sol-Gel Sci. Technol.* **2006**, *37*, 161–164. [CrossRef]
5. Darr, J.A.; Zhang, J.Y.; Makwana, N.; Weng, X.L. Continuous Hydrothermal Synthesis of Inorganic Nanoparticles: Applications and Future Directions. *Chem. Rev.* **2017**, *17*, 11125–11238. [CrossRef] [PubMed]
6. Liu, H.; Zhang, S.Y.; Liu, Y.; Yang, Z.; Feng, X.; Lu, X.H.; Huo, F.W. Well-Dispersed and Size-Controlled Supported Metal Oxide Nanoparticles Derived from MOF composites and Further Application in Catalysis. *Small* **2015**, *11*, 3130–3134. [CrossRef] [PubMed]
7. Dhall, A.; Self, W. Cerium Oxide nanoparticles: A Brief Review of Their Synthesis Methods and Biomedical Applications. *Antioxidants* **2018**, *7*, 97. [CrossRef] [PubMed]
8. Shi, W.D.; Song, S.S.; Zhang, H.J. Hydrothermal Synthetic Strategies of Inorganic Semiconducting Nanostructures. *Chem. Soc. Rev.* **2013**, *42*, 5714–5743. [CrossRef] [PubMed]
9. Taguhi, M.; Takami, S.; Naka, T.; Adschiri, T. Growth Mechanism and Surface Chemical Characteristics of dicarboxylic Acid-Modified CeO_2 Nanoparticles Produced in Supercritical water: Tailor-Made Water-Soluble CeO_2 Nanocrystals. *Cryst. Growth Des.* **2009**, *9*, 5297–5303. [CrossRef]
10. Takami, S. Hydrothermal Synthesis of Organic Modified Metal Oxide Nanoparticles. *Mater. Jpn.* **2020**, *59*, 199–206. [CrossRef]
11. Modena, M.M.; Ruhle, B.; Burg, T.P.; Wuttke, S. Nanoparticle Characterization: What to Measure? *Adv. Mater.* **2019**, *31*, 1970226. [CrossRef]
12. Mourdikoudis, S.; Pallares, R.M.; Thanh, N.T.K. Characterization Techniques for Nanoparticles: Comparison and Complementarity upon Studying Nanoparticle Properties. *Nanoscale* **2018**, *10*, 12871–12934. [CrossRef] [PubMed]
13. Kjellman, T.; Asahina, S.; Schmitt, J.; Imperor-Clerc, M.; Terasaki, O.; Alfredsson, V. Direct Observation of Plugs and Intrawall Pores in SBA-15 using Low Voltage High Resolution Scanning Electron Microscopy and the Influence of Solvent Properties on Plug-Formation. *Chem. Mater.* **2013**, *25*, 4105–4112. [CrossRef]
14. Kobayashi, M.; Susuki, K.; Otsuji, H.; Sakuda, Y.; Asahina, S.; Kikuchi, N.; Kanazawa, T.; Kuroda, Y.; Wada, H.; Shimojima, A. Direct Observation of the Outermost Surfaces of Mesoporous Silica Thin Films by High Resolution Ultralow Voltage Scanning Electron Microscopy. *Langmuir* **2017**, *33*, 2148–2156. [CrossRef] [PubMed]
15. Asahina, S.; Uno, S.; Suga, M.; Stevens, S.M.; Klingstedt, M.; Okano, Y.; Kudo, M.; Schuth, F.; Anderson, M.W.; Adschiri, T.; et al. A New HRSEM Approach to Observe Fine Structures of Novel Nanostructured Materials. *Microporous Mesoporous Mater.* **2011**, *146*, 11–17. [CrossRef]

16. Jahangiri, A.R.; Kakvani, P.R.; Shapouri, S.; Sari, A.; Talu, S.; Jalili, Y.S. Quantitative SEM Characterization of Ceramic Target Prior and After Magnetron Sputtering: A Case Study of Aluminium Zinc Oxide. *J. Microsc.* **2021**, *281*, 190–201. [CrossRef] [PubMed]
17. Asahina, S.; Takami, S.; Otsuka, T.; Adschiri, T.; Terasaki, O. Exploitation of Surface-Sensitive Electrons in Scanning Electron Microscopy Reveals the Formation Mechanism of New Cubic and Truncated Octahedral CeO_2 Nanoparticles. *ChemCatChem* **2011**, *3*, 1028–1044. [CrossRef]
18. Hojo, D.; Togashi, T.; Iwasa, D.; Arita, T.; Minami, K.; Takami, S.; Adschiri, T. Fabrication of two-Dimensional Structures of Metal Oxide Nanocrystals Using Si Substrate Modified with 3,4-Dihydroxyhydrocinnamic Acid. *Chem. Mater.* **2010**, *22*, 1862–1869. [CrossRef]
19. Lu, J.; Asahina, S.; Takami, S.; Yoko, A.; Seong, G.; Tomai, T.; Adschiri, T. Interconnected 3D Framework of CeO_2 with High Oxygen Storage Capacity: High-Resolution Scanning Electron Microscopic Observation. *ACS Appl. Nano Mater.* **2020**, *3*, 2346–2353. [CrossRef]
20. Reimer, L. *Scanning Electron Microscopy: Physics of Image Formation and Microanalysis*, 2nd ed.; Springer: Heidelberg, Germany, 1998; pp. 30–32.
21. Lee, T.S.; Patil, S.B.; Kao, Y.T.; An, J.Y.; Lee, Y.C.; Lai, Y.H.; Chang, C.K.; Cheng, Y.S.; Chuang, Y.C.; Sheu, H.S.; et al. Real-Time Observation of Anion Reaction in High-Performance Al Ion Batteries. *ACS Appl. Mater. Interfaces* **2020**, *12*, 2572–2580. [CrossRef] [PubMed]

Article

The Role of the Surface Acid–Base Nature of Nanocrystalline Hydroxyapatite Catalysts in the 1,6-Hexanediol Conversion

Asato Nakagiri, Kazuya Imamura, Kazumichi Yanagisawa and Ayumu Onda *

Research Laboratory of Hydrothermal Chemistry, Faculty of Science, Kochi University, 2-17-47 Asakurahonmachi, Kochi 780-8073, Japan; nkgrpon3@gmail.com (A.N.); imamura-kazuya@kochi-u.ac.jp (K.I.); yanagi@kochi-u.ac.jp (K.Y.)
* Correspondence: aonda@kochi-u.ac.jp; Tel.: +81-88-844-8353

Abstract: Hydroxyapatite is known to have excellent catalytic properties for ethanol conversion and lactic acid conversion, and their properties are influenced by the elemental composition, such as Ca/P ratio and sodium content. However, few reports have been examined for the surface acid–base nature of hydroxyapatites containing sodium ions. We prepared nanocrystalline hydroxyapatite (Ca-HAP) catalysts with various Ca/P ratios and sodium contents by the hydrothermal method. The adsorption and desorption experiments using NH_3 and CO_2 molecules and the catalytic reactions for 2-propenol conversion revealed that the surface acid–base natures changed continuously with the bulk Ca/P ratios. Furthermore, the new catalytic properties of hydroxyapatite were exhibited for 1,6-hexanediol conversion. The non-stoichiometric Ca-HAP(1.54) catalyst with sodium ions of 2.3 wt% and a Ca/P molar ratio of 1.54 gave a high 5-hexen-1-ol yield of 68%. In contrast, the Ca-HAP(1.72) catalyst, with a Ca/P molar ratio of 1.72, gave a high cyclopentanemethanol yield of 42%. Both yields were the highest ever reported in the relevant literature. It was shown that hydroxyapatite also has excellent catalytic properties for alkanediol conversion because the surface acid–base properties can be continuously controlled by the elemental compositions, such as bulk Ca/P ratios and sodium contents.

Keywords: hydroxyapatite; acid–base catalyst; hydrothermal synthesis; 1,6-hexanediol; nanocrystalline materials; sodium containing hydroxyapatite

1. Introduction

Hydroxyapatite (Ca-HAP) is a major component of bone and tooth enamel. Ca-HAP has a high thermal stability and biocompatibility. The stoichiometric form of Ca-HAP is shown as $Ca_{10}(PO_4)_6(OH)_2$, and its Ca/P molar ratio is 1.67. However, while maintaining the single phase of apatite crystal structure, it is possible to change the Ca/P ratio ranging from 1.5 to 1.7 and substitute the constituents of Ca^{2+}, PO_4^{3-}, and OH^- with other elements to some extents [1]. Ca-HAP exhibits various unique properties, depending on changes in the chemical compositions. In recent years, much research has been conducted on the synthesis and application of Ca-HAP, and its application is expected in the fields of adsorbents, catalysts and catalyst supports, composite materials with polymers, biomedical materials, and nanopapers.

Many research examples have been reported in recent years regarding the catalytic properties of Ca-HAP [2–9]. Ca-HAP exhibits highly stable against reaction conditions and no highly acidic and/or base properties, so it is appropriate as a catalyst. These properties are particularly effective in converting oxygen-containing organic compounds, such as alcohols and carboxylic acids, at reaction temperatures of around 300–400 °C. Previous studies revealed a high selectivity of 1-butanol production in ethanol conversion [8–10] and a high selectivity of acrylic acid production in lactic acid conversion [11–14]. Many studies have been conducted on the effects of element types and Ca/P ratios on these reactions [8–13,15]. In addition, in lactic acid conversion, the Na-containing HAP catalyst showed a particularly high acrylic acid yield [12]. It has been reported to have catalytic deactivation resistance by the other group [16].

However, although hydroxyapatite is expected to be effective for selective conversion of many oxygen-containing organic compounds, such as alkanediols, there are few reports of attractive catalytic properties. Alkanediol is now produced from petroleum and widely used as a polymer raw material. In recent years, it has been found that alkanediols, such as ethylene glycol, propylene glycol, and hexanediol, were obtained from biomass-derived compounds, such as glucose, [17–22] and it is expected as one of basic renewable raw materials for various bulk and fine chemicals in the future to realizing a sustainable society [23–26].

For use hydroxyapatite as a catalyst with high catalytic activity and high product selectivity, high specific surface area and high surface uniformity are important. In addition, it is expected to change the surface Ca/P ratios of hydroxyapatite nanocrystalline particles. In general, the hydrothermal method is suitable for synthesizing high crystalline nanoparticles. However, in the hydrothermal synthesizing methods, nanoparticles grow from the crystal nucleus vis the repeating of dissolution and precipitation. Therefore, does the surface structure and catalytic properties continuously change in response to changes in bulk Ca/P ratios, or does those surface properties not continuously change but maintain a relatively stable surface? The question about whether to take the structure is not revealed. In particular, there are few studies on the effect of the Ca/P ratio on Na-containing Ca-HAP particles which have unique catalytic properties.

In this study, nanocrystalline Na-containing hydroxyapatite catalysts with various Ca/P ratios were prepared by the hydrothermal method, and the effects of Ca/P ratios of the nanocrystalline HAPs on the catalytic properties were examined. The relationship between the acid–base adsorption properties and the catalytic properties of the HAPs with various bulk Ca/P ratio were investigated to clarify whether the surface structures and catalytic properties continuously changes with changes in the bulk Ca/P ratio. In addition, this study first reports the unique catalytic properties of nanocrystalline Ca-HAP in 1,6 hexanediol conversions (Scheme 1). By controlling Ca/P ratios of nanocrystalline Ca-HAP catalysts, the 1,6 hexanediol conversion provides both useful compounds, unsaturated alcohols and cycloalkane alcohols, with higher yields and selectivity than the previously reported catalysts.

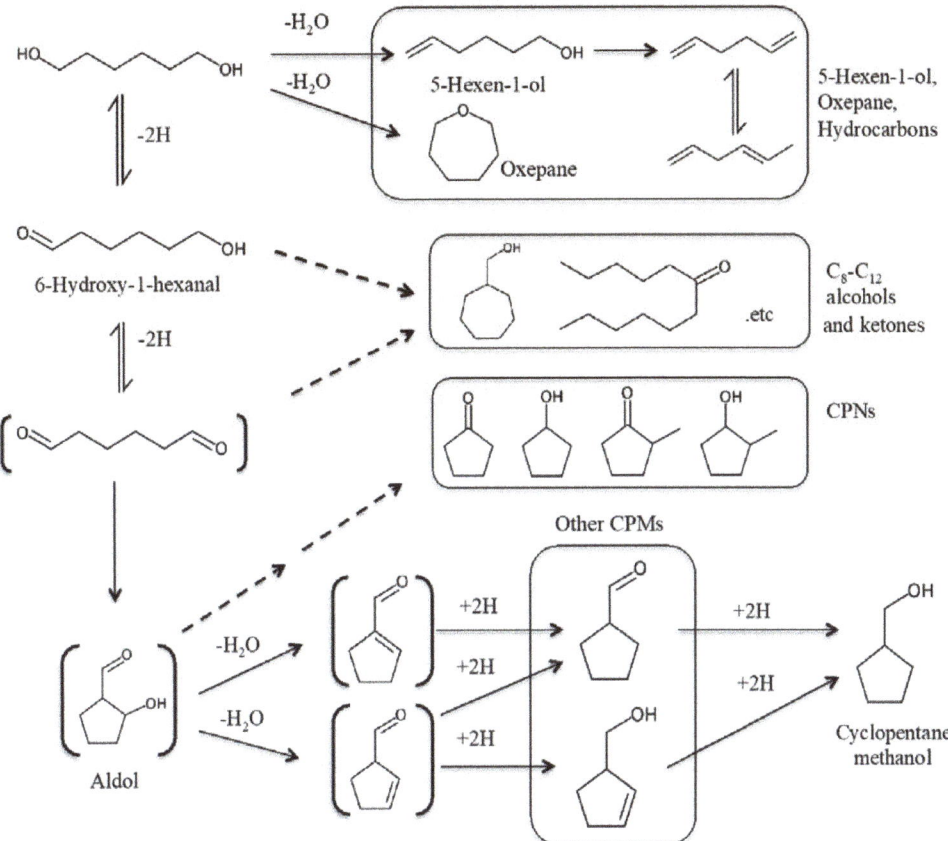

Scheme 1. Probable reaction route of 1,6-hexanediol conversion based on reference [24].

2. Materials and Methods

2.1. Materials

For use in this study, Ca(NO$_3$)$_2$·4H$_2$O, P$_2$O$_5$, NaOH, SiO$_2$, Ca(OH)$_2$ and 2-propanol were purchased from Fujifilm Wako Pure Chemical Co., Osaka, Japan. The necessary 1,6-hexanediol and ethanol were purchased from Sigma-Aldrich Co., St. Louis, MO, USA. A commercially available stoichiometric hydroxyapatite (HAP-100) was a supplied by Taihei Chemical Industrial Co. Ltd., Osaka, Japan. Furthermore, Sc$_2$O$_3$ was purchased from Kanto Chemical Co. Inc., Tokyo, Japan. In addition, ZrO$_2$ were supplied from the Catalysis Society of Japan (Tokyo, Japan) as JRC-ZRO-7. All chemicals were used as received without purification.

2.2. Catalyst Preparations

A calcium-phosphorus hydroxyapatite catalyst (Ca-HAP) was prepared using hydrothermal methods [5,8]. First, a solution containing Ca(NO$_3$)$_2$·4H$_2$O (132–238 mmol) in 564 mL of distilled water was added to a solution containing P$_2$O$_5$ (66 mmol) in 490 mL of distilled water with NaOH (460 mmol). Each preparation of the Ca/P molar ratio in the mixed solution was between 1.0 and 1.8. The resultant suspension was treated under hydrothermal conditions at 110 °C for 16 h with agitation in a Teflon-lined autoclave. After hydrothermal treatments, the resultant precipitates were washed repeatedly using centrifugal separation with distilled water; then they were dried overnight at 100 °C.

Using an impregnation method, SiO$_2$-supported P$_2$O$_5$ catalyst (P$_2$O$_5$/SiO$_2$) was prepared. The 0.9 g of silica-gel was added to 2 mL of 0.5 molL^{-1} P$_2$O$_5$ aqueous solution. Then the mixture was stirred on a water bath until dry. The obtained material was dried further at 60 °C overnight. Before catalytic tests and characterizations, the prepared HAP catalysts were typically pretreated at 500 °C under vacuum or inert gas.

2.3. Catalyst Characterizations

Crystalline phases of as-prepared and used catalysts were identified using powder X-ray diffraction (XRD, Ultima IV; Rigaku Co., Tokyo, Japan) with Cu K radiation (40 kV, 20 mA). The prepared samples were observed using transmission electron microscopy (TEM, H-7000; Hitachi Ltd., Tokyo, Japan). The elemental compositions of hydroxyapatites were ascertained using inductively coupled plasma (ICP, ICPE-9000; Shimadzu Co., Kyoto, Japan). Approximately 4 mg of sample was dissolved in 10 mL of 0.1 M HNO$_3$ solution, then it was diluted to 100 mL using distilled water. The ICP measurement error for Ca/P molar ratio was ±0.01. The specific surface areas of as-prepared and used catalysts were measured using nitrogen physisorption with the Brunauer-Emmett-Teller (BET) method (Belsorp Max; BEL Japan Inc., Osaka, Japan). The particle sizes of the prepared Ca-HAP samples were determined by the manual approach and calculated by averaging the diameters of ~150 particles in the TEM images. In addition, the particle sizes were also roughly estimated based on BET surface areas by assuming the density of 3.1 g cm^{-3} and spherical shape particles and also roughly calculated by using Scherrer's equation based on X-ray diffraction peak of (002). The temperature-programmed desorption (TPD) of CO$_2$ were carried out using a glass U-tube reactor equipped with a quadrupole mass spectrometer (CANON ANELVA M-201QA-TDM, Kawasaki, Japan). Samples (0.1 g) were preheated in He at 500 °C and exposure to a 10% CO$_2$/He (50 kPa) gas at room temperature until saturation coverages were reached. Weakly adsorbed CO$_2$ was removed by flushing with He at room temperature for 30 min. The temperature was then increased at a linear rate of 5 °C/min from 20 to 800 °C. The TPD spectra were normalized at the specific surface area of samples. The TPD of NH$_3$ was carried out by almost same method as that of TPD of CO$_2$, except for using 10% NH$_3$/He gas. The densities of acidic sites and basic sites were estimated respectively from the amounts of adsorbed NH$_3$ and CO$_2$, which had been determined by the isotherms at 250 °C using Belsorp Max instrument after the pretreatments of catalysts at 430 °C for 3 h under vacuum. The temperature of 430 °C was maximum for a heating oven attached with the instrument.

2.4. Catalytic Reaction

Catalytic conversions of 1,6-hexanediol were typically conducted at 375 °C using 0.1 g of catalysts in a fixed-bed continuous-flow glass reactor (8 mm i.d.) under atmospheric pressure. After the sample powders were pelletized and crushed to the desired size (250–500 µm) to avoid pressure gradients in reactor, they were pretreated at 500 °C for 3 h in N$_2$ flow before catalytic reactions. Then 1,6-hexanediol was diluted by ethanol (1,6-hexanediol concentration was 10 mol%). The mixed solution was introduced into the reactor using a syringe pump with flowing N$_2$ (30 mL min^{-1}). The reaction products were condensed in a cold ethanol trap at −50 °C. The collected liquid samples and gas samples at the outlet of the trap were analyzed by using a GC-FID (GC-2014; Shimadzu Co., Kyoto, Japan) With a DB-wax capillary column (30 m, 0.32 mm) and a GC-MS (6890N/5973N; Agilent Technologies, Inc., Santa Clara, CA, USA) with a DB-1 capillary column (60 m, 0.25 mm). The 1,6-hexanediol conversions, product yields, and product selectivities were calculated based on the following equations.

$$1,6-\text{Hexanediol conversion (C}-\%) = \left(1 - \frac{\text{C mmol unreacted } 1,6-\text{hexanediol}}{\text{C mmol of introduced } 1,6-\text{hexanediol}}\right) \times 100$$

$$\text{Product yield (C}-\%) = \frac{\text{C mmol of product}}{\text{C mmol of introduced } 1,6-\text{hexanediol}} \times 100$$

$$\text{Product selectivity (C}-\%) = \frac{\text{Product yield}}{1,6-\text{hexanediol conversion}} \times 100$$

The mass balance was calculated as the ratio of the weight of collected liquid to the weight of introduced solution. The mass balance in ethanol conversion over the Ca-HAP catalysts was $100 \pm 1\%$.

The catalytic conversions of 2-propanol were conducted using a fixed-bed continuous-flow glass reactor (10 mm i.d.) under atmospheric pressure to estimate the acid–base catalytic properties. Vapor of 2-propanol was introduced into the reactor with flowing N_2. The total flow rate was 30 mL min^{-1}. The partial pressure of 2-propanol was 960 Pa. The reaction products were analyzed using GC-FID (GC-2010; Shimadzu Co., Kyoto, Japan) with a Stabilwax capillary column (30 m, 0.32 mm).

$$\text{Acetone selectivity (C}-\%) = \frac{\text{C mmol of acetone}}{\text{C mmol of acetone + C mmol of propylene}}$$

3. Results

3.1. Characterization

Calcium-phosphorus hydroxyapatite catalyst (Ca-HAP) was prepared by the hydrothermal method at 110 °C using sodium hydroxide as an alkaline source. The initial Ca/P molar ratios in the mixed solution were between 1.0 and 1.8. The pH of the filtrates after the preparations was 12.2–12.8. Table 1 shows the chemical compositions and the specific surface areas of the prepared Ca-HAP catalysts. In the case of initial Ca/P ratios lower than 1.66, the resulted Ca/P ratios of particles were higher than the initial Ca/P ratios. This tendency was similar to the cases where ammonia was used as an alkali source without sodium species [15]. Whereas in the case of initial Ca/P ratios higher than 1.67, the resulted Ca/P ratios of particles were lower than the initial Ca/P ratios. The Ca/P molar ratios of particles were 1.54–1.72.

Table 1. Characterizations of prepared hydroxyapatite catalysts.

	Initial Solution	Resulted Particles			Surface Area/m^2 g^{-1}		pH
	Ca/P Molar Ratio	Ca/P Molar Ratio	Na Content wt%	(Ca+Na)/P Molar Ratio	As-Prepared	Used in 1,6-Hexanediol Conversion	After Preparation
Ca-HAP(1.54)	1	1.54	2.3	1.69	79	41	12.9
Ca-HAP(1.58)	1.5	1.58	1.7	1.69	65	46	12.8
Ca-HAP(1.62)	1.55	1.62	0.8	1.67	52	41	12.5
Ca-HAP(1.65)	1.67	1.65	0.4	1.67	52	40	12.2
Ca-HAP(1.69)	1.72	1.69	0.0	1.69	50	38	12.1
Ca-HAP(1.72)	1.8	1.72	0.0	1.72	65	55	11.0

As shown in Table 1, the specific surface areas of the prepared Ca-HAP catalysts were 51–79 m^2 g^{-1}. Figure 1a shows powder XRD patterns of prepared Ca-HAP catalysts. All XRD patterns were attributable to hydroxyapatite $Ca_5(PO_4)_3OH$ (JCPDS #01-074-0506) with a single phase. The peak intensities were almost identical, suggesting that the prepared catalysts had siTmilar crystallinity. In previous reports, $Ca(OH)_2$, which coexists in Ca-HAP, with Ca/P of 1.7 or more were considered [1,27]. In this study, hydroxyapatite particles with a Ca/P ratio of 1.7 were obtained to be a single phase by the hydrothermal method. It was considered that Ca-HAP(1.72) could be a solid solution resulting from the partial replacement of PO_4^{3-} ions with CO_3^{2-} ions [1,27]. In consequence, according to the XRD patterns, it was observed that single-phase hydroxyapatite particles were synthesized with a Ca/P molar ratio of 1.54–1.72.

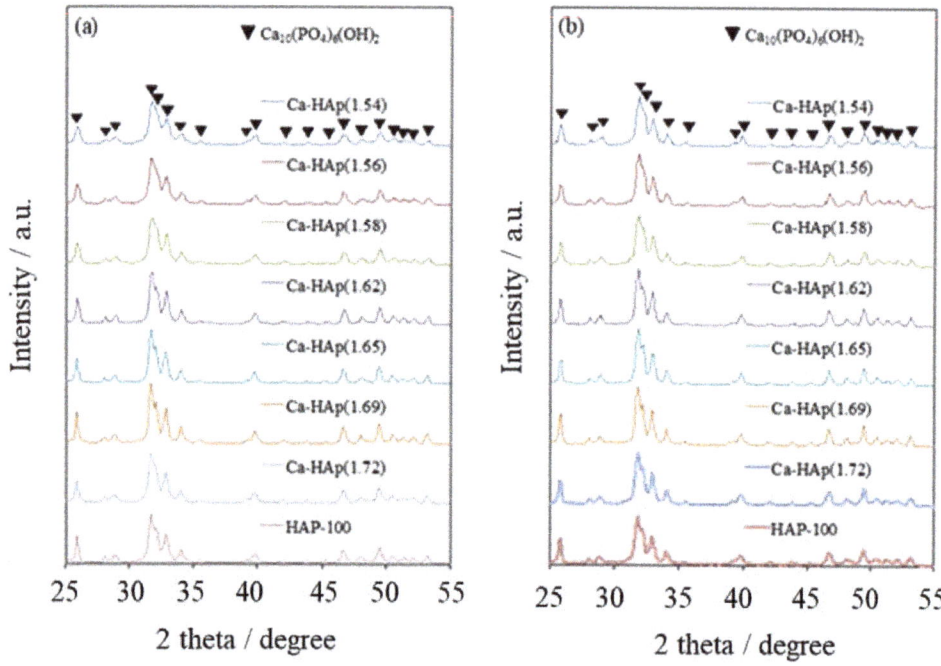

Figure 1. Powder X-ray diffraction (XRD) patterns of hydroxyapatites (**a**) as-prepared and (**b**) after conversion of 1,6-hexanediol.

Some TEM images of the Ca-HAP catalysts are presented in Figure 2. All prepared Ca-HAP hydroxyapatites were column particles with average lengths of 24–45 nm. As shown in Table 2, the average widths were 14–27 nm. The particle sizes increased concomitantly with increasing Ca/P ratios, except for Ca-HAP(1.72). Table 2 also shows the particle sizes obtained by a simple calculation assuming the density of 3.1 g cm^{-3} and spherical shape particles. The particle sizes were 25–39 nm and were closed to the particle sizes observed by TEM. These results indicated that the particles observed by TEM were average values reflecting the whole of particles. Table 2 also shows particle sizes of Ca-HAP catalysts roughly calculated by using Scherrer's equation based on XRD patterns presented in Figure 1 The particle sizes were 25–40 nm and were almost equal to those estimated with TEM, which implied that the particles observed in TEM images were single crystals. In consequence, the prepared Ca-HAP catalysts with various Ca/P molar ratios of 1.54–1.72 had almost identical crystallinity of hydroxyapatite structure to that of single-phase materials, with particle sizes of about 20–40 nm.

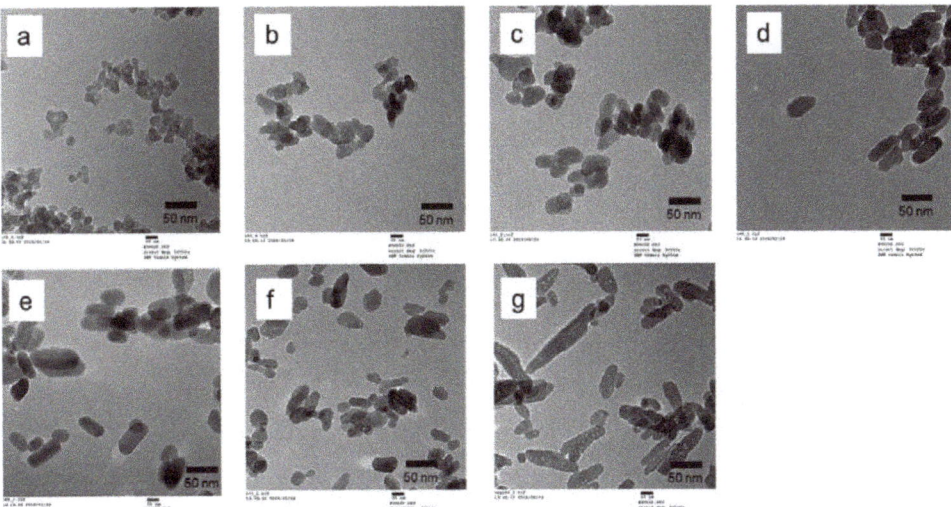

Figure 2. Transmission electron microscopy (TEM) images of the Ca-HAP catalysts. (**a**) Ca-HAP(1.54), (**b**) Ca-HAP(1.58), (**c**) Ca-HAP(1.62), (**d**) Ca-HAP(1.65), (**e**) Ca-HAP(1.69), (**f**) Ca-HAP(1.72), and (**g**) commercially available hydroxyapatite (HAP-100).

Table 2. Average particle sizes of prepared Ca-HAP particles.

	Average Particle Sizes (TEM)		Average Particle Sizes (BET)	Average Particle Sizes (XRD)
	Length/nm	Width/nm	nm	nm
Ca-HAP(1.54)	24	14	25	25
Ca-HAP(1.58)	29	19	28	30
Ca-HAP(1.62)	37	24	36	37
Ca-HAP(1.65)	39	25	33	37
Ca-HAP(1.69)	45	27	40	39
Ca-HAP(1.72)	34	20	29	30

As shown in Table 1, Na$^+$ ions were detected in the prepared Ca-HAP catalysts with Ca/P molar ratios less than 1.65. The Na$^+$ contents increased concomitantly with decreasing Ca/P molar ratios. In addition, the molar ratios of (Ca+Na)/P in the prepared resulted Ca-HAP particles except for Ca-HAP(1.72) were 1.67–1.69, which were closed to stoichiometric ratio of 1.67. These results showed agreement with earlier reports, suggesting that Na$^+$ ions might replace the Ca^{2+} deficiency in the Ca-HAP catalysts with low Ca/P ratios [13]. Because the ion radius of Na$^+$ ion is close to that of Ca^{2+} ions and there were almost the same amounts of Na$^+$ ions as those of the deficient Ca^{2+} ions in the Ca-HAP prepared by the hydrothermal method using NaOH, Na$^+$ ions might occupy the calcium-deficient vacancy-sites in the Ca-HAP particles. Matsunaga and Murata reported that the substitution of Ca^{2+} ions in Ca-HAP by Na$^+$ ions with charge-compensating interstitial H$^+$ ions was energetically favorable, and that the H$^+$ ion is attached to the OH group to form an H$_2$O group [28]. Based on this first principle calculation investigation, the Ca-HAP particles could be expressed as $Ca_{10-n}Na_n(PO_4)_6(OH)_{2-n}(H_2O)_n$.

3.2. Characterization

The adsorption amounts of NH$_3$ and CO$_2$ on the Ca-HAP catalysts at 250 °C were measured. Before the adsorption measurements, the catalyst was pretreated at 500 °C which was the same temperature as the pretreatment of catalytic reactions. The XRD measurement and BET specific surface area were almost the same as the post-reaction data shown in Figure 1b and Table 1, and no crystal phase other than hydroxyapatite was observed, the crystallite size increased slightly, and the specific surface area decreased by about 20%. As shown in Table 3, the NH_3 adsorption amounts increased slightly with the increase of Ca/P molar ratios. The CO_2 adsorption amounts increased greatly with the increase of Ca/P molar ratios. Figures S1 and S2, respectively, portray NH_3-TPD and CO_2-TPD of the Ca-HAP catalysts. The NH_3-TPD intensity increased at around 250 °C with the increase of the Ca/P molar ratios. The CO_2-TPD intensity at around 250 °C also increased concomitantly with the increase of the Ca/P molar ratios. The results corresponded with the adsorbed NH_3 and CO_2 amount data presented in Table 3, indicating that both the basic site density and acidic site density increased concomitantly with increasing Ca/P molar ratios of the Ca-HAP catalysts.

Table 3. Adsorption amounts of NH_3 and CO_2 on Ca-HAP catalysts with various Ca/P molar ratios.

Sample	Adsorption Amount of NH_3 (μmol m^{-2})	Adsorption Amount of CO_2 (μmol m^{-2})
Ca-HAP(1.72)	3.93	1.62
Ca-HAP(1.65)	3.45	1.58
Ca-HAP(1.58)	3.12	0.95
Ca-HAP(1.54)	2.65	0.25

Table 4 shows the 2-propanol conversions over the prepared Ca-HAP catalysts, P_2O_5/SiO_2, $Ca(OH)_2$ catalysts. The specific surface area of each catalyst is presented in Table S1. Acetone and propylene were formed as products. Over Ca-HAP catalysts, CO and CO_2 were not detected. The product selectivity was constant with time on stream for at least 20 h [5]. The selectivity of propylene was 100% over acidic catalysts as P_2O_5/SiO_2. In contrast, the acetone selectivities of basic catalysts as $Ca(OH)_2$ were, respectively, 90% and 99%. The acidic catalyst showed almost 100% of the propylene selectivity. The basic catalyst showed 90% or more of the acetone selectivity. However, the acid–base bifunctional Ca-HAP catalysts with low Ca/P molar ratio showed high selectivity of propylene. Moreover, the acetone selectivity increased concomitantly with increase of the Ca/P molar ratio. These results indicate that Ca-HAP catalysts with a low Ca/P molar ratio exhibited high acidity; the basicity became dominant instead of acidity as the Ca/P molar ratio increased. According to Figure 2 and Table 4, the morphology effects of Ca-HAP catalyst particles on the product selectivities were almost negligible. Table 3 shows that the specific ratios of the basic site density to the acidic site density increased concomitantly with increasing Ca/P molar ratios. The specific ratios of the basic site density to the acidic site density showed good agreement with the acetone selectivity in the 2-propanol conversion. The results suggest that the Ca-HAP catalysts with higher Ca/P molar ratios had higher basic properties to accelerate the dehydrogenation of hydroxy groups of alcohols into ketones and aldehydes preferentially over dehydration into unsaturated compounds.

Table 4. Catalytic conversion of 2-propanol into propylene and acetone.

Catalyst	Selectivity (C-%)	
	Propylene	Acetone
Ca-HAP(1.54)	96	4
Ca-HAP(1.58)	62	38
Ca-HAP(1.62)	45	55
Ca-HAP(1.65)	23	77
Ca-HAP(1.69)	14	86
Ca-HAP(1.72)	4	96
HAP-100	25	75
P_2O_5/SiO_2	100	0
$Ca(OH)_2$	1	99
Sc_2O_5	40	60
ZrO_2	95	5

With a 250 °C reaction temperature, 980 kPa; 2-propanol partial pressure, 30 mL min^{-1}; N$_2$ flow, 0.05–0.2 g; catalyst, 0.3–5.8%; Conversion (100% conversion over P_2O_5/SiO_2).

3.3. Catalytic Conversion of 1,6-Hexanediol over Ca-HAP Catalysts

Catalytic conversions of 1,6-hexanediol over Ca-HAP catalysts with various Ca/P ratios were carried out using a fixed bed reactor at 375 °C (Scheme 1). Before reaction, the catalysts were pretreated at 500 °C under flowing nitrogen. Changes in major product selectivities against the 1,6-hexanediol conversions over Ca-HAP(1.72), Ca-HAP(1.62), and Ca-HAP(1.54) catalysts are presented in Figure 3. The major products were cyclopentanemethanol and 5-hexen-1-ol, and the by-products were 6-hydroxy-1-hexanal, C$_8$-C$_{12}$ alcohols and ketones, and hydrocarbons. In addition, acetaldehyde and 1-butanol were formed as by-products derived from ethanol. Ethanol was a solvent and a hydrogen source and was introduced nine times more than 1,6-hexanediol. However, the total yields of by-products from ethanol, such as acetaldehyde and 1-butanol, were about 2 C-% or less based on 1,6-hexanediol, even when those from 1,6 hexanediol were over 90%. These results implied that 1,6-hexanediol was much more reactive than ethanol on Ca-HAP catalysts.

As portrayed in Figure 3a for Ca-HAP(1.72) catalyst, the main product was cyclopentanemethanol. It increased concomitantly with increasing of conversion up to 96%. The C$_8$-C$_{12}$ oxygenated compound selectivity also increased. In contrast, the selectivities of 6-hydroxy-1-hexanal and the undetected compounds decreased concomitantly with increase of the conversion. The selectivities of 5-hexen-1-ol and hydrocarbons, respectively, recorded as 6% and 1%, were almost constant. As portrayed in Figure 3b for Ca-HAP(1.62) catalyst, the selectivities of the C$_8$-C$_{12}$ oxygenated compounds, cyclopentanemethanol, and hydrocarbons increased concomitantly with increase of the 1,6-hexanediol conversion. In contrast, the selectivities of 6-hydroxy-1-hexanal and 5-hexene-1-ol decreased concomitantly with increasing conversion. Figure 3c shows that the product selectivities of Ca-HAP(1.54) catalyst were less dependent on the conversion than Ca-HAP(1.72) and Ca-HAP(1.62) catalysts. The main product over Ca-HAP(1.54) catalyst was 5-hexene-1-ol. The 5-hexen-1-ol selectivity decreased concomitantly with increase at conversion of more than 90%. Instead, the hydrocarbon selectivity increased.

Over Ca-HAP(1.72) and Ca-HAP(1.62) catalysts, the selectivity of cyclopentanemethanol and the C$_8$-C$_{12}$ oxygenated compounds increased concomitantly with increasing conversion. Instead, the 6-hydroxy-1-hexanal selectivity decreased. Those results indicate that 6-hydroxy-1-hexanal is an intermediate of the formations of cyclopentanemethanol and the C$_8$-C$_{12}$ deoxygenated compounds. Cyclopentanemethanol formation was found to be predominant over Ca-HAP(1.72), whereas the formations of the C$_8$-C$_{12}$ deoxygenated compounds were predominant over Ca-HAP(1.62). The selectivity difference was regarded as attributable to the stronger basic properties of Ca-HAP(1.72) to accelerate the intramolecular aldol reaction. However, over Ca-HAP(1.54) catalyst, the dehydration reaction to 5-hexen-1-ol became dominant. In addition, Table S2 presents results of catalytic conversion of 5-hexen-1-ol over Ca-HAP(1.54) under almost the same reaction conditions. The

5-hexen-1-ol conversion was 20%. This result indicated that 5-hexen-1-ol conversion was less reactive than 1,6-hexanediol conversion over Ca-HAP(1.54) catalyst. For that reason, Ca-HAP(1.54) showed high selectivity of 5-hexen-1-ol.

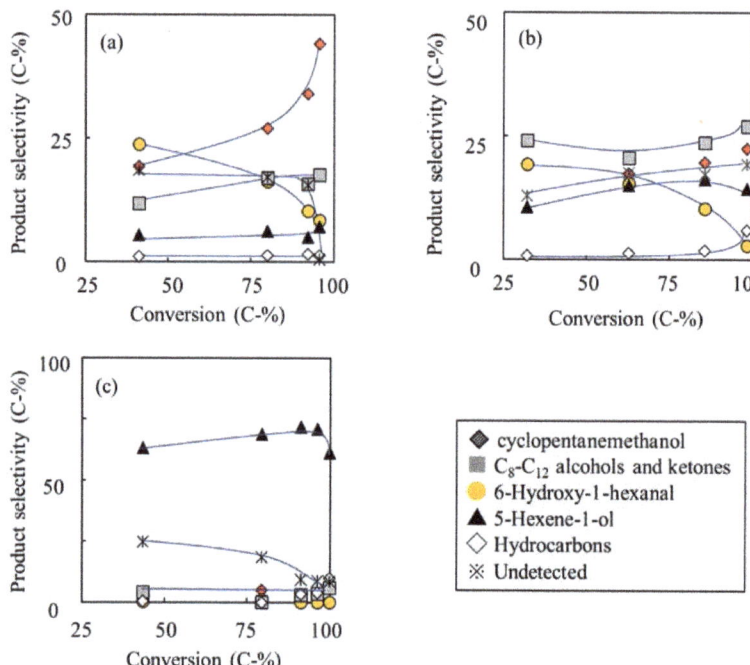

Figure 3. Relation between products selectivity and 1,6-hexanendiol conversion over Ca-HAP catalysts with various Ca/P molar ratios: (**a**) Ca-HAP(1.72), (**b**) Ca-HAP(1.62), and (**c**) Ca-HAP(1.54). Temperature 375 °C; 1,6-hexanediol 10 mol% EtOH solution, 1 mL h^{-1}; N$_2$, 30 mL min^{-1}. Conversion and selectivity were averaged for the initial 3–5 h.

Table 5 presents results of 1,6-hexanediol conversions over the Ca-HAP catalysts and the other acidic, basic, and acid–base bifunctional catalysts at 375 °C. The loading amounts of catalysts were adjusted so that the conversions were 93–98%, except for P$_2$O$_5$/SiO$_2$ catalysts. The conversion and product selectivity were approximately the same during 5 h of time on stream over the entire tested catalysts. The catalysts remained almost white after reactions, except for P$_2$O$_5$/SiO$_2$.

Table 5. 1,6-hexanediol conversion over various catalysts.

Catalyst	Catalyst Weight (g)	Conversion (C-%)	Hydro Carbons	Oxepane	5-Hexen-1-ol	Cyclopentanemethanol	Selectivity (C-%) Other CPMs [a]	CPNs [b]	6-Hydroxy-1-hexanal	C_8-C_{12} -OH, CO [c]	Others (Undetected)
Ca-HAP(1.54)	0.26	96.2	4.0	8.0	71.0	3.9	0.1	0.3	0.3	3.7	8.6
Ca-HAP(1.56)	0.20	93.5	3.9	4.9	38.7	16.2	3.5	1.0	5.6	23.5	6.6
Ca-HAP(1.58)	0.20	93.1	1.7	2.3	23.0	19.9	2.1	0.7	3.3	24.5	22.6
Ca-HAP(1.62)	0.20	98.3	5.6	3.4	14.9	22.4	3.1	1.9	2.9	27.0	18.9
Ca-HAP(1.65)	0.20	93.9	2.8	3.5	13.8	34.7	5.0	1.4	5.7	17.9	15.2
Ca-HAP(1.69)	0.15	97.5	3.8	2.6	6.5	41.7	3.6	2.7	3.1	19.7	16.4
Ca-HAP(1.72)	0.40	95.6	1.3	4.8	7.2	44.2	9.7	6.1	8.5	17.7	0.6
HAP-100	0.18	98.0	2.6	3.2	22.3	26.1	1.7	1.9	0.9	32.2	9.1
P_2O_5/SiO_2	0.20	44.5	1.0	18.8	59.6	0.0	0.0	0.0	0.0	0.0	20.6
Ca(OH)$_2$	3.00	89.7	1.3	1.1	3.0	15.7	2.5	15.4	11.7	16.4	32.8
Sc_2O_3	0.30	93.3	0.7	0.3	61.4	0.4	0.0	0.5	0.0	0.0	36.7
ZrO_2	0.20	97.8	11.9	14.8	36.7	0.4	0.0	0.0	0.0	0.0	36.2

375 °C temperature, 1.0 mL h^{-1}; 1,6-hexanediol 10 mol%; EtOH solution, 30 mL min^{-1}; N$_2$, 5 h time on stream. [a] Other CPMs; 2-cyclopenthenylmethanol+cyclopentacarbaldehyde. [b] CPNs; cyclopentanone+cyclopentanol+2-methyl cyclopentanone+2-methyl cyclopentanonol. [c] C_8-C_{12}-OH, -CO; C_8-C_{12} oxygenated compounds such as cycloheptanemethanol, 2-ethyl-1-hexanol, and 6-undecanone.

Among the prepared Ca-HAP catalysts in Table 5, the selectivity of 5-hexen-1-ol was maximized using Ca-HAP(1.54) catalyst. The selectivity decreased concomitantly with decrease of the Ca/P molar ratio. By contrast, the highest cyclopentanemethanol selectivity was obtained using Ca-HAP(1.72) catalyst. The selectivity increased concomitantly with increasing Ca/P molar ratio. Furthermore, the products formed were 6-hydroxy-1-hexanal, oxepane, and other CPMs, i.e., 2-cyclopentenylmethanol and cyclopentacarbaldehyde, and CPNs, i.e., cyclopentanone, cyclopentanol, 2-methylcyclopentanone, and 2-methylcyclopentanol, C_8-C_{12} oxygenated compounds, such as cycloheptanemethanol, 2-ethyl-1-hexanol, 6-undecanone, and C_4-C_6 hydrocarbons. The selectivities of other CPMs and CPNs increased concomitantly with increase of Ca/P molar ratio. A similar tendency was apparent: 6-hydroxy-1-hexanal selectivity was the lowest at 0.3% over Ca-HAP(1.54) and the highest at 8.5% over Ca-HAP(1.72). Conversely, the hydrocarbon selectivity was the highest at 4.0% over Ca-HAP1.54 and the lowest over Ca-HAP1.72. The selectivities of the C_8-C_{12} oxygenated compounds were higher over Ca-HAP catalysts with medium Ca/P molar ratios as Ca-HAP(1.62). The specific catalytic activities per the specific surface areas among Ca-HAP catalysts were as follows:

Ca-HAP(1.69) > Ca-HAP(1.65), Ca-HAP(1.62), Ca-HAP(1.58) > Ca-HAP(1.54) > Ca-HAP(1.72)

Ca-HAP(1.69) with almost the stoichiometric Ca/P ratio had the highest catalytic activity per specific surface area, and the catalytic activity decreased with decreasing the Ca/P ratio decreases, and the catalytic activity of Ca-HAP(1.72) with a higher Ca/P ratio than the stoichiometry decreased by about half. The cause of the different catalytic activity was complicated, but it is considered that it is mainly due to the difference in product selectivity due to the difference in surface acid–base nature of the Ca-HAP catalysts with various Ca/P ratios. In addition, the lower the aspect ratios of the morphology and the smaller the particles, the lower the activity per specific surface area tended to be, suggesting that the morphology of Ca-HAP catalysts might also have an effect on the catalytic activity.

Table 5 also shows the catalytic properties of phosphoric acid and calcium hydroxide, which were the constituents of Ca-HAP catalysts. The P_2O_5/SiO_2 catalyst showed about 60% selectivity to 5-hexene-1-ol and 20% selectivity to oxepane. $Ca(OH)_2$ catalysts with specific surface area of 16 $m^2\ g^{-1}$ (see Table S1) had low catalytic activity and required 3.0 g to obtain 90% conversion of 1,6-hexanediol. When the experiment was carried out even using 3.0 g of catalyst, we checked no pressure drop. $Ca(OH)_2$ catalysts gave cyclopentanemethanol and 5-hexene-1-ol as products, but their selectivities were as low as less than 3%. In previous representative studies, Sato et al. reported that 5-hexen-1-ol selectivities in 1,6-hexanediol conversion over Sc_2O_5 and ZrO_2 were, respectively, 62% and 33% [24]. Sc_2O_5 and ZrO_2 catalysts also showed acid–base catalytic properties as shown in Table 4. In our experiments for 1,6 hexanediol conversions over Sc_2O_5 and ZrO_2 catalysts, 5-hexen-1-ol selectivities were, respectively, 61% and 37%, which showed good agreement with reported results. The results revealed that Ca-HAP(1.54) catalyst have the both acid–base bifunctionality similar to Sc_2O_3 and ZrO_2 catalysts and the significant higher catalytic selectivity for unsaturated alcohol productions.

In consequence, over acidic catalysts, C_4–C_6 hydrocarbons and 5-hexene-1-ol were obtained as major products. Over base catalysts, many kinds of five-membered and six-membered cyclic compounds and 6-hydroxy-1-hexanal were obtained. Over Ca-P hydroxyapatite catalyst, 5-hexen-1-ol and cyclopentanemethanol were mainly obtained. Remarkably, Ca-HAP(1.72) catalyst showed the highest cyclopentanemethanol yield ever reported. Furthermore, the Ca-HAP(1.54) catalyst showed the highest 5-hexene-1-ol yield ever reported. The next section describes details of an investigation of the effects of Ca/P molar ratios on product selectivities over Ca-HAP catalysts.

3.4. Effects of Acid–Base Properties of Catalysts on the Products Selectivity

Good correlation was found between the acid–base properties and the product selectivity in the 1,6-hexanediol conversion over Ca-HAP catalysts with various Ca/P molar ratios. Figure 4 presents the relation between the acid–base properties of Ca-HAP catalysts and the product selectivity for 5-hexen-1-ol and cyclopentanemethanol. The acid–basic properties of Ca-HAP catalysts were estimated based on the selectivities of acetone and propylene in the 2-propanol conversions shown in Table 3. Among Ca-HAP catalysts, the cyclopentanemethanol selectivity increased along with the increase of the acetone selectivity. In contrast, the 5-hexen-1-ol selectivity increased concomitantly with the decrease of acetone selectivity because the higher acidity of Ca-HAP catalyst showed higher selectivity of dehydration to 5-hexen-1-ol from 1,6-hexanediol. The Ca-HAP catalysts had little catalytic activity for 5-hexen-1-ol conversion. A good correlation was found between the product selectivity of acetone and propylene in the 2-propanol conversion and the product selectivity of 5-hexen-1-ol and cyclopentanemethanol in the 1,6-hexanediol conversion. The Ca-HAP(1.54) showed high 5-hexene-1-ol yield from 1,6-hexanediol, whose catalytic properties were affected by the Ca/P ratios and the Na amounts but hardly affected by the morphology. The Ca-HAP(1.54) catalyst also showed high acrylic acid yields in the lactic acid conversion. It was probably suitable for the formation of unsaturated oxygen-containing compounds. Further details will be clarified in future research. In addition, Ca(OH)$_2$ catalysts with high basicity gave high selectivity to by-products because of insufficient activity to proceed with the hydrogenation steps of the intermediates. The Ca-HAP catalysts with middle Ca/P molar ratios as Ca-HAP(1.62), which had moderate acid–base properties, also gave high selectivity to by-products because of the stronger selectivity to 5-hexen-1-ol and insufficient activity for the hydrogenation steps.

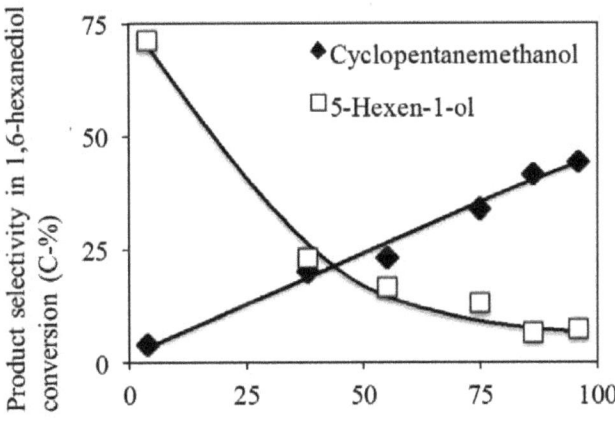

Figure 4. Effects of acid–base property on the products selectivity in the 1,6-hexanediol conversion over the Ca-HAP catalysts with various Ca/P molar ratios.

4. Conclusions

Nanocrystalline hydroxyapatite catalysts with various Ca/P ratios and sodium contents were formed by the hydrothermal method. According to adsorption and desorption experiments using NH$_3$ and CO$_2$ molecules and the catalytic reactions for 2-propenol conversions, the surface acid–base natures changed continuously with the bulk Ca/P ratios. In 1,6-hexanediol conversion over Ca-P hydroxyapatite catalysts, 5-hexen-1-ol, and cyclopentanemethanol were mainly obtained. Remarkably, those product selectivities were changed considerably with changing Ca/P molar ratios of Ca-HAP catalysts. The selectivity of 5-hexen-1-ol was maximized over Ca-HAP(1.54). The yield was 68%. However, the highest cyclopentanemethanol selectivity was obtained over Ca-HAP(1.72): the yield was 42%. Both yields were the highest ever reported in the relevant literature. The reason was presumably that Ca-HAP(1.72) mainly had basic properties and only slightly acidic properties.

Supplementary Materials: The following are available online at https://www.mdpi.com/2079-4991/11/3/659/s1. Table S1: Surface areas of solid catalysts; Table S2: Catalytic conversion of 5-hexen-1-ol over Ca-HAP(1.54); Figure S1: NH_3-TPD spectra of hydroxyapatites; Figure S2: CO_2-TPD spectra of hydroxyapatites.

Author Contributions: A.N. investigation, writing—original draft preparation, visualization; K.I. and K.Y. validation; A.O. conceptualization, supervision, validation, writing—review and editing. All authors have read and agreed to the published version of the manuscript.

Funding: This research received no external funding.

Acknowledgments: This work was supported by the Kochi University research project of the Biomass Refinery of Marin Algae and the Wood. We also acknowledge the support of Sangi Co. Ltd. Tokyo, Japan.

Conflicts of Interest: The authors declare no conflict of interest.

References

1. Elliott, J.C. *Structure and Chemistry of the Apatites and Other Calcium Orthophosphates*; Elsevier Science, B.V.: Amsterdam, The Netherland, 1994; pp. 1–404.
2. Elazarifi, N.; Ezzamarty, A.; Leglise, J.; De Ménorval, L.; Moreau, C. Kinetic study of the condensation of benzaldehyde with ethylcyanoacetate in the presence of Al-enriched fluoroapatites and hydroxyapatites as catalysts. *Appl. Catal. A* **2004**, *267*, 235–240. [CrossRef]
3. Resende, N.S.; Nele, M.; Salim, V.M.M. Effects of anion substitution on the acid properties of hydroxyapatite. *Thermochim. Acta* **2006**, *451*, 16–21. [CrossRef]
4. Hajimirzaee, S.; Chansai, S.; Hardacre, C.; Banksa, C.E.; Doyle, A.M. Effects of surfactant on morphology, chemical properties and catalytic activity of hydroxyapatite. *J. Solid State Chem.* **2019**, *276*, 345–351. [CrossRef]
5. Ogo, S.; Onda, A.; Yanagisawa, K. Hydrothermal synthesis of vanadate-substituted hydroxyapatites, and catalytic properties for conversion of 2-propanol. *Appl. Catal. A Gen.* **2008**, *348*, 129–134. [CrossRef]
6. Diallo-Garcia, S.; Osman, M.B.; Krafft, J.M.; Casale, S.; Thomas, C.; Kubo, K.; Costentin, G. Identification of Surface Basic Sites and Acid–Base Pairs of Hydroxyapatite. *J. Phys. Chem. C* **2014**, *118*, 12744–12757. [CrossRef]
7. Tsuchida, T.; Kubo, J.; Yoshioka, T.; Sakuma, S.; Takeguchi, T.; Ueda, W. Reaction of ethanol over hydroxyapatite affected by Ca/P ratio of catalyst. *J. Catal.* **2008**, *259*, 183–189. [CrossRef]
8. Ogo, S.; Onda, A.; Yanagisawa, K. Selective synthesis of 1-butanol from ethanol over strontium phosphate hydroxyapatite catalysts. *Appl. Catal. A Gen.* **2011**, *402*, 188–195. [CrossRef]
9. Fihri, A.; Len, C.; Varma, R.S.; Solhy, A. Hydroxyapatite: A review of syntheses, structure and applications in heterogeneous catalysis. *Coord. Chem. Rev.* **2017**, *343*, 48–76. [CrossRef]
10. Ogo, S.; Onda, A.; Iwasa, Y.; Hara, K.; Fukuoka, A.; Yanagisawa, K. 1-Butanol synthesis from ethanol over strontium phosphate hydroxyapatite catalysts with various Sr/P ratios. *J. Catal.* **2012**, *296*, 24–30. [CrossRef]
11. Ghantani, V.C.; Lomate, S.T.; Umbarkar, S.B. Catalytic dehydration of lactic acid to acrylic acid using calcium hydroxyapatite catalysts. *Green Chem.* **2013**, *15*, 1211–1217. [CrossRef]
12. Matsuura, Y.; Onda, A.; Yanagisawa, K. Selective conversion of lactic acid into acrylic acid over hydroxyapatite catalysts. *Catal. Commun.* **2014**, *48*, 5–10. [CrossRef]
13. Matsuura, Y.; Onda, A.; Ogo, S.; Yanagisawa, K. Acrylic acid synthesis from lactic acid over hydroxyapatite catalysts with various cations and anions. *Catal. Today* **2014**, *226*, 192–197. [CrossRef]
14. Yan, B.; Tao, L.; Liang, Y.; Xu, B. Sustainable Production of Acrylic Acid: Catalytic Performance of Hydroxyapatites for Gas-Phase Dehydration of Lactic Acid. *ACS Catal.* **2014**, *4*, 1931–1943. [CrossRef]
15. Tsuchida, T.; Kubo, J.; Yoshioka, T.; Sakuma, S.; Takeguchi, T.; Ueda, W. Influence of preparation factors on Ca/P ratio and surface basicity of hydroxyapatite catalyst. *J. Jpn. Petrol. Inst.* **2009**, *52*, 51–59. [CrossRef]
16. Li, C.; Zhu, Q.; Cui, Z.; Wang, B.; Tan, T. Insight into deactivation behavior and determination of generation time of the hydroxyapatite catalyst in the dehydration of lactic acid to acrylic acid. *Ind. Eng. Chem. Res.* **2019**, *58*, 53–59. [CrossRef]
17. Buntara, T.; Noel, S.; Phua, P.H.; Meliµn-cabrera, I.; De Vries, J.G.; Heeres, H.J. Caprolactam from Renewable Resources: Catalytic Conversion of 5-Hydroxymethylfurfural into Caprolactone. *Angew. Chem. Int. Ed.* **2011**, *50*, 7083–7087. [CrossRef]
18. Alamillo, R.; Tucker, M.; Chia, M.; Pagán-torres, Y.; Dumesic, J. The selective hydrogenation of biomass-derived 5-hydroxymethylfurfural using heterogeneous catalysts. *Green Chem.* **2012**, *14*, 1229–1546. [CrossRef]
19. Melia, I.; Heeres, H.J. From 5-Hydroxymethylfurfural (HMF) to polymer precursors: Catalyst screening studies on the conversion of 1,2,6-hexanetriol to 1,6-hexanediol. *Top. Catal.* **2012**, *55*, 612–619.
20. Chen, K.; Koso, S.; Kubota, T.; Nakagawa, Y.; Tomishige, K. Chemoselective hydrogenolysis of tetrahydropyran-2-methanol to 1,6-hexanediol over rhenium-modified carbon-supported rhodium catalysts. *ChemCatChem* **2010**, *2*, 547–555. [CrossRef]
21. Said, A.; Perez, D.D.S.; Perret, N.; Pinel, C.; Besson, M. Selective C−O hydrogenolysis of erythritol over supported Rh-ReOx catalysts in the aqueous phase. *ChemCatChem* **2017**, *9*, 2768–2783. [CrossRef]

2. Allgeier, A.M.; De Silva, W.I.N.; Korovessi, E.; Menning, C.A.; Ritter, J.C.; Sengupta, S.K.; Stauffer, C.S. Process for Preparing 1,6-Hexanediol. U.S. Patent US8889912B2, 18 November 2014.
3. Larkin, D.R. The Vapor Phase Catalytic Dehydrogenation of 1,6-Hexanediol. *J. Org. Chem.* **1965**, *30*, 335–339. [CrossRef]
4. Abe, K.; Ohishi, Y.; Okada, T.; Yamada, Y.; Sato, S. Vapor-phase catalytic dehydration of terminal diols. *Catal. Today* **2011**, *164*, 419–424. [CrossRef]
5. Akashi, T.; Sato, S.; Takahashi, R.; Sodesawa, T.; Inui, K. Catalytic vapor-phase cyclization of 1,6-hexanediol into cyclopentanone. *Catal. Commun.* **2003**, *4*, 411–416. [CrossRef]
6. Mounguengui-Diallo, M.; Vermersch, F.; Perret, N.; Pinel, C. Base free oxidation of 1,6-hexanediol to adipic acid over supported noble metal mono- and bimetallic catalysts. *Appl. Catal. A Gen.* **2018**, *551*, 88–97. [CrossRef]
7. Bonel, G.; Heughebaert, J.C.; Heughebaert, M.; Lacout, J.L.; Lebugle, A. Apatitic calcium orthophosphates and related compounds for biomaterials preparation. *Ann. N. Y. Acad. Sci.* **1988**, *523*, 115–130. [CrossRef] [PubMed]
8. Matsunaga, K.; Murata, H. Formation Energies of Substitutional Sodium and Potassium in Hydroxyapatite. *Mater. Trans.* **2009**, *50*, 1041–1045.

Article

Hydrothermal Synthesis of Various Shape-Controlled Europium Hydroxides

Hongjuan Zheng [1], Kongjun Zhu [1], Ayumu Onda [2] and Kazumichi Yanagisawa [2,*]

[1] State Key Laboratory of Mechanics and Control of Mechanical Structures, Nanjing University of Aeronautics and Astronautics, Nanjing 210016, China; zhenghj2012@126.com (H.Z.); kjzhu@nuaa.edu.cn (K.Z.)
[2] Research Laboratory of Hydrothermal Chemistry, Faculty of Science, Kochi University, Kochi 780-8520, Japan; aonda@kochi-u.ac.jp
* Correspondence: yanagi@kochi-u.ac.jp; Tel.: +81-26-269-5389

Abstract: $Eu(OH)_3$ with various shape-controlled morphologies and size, such as plate, rod, tube, prism and nanoparticles was successfully synthesized through simple hydrothermal reactions. The products were characterized by XRD (X-Ray Powder Diffraction), FE-SEM (Field Emission- Scanning Electron Microscopy) and TG (Thermogravimetry). The influence of the initial pH value of the starting solution and reaction temperature on the crystalline phase and morphology of the hydrothermal products was investigated. A possible formation process to control morphologies and size of europium products by changing the hydrothermal temperature and initial pH value of the starting solution was proposed.

Keywords: rare earth; europium hydroxides; hydrothermal; morphology control; nanoparticles

1. Introduction

Rare earth (RE) compounds have drawn continuous research attention for many years because of their unique optical, magnetic, electric, and catalytic properties resulting from their unique 4f-electron configuration. The several frequently used RE compounds currently include RE oxides [1–3], hydroxides [4–6], fluorides [7–9], phosphates [10,11], and so on. Certainly, RE hydroxides are of great importance because RE oxides can be directly formed through dehydration of their corresponding hydroxides, and RE sulfides can be directly obtained through sulfuration of the hydroxides. The hydroxyl group may also act as active sites for grafting other organic or inorganic functional groups [12]. Therefore, RE hydroxides are an important intermediate to synthesize oxides, sulfides, or other functional materials.

The properties of RE compounds depend strongly on their morphologies. Many kinds of morphological RE compounds depending on their inherent anisotropy have been explored in the recent decades, such as one-dimensional (1D) nanostructures (nanowires, nanorods, nanotubes), two-dimensional (2D) (nanosheet, nanobelts) and three-dimensional (3D) architectures (flower-like [13], spindle-like [14], hierarchical architectures [15]). In order to achieve the specific properties for their further applications, the selection of well-defined synthetic methods is required.

Among all the RE ions, Eu^{3+} can be effectively activated by ultraviolet rays or cathode rays and emits high purity red light because of its unique $4f^6$ configuration [16]. The major emission band of Eu^{3+} is centered near 611 nm (red), which is one of the primary colors [17]. Therefore, Eu^{3+} is a good activator with sharp and intense luminescence in the red region of the visible spectrum. Monodisperse hexagonal $Eu(OH)_3$ submicrospindles with a diameter of 80−200 nm and a length of 500−900 nm have been synthesized in a large scale via a facile aqueous solution route by Xu et al. [18]. The morphology of Eu_2O_3 obtained by the calcination of $Eu(OH)_3$ submicrospindles maintains the same morphology as $Eu(OH)_3$. The similar method of annealing the $Eu(OH)_3$ precursor to obtain Eu_2O_3 with the same morphology as

Citation: Zheng, H.; Zhu, K.; Onda, A.; Yanagisawa, K. Hydrothermal Synthesis of Various Shape-Controlled Europium Hydroxides. *Nanomaterials* **2021**, *11*, 529. https://doi.org/10.3390/nano11020529

Academic Editor: Drazic Goran
Received: 13 January 2021
Accepted: 2 February 2021
Published: 19 February 2021

Publisher's Note: MDPI stays neutral with regard to jurisdictional claims in published maps and institutional affiliations.

Copyright: © 2021 by the authors. Licensee MDPI, Basel, Switzerland. This article is an open access article distributed under the terms and conditions of the Creative Commons Attribution (CC BY) license (https://creativecommons.org/licenses/by/4.0/).

the former have been reported by others in succession in recent years [19–21]. Xu et al. [22] obtained the monodisperse and well-defined 1D rare earth fluorides (β-NaREF$_4$) (RE = Y, Sm, Eu, Gd, Tb, Dy, and Ho) nanowires/nanorods using RE(OH)$_3$ as precursors via a facile hydrothermal route and characterized their photoluminescence properties. Among them, the diameter and length of Eu(OH)$_3$ nanowires are about 10−20 nm and 0.1–0.2 μm, and the final β-NaEuF$_4$ samples inherit their Eu(OH)$_3$'s morphology by this conversion process. Zhang et al. [23] synthesized the biocompatible Eu(OH)$_3$ nanoclusters composed of approximately 5 nm nanoparticles with a modified microwave-assisted hydrothermal method and showed the as-synthesized Eu(OH)$_3$ nanoclusters exhibited excellent physiological stability and biocompatibility both in vitro and in vivo, and possessed considerable pro-proliferative activities in human umbilical vein endothelial cells. This study developed the application of Eu(OH)$_3$ to biofunctional nanomaterials.

Because of the unique nature of RE ions, the assembly of Eu complexes offers great challenges and opportunities in terms of controlling fascinating frameworks and specific properties. To date, various approaches have been used to prepare Eu(OH)$_3$ particles, such as homogeneous precipitation [24], solvothermal [25], microwave [23], and the hydrothermal method [21], and so on. The hydrothermal method has drawn tremendous attention owing to its advantages, such as simplicity, low energy consumption, environmental friendliness, and well-defined product morphology.

Eu(OH)$_3$ will be continuously studied for development of their various applications. At this point, it is particularly important to exploit unique products with a variety of morphologies. In this study, we proposed a simple hydrothermal route for the synthesis of Eu(OH)$_3$ nanoparticles with various controlled morphologies. The influence of the pH value of the starting solutions and reaction temperature on the crystallized phase, particle size, and morphology of europium compounds was systematically investigated. In addition, the mechanism of morphology evolution under hydrothermal conditions was proposed.

2. Experimental

2.1. Raw Materials

The starting materials, europium oxide (Eu$_2$O$_3$) (99.9%), nitric acid (HNO$_3$) (60%) and ammonia (NH$_3$·H$_2$O) (28%), were received from Wako Pure Chemical Industries, Ltd. (Osaka, Japan). All chemicals were used as received without further purification.

2.2. Preparation of Eu(OH)$_3$ Powders

In a typical synthesis process, 0.45 g of Eu$_2$O$_3$ was dissolved in 6ml of 3.0 M HNO$_3$ solution through hydrothermal treatment at 120 °C for 2 h. Then, NH$_3$·H$_2$O was added to adjust the pH of the solution to a designated value for getting the precipitates under vigorous agitation. When NH$_3$·H$_2$O was added to the Eu solutions, the transparent solutions were changed to opaque colloidal solutions consisting of amorphous particles. The range of pH value was between 7 and 12 (the maximum pH value adjusted with a concentrated ammonia solution is about 12). The final volume of the resulting solution was adjusted to reach 15 mL. The as-obtained colloidal solution was immediately transferred into a 25 mL Teflon-lined autoclave (homemade, Kochi, Japan), followed by hydrothermal treatment at temperatures from 80 to 220 °C for 24 h. After cooling, the received white precipitate was collected by a centrifuge, washed with distilled water several times and dried at 80 °C overnight.

2.3. Characterization

The as-prepared powders were characterized by X-ray powder diffraction (XRD) to determine the crystal structure under a Rigaku RTP-300RC X-ray diffractometer (Tokyo, Japan) with Cu Kα radiate on (λ = 1.5418 Å). Specific scan parameters were a tube voltage of 40 kV and tube current of 20 mA. The patterns were collected in the range of 5 to 70° with a 0.02° step and scanning speed of 4°/min. Field emission scanning electron microscopic (FE-SEM) observation was carried out on a JEOL JSM-6500F instrument (Tokyo, Japan) at 15 kV to collect the structure information of the powders. Thermogravimetry-Differential Thermal Analysis (TG-DTA) study was performed using a TG/DTA System (Mac Science Co. Tokyo, Japan, TG-DTA2020S) in air. The temperature was raised from room temperature to 900 °C at a heating rate of 10 °C/min, and held at 900 °C for 10 min.

3. Results and Discussion
3.1. Crystalline Phases

The crystal structure of the hydrothermally synthesized europium products was characterized by XRD, as shown in Figure 1. First, the lowest reaction temperature was fixed to 80 °C, and the pH value of the starting solution was increased from 7 to 12. The positions of diffraction peaks of the products obtained with the pH value of 7.34 and 9.24 are the same. The diffraction patterns exhibit a series of strong (001) and sharp (220) diffractions, suggesting that all the peaks can be well indexed to a layered phase [26], as shown in Figure 1a,b. These diffraction peaks show exactly the same patterns with Eu-doped $Y_2(OH)_5NO_3 \cdot 2H_2O$ according to the result reported by Zhang et al. [27] and Wu et al. [28] Therefore, the products can be temporarily identified as the pure phase of $Eu_2(OH)_5NO_3 \cdot 2H_2O$. As the pH value was increased to 10.82, the pure phase of $Eu(OH)_3$ was successfully formed, as shown in Figure 1c. The diffraction peaks matched the standard data of a hexagonal phase with a P63/m space group, according to the JCPDS Card No. 83-2305. When the pH value was higher than 12, the pure hexagonal phase of $Eu(OH)_3$ was also synthesized, as shown in Figure 1d. It is important to note that the relative intensity ratio of (110) and (101) diffraction was changed with the increase in pH value of the starting solution, suggesting that the preferential crystal growth direction of two kinds of $Eu(OH)_3$ products might be different. In other words, the crystals of these products might show different morphologies. The intensity of the main diffraction peak (100) of the product obtained in the solution with the highest pH value was lowest, which indicated that the crystallinity of $Eu(OH)_3$ was decreased with the increase in pH value of the starting solution.

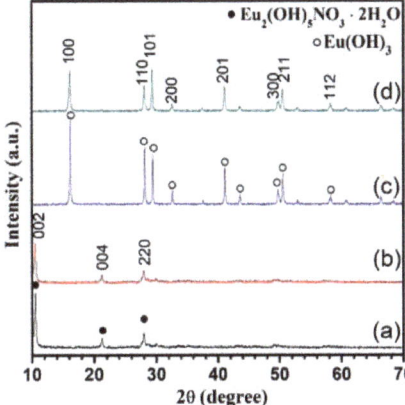

Figure 1. XRD patterns of the products obtained at 80 °C for 24 h as a function of the pH value of the starting solution. The pH value used is 7.34 (**a**), 9.24 (**b**), 10.82 (**c**), 12 (**d**).

Figure 2 shows the TG curve of the product synthesized at 80 °C for 24 h in the solution with a pH value of 7.34. It can be concluded that the weight loss of 17.36% below 450 °C corresponds to the evaporation of water and the release of OH species. The weight loss at temperatures higher than 450 °C is 10.14%, which associated with the release of NO species. The overall weight loss of 27.5% is consistent with the theoretical value of 27.7% of the transition from $Eu_2(OH)_5NO_3 \cdot 2H_2O$ to Eu_2O_3. The above analysis is consistent with the thermal decomposition of $Y_2(OH)_{5.14}(NO_3)_{0.86} \cdot H_2O$ reported by Li et al. [2] Therefore, the product obtained in the solution with a pH value of 7.34 was proved to be $Eu_2(OH)_5NO_3 \cdot 2H_2O$.

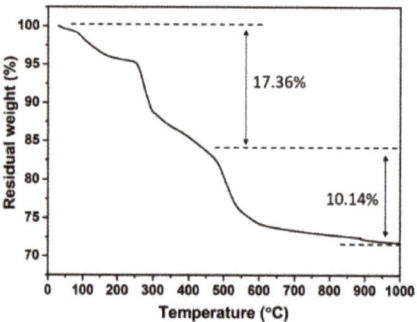

Figure 2. TG curves of the product obtained at 80 °C for 24 h in the solution with a pH value of 7.34.

To obtain a further understanding of the effect of the pH valve on the crystal forms of hydrothermally synthesized europium products, Figure 3 exhibits the XRD patterns of hydrothermally synthesized europium products at 160 °C for 24 h as a function of the pH value of the initial hydrothermal solution. According to the XRD patterns, when the pH value was 7.26, the diffraction peaks can be indexed to be the layered phase of $Eu_2(OH)_5NO_3 \cdot 2H_2O$, as shown in Figure 3a. When the pH value was higher than 8.99, the pure hexagonal phase of $Eu(OH)_3$ can be formed, as shown in Figure 3b–d. However, the crystallinity of $Eu(OH)_3$ was gradually decreased with the increase in pH value of the starting solution, suggested by the decrease in the intensity of the main diffraction peak (100) of the product.

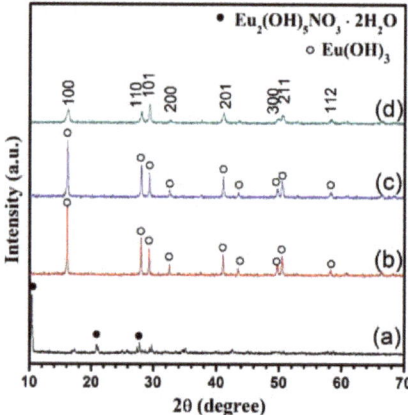

Figure 3. XRD patterns of the products obtained at 160 °C for 24 h as a function of the pH value of the starting solution. The pH value used is 7.26 (a), 8.99 (b), 10.48 (c), 12 (d).

Finally, the effect of the pH valve on the crystals form of hydrothermally synthesized europium products at high temperature of 220 °C was further investigated. The crystallized phases of the synthesized europium products were similar to the products obtained at 160 °C, as shown Figure 4. Unlike the Figure 1a,b and Figure 3a, the (220) diffraction of layered $Eu_2(OH)_5NO_3 \cdot 2H_2O$ crystals prepared at 220 °C were disappeared, which suggests that the crystals tend to grow oriented in the c axis direction. The crystallinity of $Eu(OH)_3$ crystals synthesized in the solution with a pH value of 9.15 (Figure 4b) was obviously higher than in the solution with a pH value of 10.90 (Figure 4c) and 12 (Figure 4d) according to their intensity of diffractions.

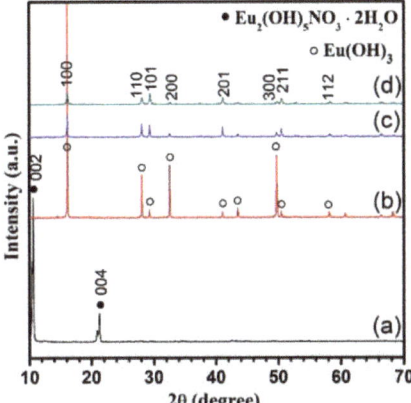

Figure 4. XRD patterns of the products obtained at 220 °C for 24 h as a function of the pH value of the starting solution. The pH value used is 7.63 (**a**), 9.15 (**b**), 10.90 (**c**), 12 (**d**).

3.2. Morphology

In addition to the crystalline phases obtained, the pH value of the initial solutions and reaction temperature also have an impact on the morphology. The morphology of the products synthesized at 80 °C for 24 h was shown in Figure 5 as a function of the pH value of the starting solutions. When the pH of the starting solution was 7.34, the $Eu_2(OH)_5NO_3 \cdot 2H_2O$ crystals comprised flower-like agglomerates, which contained the ultrathin plate-like crystals with a size of above 1 µm, as shown in Figure 5a. As the pH was increased to 9.24, $Eu_2(OH)_5NO_3 \cdot 2H_2O$ crystals were composed of nanorods with a diameter of 100 nm and a length of 400–500 nm, as shown in Figure 5b. $Eu(OH)_3$ crystals synthesized in the solution with a pH value of about 11 is shown in Figure 5c. $Eu(OH)_3$ crystals consisted of nanotubes with an outer diameter of 200–300 nm, an inner diameter of 80–100 nm, and a length of 500 nm. Figure 5d reveals that the fine $Eu(OH)_3$ nanorods with diameter of 30–40 nm and length of 100 nm were formed in the solution with a pH value of about 12.

Figure 5. SEM images of the products obtained at 80 °C for 24 h as a function of the pH value of the starting solution. The pH value used is 7.34 (**a**), 9.24 (**b**), 10.82 (**c**), 12 (**d**).

Figure 6 shows SEM images of the products synthesized obtained at 160 °C for 24 h as a function of pH value of the starting solution. $Eu_2(OH)_5NO_3 \cdot 2H_2O$ crystals obtained in the starting solution with pH 7.26 were composed of small disk-like crystals with a diameter of 1.5–2 µm, as shown in Figure 6a. This disk-like morphology was very close to $Y_2(OH)_{5.14}(NO_3)_{0.86} \cdot H_2O$ in the shape of a sheet that was reported by Li, et al. [2] As the pH was increased to 8.99, $Eu(OH)_3$ crystals showed micro-cylinders with a diameter of 1–2 µm and a length up to 6 µm, as shown in Figure 6b. When the pH was changed to 10.48, the synthesized $Eu(OH)_3$ crystals showed the uniform nanorods with a diameter of 100 nm and a length of 400–500 nm, as shown in Figure 6c. Similar nanorod-like $Eu(OH)_3$ crystals with a width of 50–150 nm and a length of 500 nm–2 µm were prepared by adding ammonia by Kang et al. [21] However, they neither gave the specific pH of the starting solutions nor the added amount of ammonia. The fine $Eu(OH)_3$ nanoparticles with a diameter less than 50 nm were obtained in the solution with a pH value of about 12, as shown in Figure 6d.

Figure 6. SEM images of the products obtained at 160 °C for 24 h as a function of the pH value of the starting solution. The pH value used is 7.26 (**a**), 8.99 (**b**), 10.48 (**c**), 12 (**d**).

Figure 7 shows SEM images of the products obtained at 220 °C for 24 h as a function of the pH value of the starting solution. When the pH of the starting solution was 7.63, the plate-like $Eu_2(OH)_5NO_3 \cdot 2H_2O$ crystals were changed to a long board shape with an average aspect ratio of five, as shown in Figure 7a, which means that the crystals grew along the c axis direction. These results were consistent with those of their XRD patterns. The hexagonal prism $Eu(OH)_3$ crystals with a diameter of 8–10 µm and a length of 30–40 µm (Figure 7b) were obtained when the pH was changed to 9.15. Ji et al. [29] obtained the short hexagonal nano-prism $Eu(OH)_3$ at 120 °C and pH 8.8 by adjusting it with a NaOH aqueous solution. When the precursor pH was increased from 8.8 to 9.5 at the same temperature, the hexagonal prisms became more slender and their aspect ratio changed from 1.1 to 2.1 [29]. Thus, smaller and shorter hexagonal prisms could be synthesized at a lower temperature when ammonia was replaced by NaOH. When the pH was further increased to 10.90, $Eu(OH)_3$ crystals composed of nanorods with a diameter of 150 nm and a length of 700 nm were synthesized, as shown in Figure 7c. The fine $Eu(OH)_3$ nanoparticles with a diameter less than 50 nm were obtained in the solution with a pH value of about 12, as shown in Figure 7d.

Figure 7. SEM images of the products obtained at 220 °C for 24 h as a function of the pH value of the starting solution. The pH value used is 7.63 (**a**), 9.15 (**b**), 10.90 (**c**), 12 (**d**).

In this study, two crystalline phases, Eu(OH)$_3$ and Eu$_2$(OH)$_5$NO$_3$·2H$_2$O, were obtained as a single phase depending on reaction conditions, and they are shown in Figure 8 with their characteristic morphologies. The left and right parts of the dashed line represent different phases, i.e., Eu(OH)$_3$ (marked with a circle) on the right and Eu$_2$(OH)$_5$NO$_3$·2H$_2$O (marked with a star) on the left. To a certain degree, the dashed line represents the boundary of the two phases. A proposed formation process seems to be responsible for the observed development of the morphologies of Eu$_2$(OH)$_5$NO$_3$·2H$_2$O and Eu(OH)$_3$ crystals synthesized with various pH values and temperatures. The presence of NO$_3^-$ and the concentration of OH$^-$ are the essential factors for the formation of layered Eu$_2$(OH)$_5$NO$_3$·2H$_2$O compounds. Layered Eu$_2$(OH)$_5$NO$_3$·2H$_2$O crystals were easily formed in the solution with low pH of 7 and in the solution with a pH of 9 at a low temperature. When the temperature increased in the solution with pH 9, Eu(OH)$_3$ was formed.

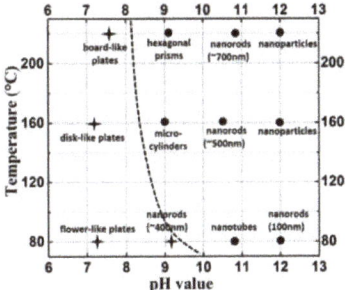

Figure 8. Morphology distribution in a temperature–pH diagram. (Closed circle: Eu(OH)$_3$; star: Eu$_2$(OH)$_5$NO$_3$·2H$_2$O).

When the reaction temperature increased, Eu$_2$(OH)$_5$NO$_3$·2H$_2$O crystals grew larger in the solution with a pH value of about 7. This can be attributed to the Ostwald ripening process under hydrothermal conditions. The reaction temperature plays a key role in the dissolution of smaller grains and the growth of larger grains [30], as observed in Figures 5a and 6a. A higher temperature of 220 °C not only results in the growth of grains, but also leads to the directional growth of grains along the c axis direction.

In the solution with pH 9, Eu$_2$(OH)$_5$NO$_3$·2H$_2$O crystallized at 80 °C but Eu(OH)$_3$ crystallized at 160 °C. Eu(OH)$_3$ is a stable phase at a higher temperature than Eu$_2$(OH)$_5$NO$_3$·2H$_2$O. The Eu(OH)$_3$ tubes were obtained only by the experiment at 80 °C with pH 11.

$Eu_2(OH)_5NO_3 \cdot 2H_2O$ precursor may be formed in the solution with pH 11 at lower temperature than 80 °C and it transforms to $Eu(OH)_3$ quickly while heating to 80 °C. Nanotubes of $Eu(OH)_3$ might be formed by rapid preferential growth along the circumferential edges in the c axis direction, similar to the formation of $Y(OH)_3$ nanotubes [2]. At 220 °C, $Eu(OH)_3$ crystals grew larger than at 160 °C. In the solution with pH 9, the temperature is still the main factor affecting the morphology and size of the grains. A high temperature promoted more quickly the dissolution of the smaller grains, and the larger grains grew up in order to reduce the surface energy. [31]

The size of $Eu(OH)_3$ grains obtained in the solution with pH 11 were small. That might be attributed to the increase in pH value which leads to the decrease in the solubility of $Eu(OH)_3$ precursors. However, $Eu(OH)_3$ crystals grew larger with the increase in reaction temperature by the Ostwald ripening process. A large amount of ammonia was added to the solution to obtain the solution with pH 12. The solubility of hydroxides must be very low in such a solution with a high pH value. Many $Eu(OH)_3$ crystal nucleus were rapidly formed when the starting solution was prepared, but they could not grow larger by the Ostwald ripening process due to the low solubility even at a high temperature. Thus the size of $Eu(OH)_3$ crystals did not change, even at the highest temperature. Based on this mechanism discussed above, $Eu(OH)_3$ crystals with more different morphologies, such as micro-cylinders, hexagonal prisms, nanotubes, nanorods and nanoparticles, are synthesized by adjusting the pH with ammonia and the temperature, even without the addition of surfactants [32].

4. Conclusions

The formation conditions of $Eu(OH)_3$ crystals with controlled morphology and size were systematically investigated through simple hydrothermal reactions by adjusting the pH value of the starting solution and reaction temperature. When the pH value of the starting solution was about 7, $Eu_2(OH)_5NO_3 \cdot 2H_2O$ crystals rather than $Eu(OH)_3$, with the flower-like agglomerates, disk-like plates and long boards, were obtained at the reaction temperature of 80 °C, 160 °C and 220 °C, respectively. In order to obtain $Eu(OH)_3$, it is necessary to further increase the reaction temperature and pH value of the starting solutions. The characteristic morphology of $Eu(OH)_3$ crystals such as micro-cylinders and hexagonal prisms could be formed in the solution with pH 9 at 160 °C and 220 °C, respectively. Nanotube and nanorod $Eu(OH)_3$ with different lengths were formed at different temperatures in the solution with pH 11. Nanoparticles were obtained at above 160 °C in the solution with pH 12, due to the decrease in the solubility of $Eu(OH)_3$ precursors in high pH solutions. The method utilized in this study to fabricate $Eu(OH)_3$ crystals could be extended to synthesize the other RE hydroxides with tunable morphologies by simply adjusting the pH value and reaction temperature.

Author Contributions: Conceptualization, K.Z. and K.Y.; methodology, H.Z. and A.O.; formal analysis, H.Z. and A.O.; investigation, H.Z. original draft preparation, review and editing, H.Z. and K.Y.: funding acquisition, K.Z. and K.Y. All authors have read and agreed to the final version of this manuscript.

Funding: This work was supported by the National Nature Science Foundation of China (NSFC No. U1904213), the Key Research and Development Program of Jiangsu Province (Grant No. BE2018008-2), the Research Fund of State Key Laboratory of Mechanics and Control of Mechanical Structures (Nanjing University of Aeronautics and astronautics) (Grant No. MCMS-0518K01), and A Project Funded by the Priority Academic Program Development of Jiangsu Higher Education Institutions (PAPD).

Institutional Review Board Statement: Not applicable.

Informed Consent Statement: Not applicable.

Data Availability Statement: The data presented in this study are available in this article.

Conflicts of Interest: The authors declare no conflict of interest.

References

1. Huang, P.X.; Wu, F.; Zhu, B.L.; Li, G.R.; Wang, Y.L.; Gao, X.P.; Zhu, H.Y.; Yan, T.Y.; Huang, W.P.; Zhang, S.M.; et al. Praseodymium Hydroxide and Oxide Nanorods and Au/Pr_6O_{11} Nanorod Catalysts for CO Oxidation. *J. Phys. Chem. B* **2006**, *110*, 1614–1620. [CrossRef] [PubMed]
2. Li, N.; Yanagisawa, K. Controlling the morphology of yttrium oxide through different precursors synthesized by hydrothermal method. *J. Solid State Chem.* **2008**, *181*, 1738–1743. [CrossRef]
3. Vishnyakov, A.V.; Korshunova, I.A.; Kochurikhin, V.E.; Sal'nikova, L.S. Catalytic activity of rare earth oxides in flameless methane combustion. *Kinet. Catal.* **2010**, *51*, 273–278. [CrossRef]
4. Li, N.; Yanagisawa, K.; Kumada, N. Facile Hydrothermal Synthesis of Yttrium Hydroxide Nanowires. *Cryst. Growth Des.* **2009**, *9*, 978–981. [CrossRef]
5. Zeng, Q.G.; Ding, Z.J.; Zhang, Z.M.; Sheng, Y.Q. Photoluminescence and Raman Spectroscopy Studies of $Eu(OH)_3$ Rods at High Pressures. *J. Phys. Chem. C* **2010**, *114*, 4895–4900. [CrossRef]
6. Wang, P.-P.; Bai, B.; Huang, L.; Hu, S.; Zhuang, J.; Wang, X. General synthesis and characterization of a family of layered lanthanide (Pr, Nd, Sm, Eu, and Gd) hydroxide nanowires. *Nanoscale* **2011**, *3*, 2529–2535. [CrossRef]
7. Wen, C.; Sun, L.; Yan, J.; Liu, Y.; Song, J.; Zhang, Y.; Lian, H.; Kang, Z. Mesoporous rare earth fluoride nanocrystals and their photoluminescence properties. *J. Colloid Interface Sci.* **2011**, *357*, 116–120. [CrossRef]
8. Gai, S.; Yang, G.; Li, X.; Li, C.; Dai, Y.; He, F.; Yang, P. Facile synthesis and up-conversion properties of monodisperse rare earth fluoride nanocrystals. *Dalton Trans.* **2012**, *41*, 11716–11724. [CrossRef]
9. Peng, C.; Li, C.; Li, G.; Li, S.; Lin, J. $YF3:Ln^{3+}$ (Ln = Ce, Tb, Pr) submicrospindles: Hydrothermal synthesis and luminescence properties. *Dalton Trans.* **2012**, *41*, 8660–8668. [CrossRef]
10. Yang, M.; You, H.; Song, Y.; Huang, Y.; Jia, G.; Liu, K.; Zheng, Y.; Zhang, L.; Zhang, H. Synthesis and Luminescence Properties of Sheaflike $TbPO_4$ Hierarchical Architectures with Different phase Structures. *J. Phys. Chem. C* **2009**, *113*, 20173–20177. [CrossRef]
11. Kim, E.; Osseo-Asare, K. Aqueous stability of thorium and rare earth metals in monazite hydrometallurgy: Eh–pH diagrams for the systems Th, Ce, La, $Nd(PO_4)(SO_4)H_2O$ at 25 °C. *Hydrometallurgy* **2012**, *113–114*, 67–78. [CrossRef]
12. Eliseeva, S.V.; Bünzli, J.-C.G. Rare earths: Jewels for functional materials of the future. *New J. Chem.* **2011**, *35*, 1165–1176. [CrossRef]
13. Yang, J.; Li, C.; Quan, Z.; Zhang, C.; Yang, P.; Li, Y.; Yu, C.; Lin, J. Self-Assembled 3D Flowerlike Lu_2O_3 and $Lu_2O_3:Ln^{3+}$ (Ln = Eu, Tb, Dy, Pr, Sm, Er, Ho, Tm) Microarchitectures: Ethylene Glycol-Mediated Hydrothermal Synthesis and Luminescent Properties. *J. Phys. Chem. C* **2008**, *112*, 12777–12785. [CrossRef]
14. Zheng, H.; Yanagisawa, K.; Onda, A.; Zhu, K. Hydrothermal synthesis of spindle-like architectures of terbium hydroxide. *J. Ceram. Soc. Jpn.* **2015**, *123*, 672–676. [CrossRef]
15. Zhou, L.; Yang, J.; Hu, X.; Luo, Y.; Yang, J. Synthesis of 3D hierarchical architectures of $Tb_2(CO_3)_3$: Eu^{3+} phosphor and its efficient energy transfer from Tb^{3+} to Eu^{3+}. *J. Mater. Sci.* **2015**, *50*, 4503–4515. [CrossRef]
16. Gonçalves, R.F.; Moura, A.P.; Godinho, M.J.; Longo, E.; Machado, M.A.C.; de Castro, D.A.; Siu Li, M.; Marques, A.P.A. Crystal growth and photoluminescence of europium-doped strontium titanate prepared by a microwave hydrothermal method. *Ceram. Int.* **2015**, *41*, 3549–3554. [CrossRef]
17. Zhang, D.; He, X.; Yang, H.; Shi, L.; Fang, J. Surfactant-assisted reflux synthesis, characterization and formation mechanism of carbon nanotube/europium hydroxide core–shell nanowires. *Appl. Surf. Sci.* **2009**, *255*, 8270–8275. [CrossRef]
18. Xu, Z.; Li, C.; Yang, P.; Hou, Z.; Zhang, C.; Lin, J. Uniform $Ln(OH)_3$ and Ln_2O_3 (Ln = Eu, Sm) Submicrospindles: Facile Synthesis and Characterization. *Cryst. Growth Des.* **2009**, *9*, 4127–4135. [CrossRef]
19. Du, N.; Zhang, H.; Chen, B.; Wu, J.; Li, D.; Yang, D. Low temperature chemical reaction synthesis of single-crystalline $Eu(OH)_3$ nanorods and their thermal conversion to Eu_2O_3 nanorods. *Nanotechnology* **2007**, *18*, 065605. [CrossRef]
20. Zhang, D.; Yan, T.; Shi, L.; Li, H.; Chiang, J.F. Template-free synthesis, characterization, growth mechanism and photoluminescence property of $Eu(OH)_3$ and Eu_2O_3 nanospindles. *J. Alloys Compd.* **2010**, *506*, 446–455. [CrossRef]
21. Kang, J.-G.; Jung, Y.; Min, B.-K.; Sohn, Y. Full characterization of $Eu(OH)_3$ and Eu_2O_3 nanorods. *Appl. Surf. Sci.* **2014**, *314*, 158–165. [CrossRef]
22. Xu, Z.; Li, C.; Yang, P.; Zhang, C.; Huang, S.; Lin, J. Rare Earth Fluorides Nanowires/Nanorods Derived from Hydroxides: Hydrothermal Synthesis and Luminescence Properties. *Cryst. Growth Des.* **2009**, *9*, 4752–4758. [CrossRef]
23. Zhang, L.; Hu, W.; Wu, Y.; Wei, P.; Dong, L.; Hao, Z.; Fan, S.; Song, Y.; Lu, Y.; Liang, C.; et al. Microwave-Assisted Facile Synthesis of $Eu(OH)_3$ Nanoclusters with Pro-Proliferative Activity Mediated by miR-199a-3p. *ACS Appl. Mater. Interfaces* **2018**, *10*, 31044–31053. [CrossRef]
24. Lian, J.; Liang, P.; Wang, B.; Liu, F. Homogeneous precipitation synthesis and photoluminescence properties of $La_2O_2SO_4$: Eu^{3+} quasi-spherical phosphors. *J. Ceram. Process. Res.* **2014**, *15*, 382–388.
25. Arunachalam, S.; Kirubasankar, B.; Murugadoss, V.; Vellasamy, D.; Angaiah, S. Facile synthesis of electrostatically anchored $Nd(OH)_3$ nanorods onto graphene nanosheets as a high capacitance electrode material for supercapacitors. *New J. Chem.* **2018**, *42*, 2923–2932. [CrossRef]
26. Zhu, Q.; Li, J.-G.; Li, X.; Qi, Y.; Sun, X. $[(Y_{1−x}Gd_x)_{0.95}Eu_{0.05}]_2(OH)_5NO_3 \cdot nH_2O$ ($0 \leq x \geq 0.50$) layered rare-earth hydroxides: Exfoliation of unilamellar and single-crystalline nanosheets, assembly of highly oriented and transparent oxide films, and greatly enhanced red photoluminescence by Gd^{3+} doping. *RSC Adv.* **2015**, *5*, 64588–64595.

27. Zhang, L.; Jiang, D.; Xia, J.; Zhang, N.; Li, Q. Enhanced fluorescence of europium-doped yttrium hydroxide nanosheets modified by 2-thenoyltrifluoroacetone. *RSC Adv.* **2014**, *4*, 17856–17859. [CrossRef]
28. Wu, X.; Li, J.-G.; Zhu, Q.; Liu, W.; Li, J.; Li, X.; Sun, X.; Sakka, Y. One-step freezing temperature crystallization of layered rare-earth hydroxide ($Ln_2(OH)_5NO_3 \cdot nH_2O$) nanosheets for a wide spectrum of Ln (Ln = Pr–Er, and Y), anion exchange with fluorine and sulfate, and microscopic coordination probed via photoluminescence. *J. Mater. Chem. C* **2015**, *3*, 3428–3437. [CrossRef]
29. Ji, X.; Hu, P.; Li, X.; Zhang, L.; Sun, J. Hydrothermal control, characterization, growth mechanism, and photoluminescence properties of highly crystalline 1D $Eu(OH)_3$ nanostructures. *RSC Adv.* **2020**, *10*, 33499–33508. [CrossRef]
30. Wu, N.-C.; Shi, E.-W.; Zheng, Y.-Q.; Li, W.-J. Effect of pH of Medium on Hydrothermal Synthesis of Nanocrystalline Cerium(IV) Oxide Powders. *J. Am. Ceram. Soc.* **2002**, *85*, 2462–2468. [CrossRef]
31. Voorhees, P.W. The theory of Ostwald ripening. *J. Stat. Phys.* **1985**, *38*, 231–252. [CrossRef]
32. Mu, Q.; Wang, Y. A simple method to prepare $Ln(OH)_3$ (Ln = La, Sm, Tb, Eu, and Gd) nanorods using CTAB micelle solution and their room temperature photoluminescence properties. *J. Alloys Compd.* **2011**, *509*, 2060–2065. [CrossRef]

Article

One-Pot Hydrothermal Synthesis of Victoria Green (Ca₃Cr₂Si₃O₁₂) Nanoparticles in Alkaline Fluids and Its Colour Hue Characterisation

Juan Carlos Rendón-Angeles [1,*], Zully Matamoros-Veloza [2], Jose Luis Rodríguez-Galicia [1], Gimyeong Seong [3], Kazumichi Yanagisawa [4], Aitana Tamayo [5], Juan Rubio [5] and Lluvia A. Anaya-Chavira [1]

1. Centre for Research and Advanced Studies of the National Polytechnic Institute, Saltillo Campus, Ramos Arizpe, Coahuila 25900, Mexico; jose.rodriguez@cinvestav.edu.mx (J.L.R.-G.); lluvia.anaya@cinvestav.mx (L.A.A.-C.)
2. Tecnológico Nacional de México/(I.T. Saltillo), Technological Institute of Saltillo, Graduate Division, Saltillo 25280, Mexico; zully.mv@saltillo.tecnm.mx
3. New Industry Creation Hatchery Center, Tohoku University, 6-6-10 Aoba, Aramaki, Aoba-Ku, Sendai 980-8579, Japan; kimei.sei.c6@tohoku.ac.jp
4. Research Laboratory of Hydrothermal Chemistry, Faculty of Science, Kochi University, Kochi 780-8073, Japan; yanagi@kochi-u.ac.jp
5. Institute of Ceramics and Glass, CSIC, 28049 Madrid, Spain; aitanath@icv.csic.es (A.T.); jrubio@icv.csic.es (J.R.)
* Correspondence: jcarlos.rendon@cinvestav.edu.mx; Tel.: +52-(844)-438-9600

Abstract: One-pot hydrothermal preparation of $Ca_3Cr_2Si_3O_{12}$ uvarovite nanoparticles under alkaline conditions was investigated for the first time. The experimental parameters selected for the study considered the concentration of the KOH solvent solution (0.01 to 5.0 M), the agitation of the autoclave (50 rpm), and the nominal content of Si^{4+} (2.2–3.0 mole). Fine uvarovite particles were synthesised at 200 °C after a 3 h interval in a highly concentrated 5.0 M KOH solution. The crystallisation of single-phase $Ca_3Cr_2Si_3O_{12}$ particles proceeded free of by-products via a one-pot process involving a single-step reaction. KOH solutions below 2.5 M and water hindered the crystallisation of the $Ca_3Cr_2Si_3O_{12}$ particles. The hydrothermal treatments carried out with stirring (50 rpm) and non-stirring triggered the crystallisation of irregular anhedral particles with average sizes of 8.05 and 12.25 nm, respectively. These particles spontaneously assembled into popcorn-shaped agglomerates with sizes varying from 66 to 156 nm. All the powders prepared by the present method exhibited CIE-L*a*b* values that correspond to the Victoria green colour spectral space and have a high near infrared reflectance property. The particle size and structural crystallinity are factors affecting the Victoria pigment optical properties, such as CIE-L*a*b* values, green tonality, and near-infrared reflectance.

Keywords: $Ca_3Cr_2Si_3O_{12}$; Victorian green pigment; hydrothermal synthesis; one-pot reaction; hierarchical architecture; nanosized dyes

1. Introduction

The traditional garnet structured $Ca_3Cr_2Si_3O_{12}$, commercially known as Victorian green, has been used as the source of colour in paintings and decoration, amongst other applications. This inorganic pigment had been available in the Earth's crust as a mineral, as the green garnet uvarovite, but only a limited number of green mineral pigments were accessible [1–3]. The uvarovite garnet is a member of a broad group of mineral silicates containing Ca^{2+} and Cr^{3+} ions as crystalline structural constituents. The calcic varieties are formed on the earth crust at lower pressures but also synthetically [3,4]. The mineral uvarovite has a dark-green tonality, and it belongs to the crystalline cubic structure (space group Ia-3d, No. 230) with unit cell 11.5–12.5 Å with chemical formula $Ca_3Cr_2Si_3O_{12}$ [4–8]. The silicate garnet crystalline structure consists of a homogeneous distributed corner-sharing SiO_4 tetrahedral units and CrO_6, together with twisted CaO_8 cubes allocated in between. The Cr^{3+} ions forming the CrO_6 octahedra unit is the chromophore that produces the dark-green hue. This particular garnet and its solid solutions have been the research subject due to its optical, magnetic, and electrical properties [3,6].

The pioneering work aiming at the preparation of the uvarovite pigment powders was conducted in 1885 with the Victorian green pigment powder obtained by solid-state reaction at high temperature 1000–1370 °C using a powder mixture containing $K_2Cr_2O_7$ $CaCO_3$, and SiO_2 [9]. Various research works have been conducted since the 1950s. Efforts have been devoted to improving the solid-state reaction synthesis conditions, focusing on the raw precursor selection and the treatment temperature (855–1490 °C) [9–12]. In addition, the Cr^{3+} loss caused by Cr_2O_3 high volatility above 1200 °C, coupled with the electric valence change of chromium (Cr^{2+}, Cr^{3+}, Cr^{6+}, and occasionally Cr^{4+} and Cr^{5+} are other processing disadvantages of this method. These factors are crucial to control the chemical compositional homogeneity of uvarovite garnets [12,13]. More recently, soft chemical processing (sol–gel), coupled with a calcination stage, has been under exhaustive investigation. This process had favoured the preparation of uvarovite and its solid solutions at even lower temperatures ranging from 700 to 900 °C in air atmosphere [13–15].

However, among the soft chemical methods available for inorganic material production, hydrothermal processing has delivered high processing efficiency for the preparation of nano-sized inorganic silicate pigments. In particular, pyroxene silicates, namely $XYSi_2O_6$ (where X and Y are both divalent metal cations), were successfully crystallised under mildly alkaline hydrothermal conditions between 180 and 240 °C [16]. The crystallisation of 3D hierarchical morphologies occurred by transforming the precursor colloidal gel by a controlled dissolution–crystallisation mechanism. The tuning of the colour tonality of $BaCuSi_2O_6$ violet stain occurred preferentially in alkaline hydrothermal media. The colour hue control was promoted by the use of NO_3^- anions, which controlled the nanometric size particle assembly process to form the fine nest-like shaped particles [16]. A literature survey suggests that natural green uvarovite crystals were likely formed under hydrothermal conditions in the Earth's crust. On the contrary, the crystallisation of the uvarovite garnet species has not been yet investigated in alkaline hydrothermal media under controlled laboratory conditions, which is the main subject in the present study.

In the work reported here, the efforts were devoted to investigating the optimum hydrothermal reaction conditions for the crystallisation of Victorian green pigment ($Ca_3Cr_2Si_3O_{12}$). In particular, we have systematically investigated the effect of various critical processing parameters that might trigger the crystallisation of the single-phase $Ca_3Cr_2Si_3O_{12}$ and the control of particle growth and morphology. The alkaline hydrothermal media concentration, stirring of the mother liquor, and the nominal mixed Si^{4+} molar content were the main parameters varied. Hypothetically, the use of KOH solutions coupled with the control of the Si^{4+} ions content deficiency might lead to promoting controlled hydrothermal conditions that would hinder the oxidation of Cr^{3+} to other valences (Cr^{4+} or Cr^{6+}). This approach might lead to the fine-tuning of a one-pot hydrothermal process for preparing nanosized $Ca_3Cr_2Si_3O_{12}$ particles free of reaction by-products. Likewise, the control of the $Ca_3Cr_2Si_3O_{12}$ particle morphology found in the precursor co-precipitated gel containing a Si^{4+} deficiency in alkaline solution could result in the tuning of the green tonalities of the pigment; the effectiveness of these approaches is discussed in detail below.

2. Materials and Methods

2.1. Materials and Ca:Cr:Si Precursor Preparation

Hydrated chemical-grade reagents (Wako, Osaka, Japan, 99.0% purity) calcium nitrate ($Ca(NO_3)_2 \cdot 4H_2O$), chromium nitrate ($Cr(NO_3)_3 \cdot 9H_2O$), and sodium metasilicate ($Na_2SiO_3 \cdot 9H_2O$) were used for the precursor gel preparation. Metal precursor solutions were prepared with deionised water at a concentration of 0.375 M for $Ca(NO_3)_2$ and Na_2SiO_3, and 0.25 M for $Cr(NO_3)_3$. The concentration of the KOH solutions selected varied in the range of 0.01–5.0 M. Equivalent volume solution mixing Ca:Cr:Si ratios of 5.5:5.5:5.5 mL (16.5 mL) and 7.5:7.5:7.5 mL (22.5 mL) were selected. Theses solution mixture ratios provide the Victorian green pigment stoichiometric Ca:Cr:Si molar ratio of 3:2:3. Initially, the Ca and Cr solutions were mixed in the Teflon chamber bottom; then, the Na_2SiO_3 solution was poured, causing spontaneously the precipitation of a milky light green colour (Gel Ca:Cr:Si) colloid. The colloidal precursor is made up of an amorphous phase and crystalline SiO_2 as shown in the X-ray diffraction plots in the Section 3.1. Subsequently, the KOH solution volume (12.5 or 18.5 mL) was added, all the experiments were carried at a total mother liquor solution volume of 35 mL with the autoclave 50% full.

2.2. Hydrothermal Treatments

2.2.1. Treatments Conducted under Stirring Conditions

The selected parameters to investigate the Victorian green pigment synthesis are as follows: the micro-autoclave agitation speed, the alkaline media concentration, and the KOH solution volume. The study also includes evaluating the effect of the kinetic parameters (temperature and reaction time). These experiments aimed to determine the effect of the alkalinity and the stirring of the hydrothermal medium to produce the $Ca_3Cr_2Si_3O_{12}$ phase. After mixing the precursor solutions, a volume of 18.5 mL of KOH solution with concentrations between 0 and 5.0 M was added. The autoclaves were sealed and placed in the rotatory device assembled within the convection oven (Figure 1). Then, the vessels were heated at a constant 5 °C speed to reach 220 °C, and the treatments were conducted for 24 h at a constant stirring speed of 50 rpm.

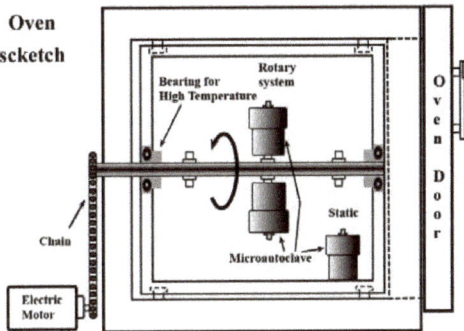

Figure 1. Scheme of the experimental oven used for carrying out the hydrothermal treatments under stirring and static conditions.

2.2.2. Treatments Conducted under Non-Stirring Conditions

The second experimental set was conducted to evaluate the feasibility for promoting the one-pot reaction without stirring. The treatments were carried using a 5.0 M KOH solution with two volumes (12.5 and 18.5 mL) for different reaction intervals (3–72 h) and temperatures (200–240 °C). Additionally, some experiments were conducted with different Si^{4+} molar contents (2.2–2.8 mole). According to the nominal 3.0 mole Si^{4+} (7.5 mL), the lower contents of 2.2, 2.4, 2.6, and 2.8 Si^{4+} mole were added proportionally to the volume ratio $Si^{4+}:H_2O$ of 5.5:2.0, 6.0:1.5, 6.5:1.0, and 7.0:0.5, respectively. After the solution mixing, the autoclave vessel was sealed and heated inside a convection oven to reach the desired temperature. After each treatment, the reaction products were gravimetrically separated and then washed vigorously, four times, with hot water (60 °C) to clean the remaining alkaline medium. Then, the resultant powders were dried overnight in an oven at 80 °C.

2.3. Characterisation

Powder X-ray diffraction (PXRD). The analyses were conducted in a Rigaku Ultima IV diffractometer (Akishima, Tokyo, Japan) operated at 40 kV and 20 mA, using Cu-Kα radiation (λ = 1.54056 Å). XRD analyses were collected in the 2θ range of 5–80° at a constant scanning speed of 20°/min with a 0.02° step. Rietveld refinement analyses were carried out on the diffraction patterns collected in the 2θ range of 15–130°, under standard conditions at a scanning speed of 0.01°/min and 0.002° step. The refinement calculations were performed using TOPAS 4.2 (Bruker AXS: Karlsruhe, Germany, 2009) software. The space group, together with the atomic position (Wyckoff number and coordinates) correspond to the COD card 96-900-7150. Additional details associated with the Rietveld refinement analysis are given in the Supplementary Supporting Information File, Section S2.3.

Fourier transform infrared spectroscopy (FT-IR). Additional structural details associated with the presence of water molecules were investigated using a FT-IR apparatus JASCO 4000, Hachioji, Tokyo, Japan. The analyses were conducted using pelletised samples prepared with 5 mg of pigment and 200 mg KBr. The powder samples were dried overnight at 60 °C previously the pellet preparation.

Morphology and microstructural observations. The microstructural aspects of the $Ca_3Cr_2Si_3O_{12}$ particles were observed by (FE-SEM JEOL JSM-7100F, Akishima, Tokyo, Japan) operated at 10 kV and 69 µA. The particle size distribution was measured from SEM images of 50 particles. Crystalline structural details of Victoria pigment particles were revealed by (HR-TEM, Philips Titan 300) operated at 300 kV.

Differential thermal analysis (DTA). $Ca_3Cr_2Si_3O_{12}$ pigment thermal stability was evaluated via thermogravimetric and differential thermal analysis (Perkin Elmer Pyris Diamond TG/DTA, Waltham, MA, USA) from 30 to 1000 °C. The analyses were carried at 10 °C/min constant heating rate in air atmosphere corresponding to the apparatus furnace chamber volume.

Optical properties. The Victoria green powders optical properties, colour CIEL*a*b*, and reflectance spectra were measured in a UV-vis near infrared (NIR) spectrometer (Perkin-Elmer Lambda 25, 800–2500 nm, Waltham, MA, USA). The $BaSO_4$ fine powder was used to calibrate the colour space parameters, as suggested by the standard CIE-L*a*b* colourimetry method.

3. Results and Discussion

A proposed one-pot hydrothermal processing scheme was investigated as a new and potentially feasible route for preparing synthetic Victorian green pigment particles ($Ca_3Cr_2Si_3O_{12}$). Initial effort was directed toward establishing the appropriate alkaline concentration over the range 0.0–5.0 M that triggers the crystallisation of $Ca_3Cr_2Si_3O_{12}$, since this had not yet been found for this pigment in hydrothermal fluids. Table 1 summarises the conditions of the selected experiments conducted in various KOH solutions that produced $Ca_3Cr_2Si_3O_{12}$ particles under alkaline hydrothermal conditions with interesting green tonalities. The major crystalline phases identified are also included together with the amount of the secondary phases calculated by the Rietveld refinement.

3.1. The Effect of the Alkalinity on the $Ca_3Cr_2Si_3O_{12}$ Chemical Stability under Hydrothermal Synthesis

Figure 2 shows the typical XRD patterns of residual powders produced at 240 °C for 24 h under stirring at 50 rpm, employing water and various KOH alkaline solutions (0–5.0 M). These experiments aimed to determine the potential feasibility to produce $Ca_3Cr_2Si_3O_{12}$ uvarovite-structured in alkaline hydrothermal medium for the first time. Generally, the synthesis of $Ca_3Cr_2Si_3O_{12}$ particles did not occur when water and low concentrated KOH (0.01–0.05 M) solutions were used as hydrothermal medium. The PXRD pattern of the reaction products could not be indexed with analogous inorganic compounds in the $CaO–Cr_2O_3–SiO_2$ system. On the other hand, using KOH media with concentrations of 0.5 and 1.0 M resulted in the formation of secondary phases of SiO_2 (Quartz high (●)), Ca_2SiO_4 (▽), and $CaCr_2O_4$ (►). The Ca:Cr:Si gel chemical reactivity was enhanced in a mildly concentrated 2.5 M KOH solution, resulting in the formation of $Ca_3Cr_2Si_3O_{12}$ (69.63 ± 2.5 wt %) together with SiO_2 (Quartz high (●), Coesite (■)) and Ca_2SiO_5 (□) crystalline phases. The Rietveld refinement algorithm calculated the content of each phase identified, and the schematic results are shown in Figure S1c,d (Supplementary Supporting Data File).

Table 1. Summary of the relevant experiments selected to investigate the hydrothermal synthesis of single-phase $Ca_3Cr_2Si_3O_{12}$ particles and relevant Rietveld refinement results. The experiments were carried out with two volumes of the KOH solution 18.5 * and 12.5 ** mL.

Sample ID	KOH Solution [M]	Temperature (°C)	Time (h)	Stirring Speed (rpm)	Si^{4+} Nominal (mole)	Crystalline Phase	Phase Content (wt %)	Average Agglomerate Size (nm)	Crystallite Size (nm)	"a_0" (Å)	Cell Volume (Å^3)	Lattice Strain	R_{wp}	GOF (χ^2)
CCS1	5.0 *	220	24	50	3.0	Uvarovite	100.0	87.0 ± 17.0	18.91 (0.31)	12.3377 (63)	1878.06 (2.9)	0.49 (0.02)	7.14	2.97
CCS2	5.0 **	220	24	50	3.0	Uvarovite	100.0		22.31 (1.29)	12.2579 (221)	1841.82 (9.9)	1.68 (0.05)	11.60	6.56
CCS4	2.5 *	220	24	50	3.0	Uvarovite Quartz, SiO_2 Ca_2SiO_5 Ca_2SiO_4	65.01 1.66 16.93 16.39	-	12.68 (0.46)	12.2379(18)	1832.82 (0.8)	1.05 (0.08)	6.73	2.43
CCS3	1.0 *	220	24	50	3.0	Quartz, SiO_2 $CaCr_2O_4$ Ca_2SiO_4	10.15 41.84 48.01	-	-	-	-	-	11.48	6.55
CCS7	0.01 *	220	24	50	3.0	Amorphous	100.0	-	-	-	-	-	-	-
CCS8	0.0 *	220	24	50	3.0	Amorphous	100.0	-	-	-	-	-	-	-
CCS9	5.0 *	240	3	50	3.0	Uvarovite	100.0	145.0 ± 22.0	22.65 (0.93)	12.3332 (148)	1876.00 (6.8)	0.76 (0.04)	13.62	7.25
CCS17	5.0 **	240	12	50	3.0	Uvarovite	100.0	137.0 ± 25.0	27.38 (1.56)	12.2468 (112)	1836.86 (5.1)	1.43 (0.04)	12.04	6.74
CCS19	5.0 **	240	6	0	3.0	Uvarovite	100.0	156.0 ± 3.0	55.86 (5.05)	12.2753 (157)	1849.71 (7.1)	1.85 (0.04)	14.48	6.83
CCS21	5.0 **	240	24	0	3.0	Uvarovite	100.0	99.0 ± 20.0	29.33 (2.24)	12.2575 (152)	1841.64 (4.8)	1.42 (0.03)	15.11	6.89
CCS23	5.0 **	240	72	0	3.0	Uvarovite	100.0	173.0 ± 29.0	23.87 (0.41)	12.2429 (39)	1835.08 (1.8)	0.78 (0.01)	6.96	2.82
CCS15	5.0 **	220	6	0	3.0	Uvarovite	100.0	112.0 ± 15.0	32.14 (4.93)	12.3016 (347)	1861.63 (15.7)	1.96 (0.10)	10.72	6.16
CCS14	5.0 **	220	12	0	3.0	Uvarovite	100.0	148.0 ± 3.0	19.96 (0.48)	12.2750 (87)	1849.55 (4.0)	1.05 (0.03)	6.90	2.84
CCS13	5.0 **	220	24	0	3.0	Uvarovite	100.0	114.0 ± 17.0	13.07 (0.25)	12.2475 (106)	1837.18 (4.8)	0.68 (0.05)	6.15	2.74
CCS31	5.0 **	220	12	0	2.6	Uvarovite	100.0	104.0 ± 23.0	14.45 (0.29)	12.2936 (67)	1858.00 (3.0)	0.49 (0.05)	6.20	2.79
CCS32	5.0 **	220	12	0	2.4	Uvarovite	100.0	90.0 ± 28.0	23.57 (1.13)	12.3277 (133)	1873.47 (6.0)	0.48 (0.04)	16.61	7.19
CCS33	5.0 **	220	12	0	2.2	Uvarovite	100.0		28.26 (0.80)	12.3372 (75)	1877.82 (0.3)	0.73 (0.01)	6.51	2.88
CCS28	5.0 **	200	6	0	3.0	Uvarovite	100.0		12.79 (1.83)	12.3340 (984)	1876.35 (45.0)	2.0 (0.39)	12.52	5.88
CCS26	5.0 **	200	24	0	3.0	Uvarovite	100.0		12.66 (0.28)	12.2767 (136)	1850.33 (6.1)	0.88 (0.04)	6.63	2.76
CCS24	5.0 **	200	72	0	3.0	Uvarovite	100.0	66.0 ± 23.0	19.06 (0.44)	12.2708 (98)	1847.66 (4.4)	1.03 (0.03)	6.71	2.81

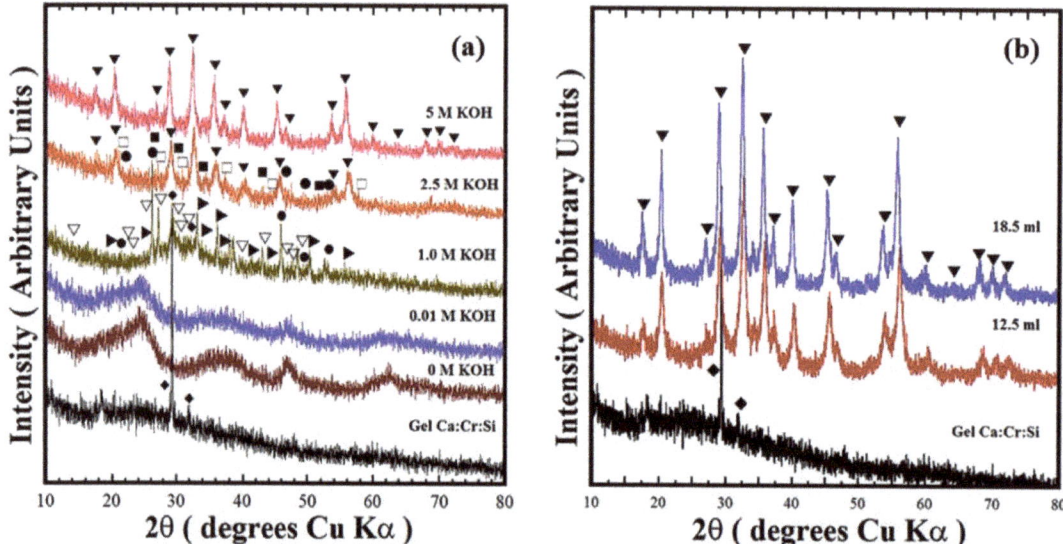

Figure 2. XRD patterns of the reaction products produced at 220 °C for 24 h with different KOH solvent solution (18.5 mL) concentrations. The experiments were carried out with stoichiometrically mixed Si^{4+} content of 3.0 mole (**a**). Those (**b**) of the powders prepared with two different volumes of the 5.0 M KOH solution; these experiments were all conducted at an autoclave stirring speed of 50 rpm. Indexed crystalline phases: (▼) Uvarovite structure $Ca_3Cr_2Si_3O_{12}$, COD card no. 96-900-7150, crystalline secondary phases: (◆) SiO_2, (●) SiO_2 Quartz high COD card 96-101-1201, (▽) Ca_2SiO_4 COD card no. 96-210-3317, (▶) $CaCr_2O_4$ COD card no. 96-200-2211 phase, (■) SiO_2 Coesite COD card no. 96-900-0805, (□) Ca_2SiO_5 COD card 96-200-1356.

The results show that $Ca_3Cr_2Si_3O_{12}$ single-phase free of reaction by-products occurs only in highly concentrated KOH media (5.0 M). The X-ray pattern of this sample was indexed with that of the cubic crystalline phase (COD Card No. 96-900-7150, space group $Ia\text{-}3d$, No. 230 (▼)), as seen in Figure 2. Likewise, the KOH nominal volume did not hinder the crystallisation of $Ca_3Cr_2Si_3O_{12}$; this compound solely was obtained either of the 5.0 M KOH solution of 12.5 mL (Sample ID CCS2) or 18.5 mL (Sample ID CCS1); see Figure 2b. These results bear evidence indicating that the one-pot hydrothermal reaction occurs in a single-step chemical reaction and is preferentially triggered under strongly alkaline conditions. The concentrated alkaline media dissolves the amorphous colloid precursor and the SiO_2, and as a consequence, solute supersaturation is reached in the system and it achieves the chemical equilibrium associated with Equation (1). The chemical reaction (Equation (1)) is conducive to Victorian green pigment powder crystallisation under the proposed one-pot hydrothermal reaction. It deserves emphasising that based on the reaction pathway elucidated by the XRD results, we infer that the KOH media is crucial to mitigate the acidic capability the anionic species, namely the high oxidant NO^{3-} ions, which might trigger the valence oxidation of Cr^{3+} to Cr^{6+}. The reaction trend is similar to that determined on silicate pigments, $BaCuSi_2O_6$ [16] and $BaCu_2Si_2O_7$ [17], where the reduction of Cu^{2+} to Cu^+ did not take place under alkaline hydrothermal conditions. Chromium oxidation might hinder the $Ca_3Cr_2Si_3O_{12}$ crystallisation due to the formation of more stable crystalline phases containing chromium (VI). This processing advantage is one of the factors associated to the one-pot hydrothermal processing efficiency in comparison with the solid-state reaction method widely used to prepare the Victorian green pigment [9–13]. Additionally, small crystalline structural differences were determined on the experiments carried at various temperatures and reaction intervals in 5 M KOH solution under hydrothermal treatments conducted without stirring. These results are reported in the Supplementary Supporting Information Data File, Section S3.

$$Ca_3Cr_2Si_{3-y}(OH)_{24-y(gel)} + ySiO_{2(s)} + 5NO_3^-{}_{(aq)} + 2Na^+{}_{(aq)}$$
$$Ca_3Cr_2Si_3O_{12(s)} + 5NO_3^-{}_{(aq)} + 2Na^+{}_{(aq)} + (24-y)OH^-{}_{(aq)} \quad (1)$$

3.2. Structural Features of the Victorian Green $Ca_3Cr_2Si_3O_{12}$ Powders Prepared under Alkaline Hydrothermal Conditions

The crystalline structural features of the cubic uvarovite particles prepared under the relevant treatment conditions are given in Table 1. Table 1 shows the results calculated by the Rietveld refinement analyses of parameters such as crystallite size, unit cell parameters, lattice strain, residual parameter R_{wp} (%) and goodness-of-fit (GOF) (χ^2) fitting refinement parameters. The refinement parameters are conducive of sufficient accuracy for calculating the structural features of the Victorian green pigment samples. The values of the R_{wp} and goodness-of-fit factor (GOF) (χ^2) (Table 1) are consistent with the $Ca_3Cr_2Si_3O_{12}$ residual line for the Victorian green pigment in the Rietveld refinement plots (Figure S2); these results reveal the high accuracy of the refinement approach used, leading to small residual differences between the experimental and calculated XRD patterns. The plots of selected pigments are shown in the Supporting Supplementary Information (see Figure S2). The unit cell lattice parameters of $Ca_3Cr_2Si_3O_{12}$ calculated are within the broad range of the "a_0" lattice parameter, 11.5–12.5 Å, which is determined in various uvarovite minerals and its synthetic parents [4–8]. Interestingly, the "a_0" lattice parameter varied within the range 12.2429–12.3372 Å (see Table 1) on the Victorian pigment powders. However, this variation is likely associated with the residual lattice strain induced in the crystallisation process as shown in Table 1, for the samples treated for short reaction intervals. However, the partial incorporation of water molecules in the $Ca_3Cr_2Si_3O_{12}$ under the one-pot hydrothermal process is another factor that might provoke the large lattice parameter values [4,18,19]. Natural mineral uvarovite is amongst the species formed at the upper mantle and transition zone. It usually incorporates water molecules that partially substitute the SiO_4 tetrahedra units by OH^- in the form of H_4O_4 [1,18,19]. Hence, the residual lattice strain coupled with the presence of H_4O_4 caused the crystalline structural variation in the Victorian green pigments produced by the new route investigated.

Additionally, $Ca_3Cr_2Si_3O_{12}$ crystalline features were also studied using FT-IR spectroscopic analyses over the wavenumber range of 400–4000 cm^{-1}. These analyses were conducted to reveal additional features of the chemical bonds in the $Ca_3Cr_2Si_3O_{12}$ pigment. The FT-IR spectra of the samples prepared at 240 °C in the 5.0 M KOH solution (with a nominal Si^{4+} content of 3.0 mole) for various intervals are shown in Figure 3. Generally, the FT-IR results revealed no further chemical compositional differences between the samples because the same bands were observed irrespective of the reaction interval. Thus, the sharp band at 528.7 cm^{-1} is assigned to the Si-O bond symmetric bending mode ($3v_4$), while two overlapped peaks at 861.1 and 920.8 cm^{-1} constituted the broad peak in the wavelength range of 750–1200 cm^{-1}. These bands correspond to the asymmetric Si-O stretching mode ($3v_3$) bands of the SiO_4 units. The new signals at 1389.6, 1483.4, and 1636.9 cm^{-1}, together with the broad one between 3000 and 3700 cm^{-1}, are attributed to the presence of O-H bonding. These bands correspond to water molecules absorbed on the pigment powders. The present results provide clear confirmatory evidence of hydro-garnet formation in highly saturated alkaline hydrothermal media, which has not previously been reported [1,18,19]. Based on these results, we argued that the water molecules absorption occurs in the OH^- supersaturated media, due to OH^- accelerating the dissolution–crystallisation mechanism that transforms the precursor gel into the $Ca_3Cr_2Si_3O_{12}$. Water absorption onto the uvarovite particles surface is likely promoted via a hydro-substitution mechanism [18]. The water (wt %) content determined by thermal gravimetric analyses was 17.0 ± 2.5 wt %; this content was measured with various uvarovite powders prepared under different conditions as shown in Figure S4 (Supplementary Supporting Data). Furthermore, the water absorption is likely to provoke the Si-O band displacement in all the samples and induce the residual strain calculated by the Rietveld refinement analyses that disturb the lattice arrangement in the cubic uvarovite structure.

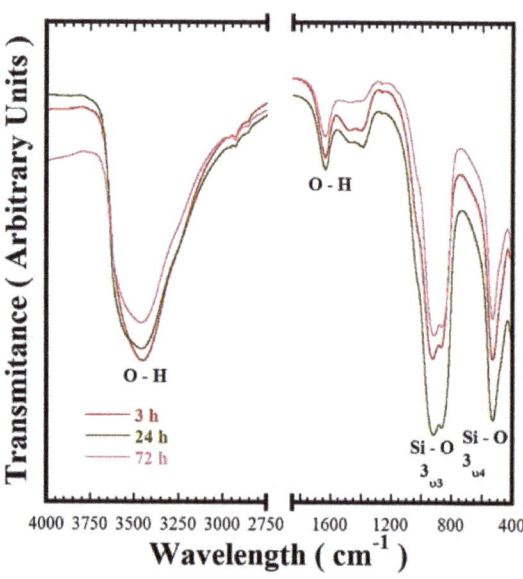

Figure 3. Fourier transform infrared (FT-IR) spectrum of $Ca_3Cr_2Si_3O_{12}$ pigment samples hydrothermally crystallised at 200 °C using a 5.0 M KOH solution (12.5 mL), the stoichiometric mixed Si^{4+} content used was 3.0 mole, and all the treatments were conducted without stirring for various reaction intervals as shown.

3.3. Tailoring the Synthesis of the $Ca_3Cr_2Si_3O_{12}$ by Controlling the Nominal Si^{4+} Content

Figure 4a shows typical PXRD patterns of Victorian green samples prepared at 220 °C for 12 h in the 5.0 M KOH solution without stirring. This experimental set aimed to investigate the effect of the nominal Si^{4+} content below the stoichiometric value of 3.0 mole required to produce the $Ca_3Cr_2Si_3O_{12}$. Crystallisation of the cubic structured $Ca_3Cr_2Si_3O_{12}$ occurred under alkaline hydrothermal conditions and was irrespective of the Si^{4+} deficiency induced in the reaction system. Thus, the Si^{4+} deficiency did not affect the equilibrium associated with the chemical reaction (Equation (2)) proposed for this reaction system that triggers the $Ca_3Cr_2Si_3O_{12}$ preparation by a single-step reaction. In addition, the nominal Si^{4+} deficiency did not produce any marked differences in the initial gel co-precipitation reaction. This inference is also supported by the dissolution of the Si-deficient precursor gel Ca:Cr:Si$_x$, which was found to proceed rapidly as in the experiments conducted with the uvarovite precursor gel containing 3.0 mole of Si^{4+} described in Section 3.1. No secondary crystalline phases containing Ca^{2+} or Cr^{3+}, namely calcium chromate, were produced during the experiments with the Si^{4+} molar deficiency (2.2–2.8 mole). Both metal ions were hydrolysed in the alkaline media as complex ions [16,17] (Equation (2)). The formation of hydro-garnet uvarovite on the residual powders prepared with various contents of Si^{4+} was confirmed by the FT-IR analyses, which are shown in Figure 4b. These results suggested that water absorption on uvarovite particles took place spontaneously during the particle crystallisation, even though the chemical equilibrium (Equation (2)) is reached with the Si^{4+} molar deficiency within the hydrothermal alkaline reaction media. These particular controlled set of experiments demonstrated that the Si^{4+} deficiency does not affect the $Ca_3Cr_2Si_3O_{12}$ synthesis, because the oxidation of Cr^{3+} to other metastable (Cr^{4+}) or Cr^{6+} stable valences was hindered on the hydrothermal alkaline media.

$$Ca_3Cr_2Si_{3-(y+x)}(OH)_{24-(y+x)(gel)} + ySiO_{2(s)} + 5NO_3^-{}_{(aq)} + 2Na^+{}_{(aq)} \rightarrow Ca_3Cr_2Si_3O_{12(s)} + 3xCa(OH)_z^{n+}{}_{(aq)} + 2xCr(OH)_m^{n+}{}_{(aq)} + 5NO_3^-{}_{(aq)} + 2Na^+{}_{(aq)} + 24 - (y+x)\ OH^-{}_{(aq)} \quad (2)$$

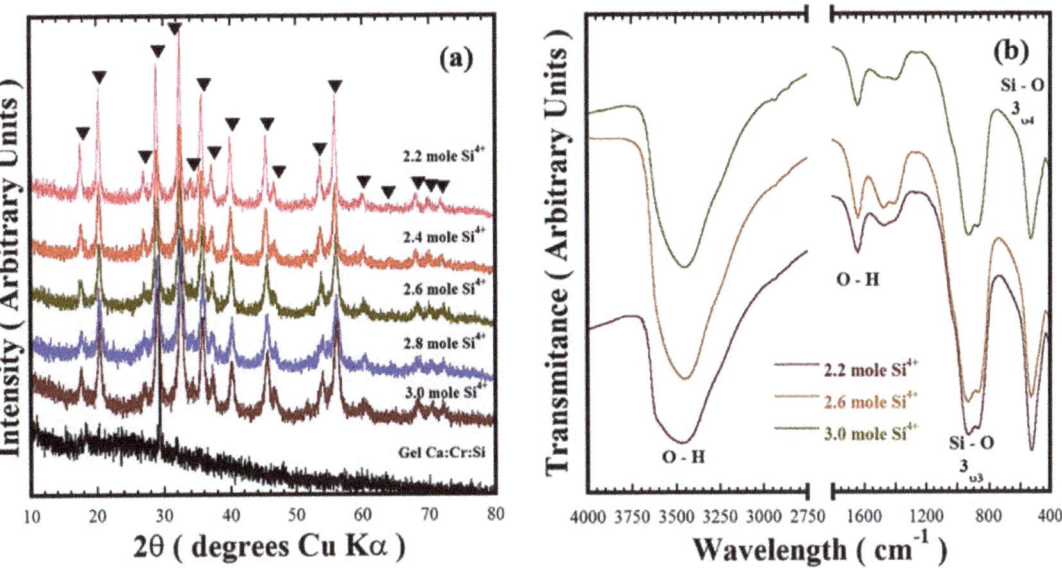

Figure 4. (a) XRD patterns and (b) FT-IR spectrum of the residual powders produced at 220 °C for 12 h without stirring in the 5.0 M KOH solvent (12.5 mL) using different Si^{4+} precursor molar contents. Crystalline phases (▼) $Ca_3Cr_2Si_3O_{12}$ uvarovite structure, COD Card No. 96-900-7150.

3.4. Morphology Evolution of $Ca_3Cr_2Si_3O_{12}$ Particles Prepared under Alkaline Hydrothermal Conditions

The morphology and particle size of the $Ca_3Cr_2Si_3O_{12}$ particles synthesised at 220 °C for 24 h in a 5.0 M KOH solution with an Si^{4+} content of 3.0 mole, under both stirred (at 50 rpm) and static conditions, are shown in Figure 5a,b: respectively. SEM observations revealed that monodispersed $Ca_3Cr_2Si_3O_{12}$ agglomerates with a popcorn quasi-spherical shape were the dominant morphology under the alkaline hydrothermal conditions. These agglomerates are constituted of tiny anhedral crystals, which self-assembled to produce the 3D hierarchical "popcorn" architecture [20,21]. The anhedral particle size varies from 12 to 34 nm, as suggested by the crystallite sizes calculated from SEM micrographs (Figure 5a,b) and the results given in Table 1. Solvent convection provoked by the autoclave rotation limited the agglomerate growth. The popcorn agglomerate size produced with a 50 rpm rotation speed was 87 ± 17 nm. By way of contrast, coarser agglomerates up to 148 ± 3 nm were produced by maintaining the autoclave static inside the oven during the hydrothermal treatment (Figure 5b). Increasing the reaction temperature to 240 °C under static treatment conditions only resulted in a slight increase of the agglomerate size up to 156 ± 3 nm (Figure 5c). We surmise that a homogeneous colloid dispersion provoked by the autoclave agitation caused the reduction of the popcorn-shaped agglomerates. Agitation breaks up the colloid, accelerating its rapid dissolution in the solvent; as a consequence, a greater quantity of embryos is precipitated compared to that produced without stirring. The large molar volume of embryos homogeneously dispersed reduces the local solute saturation, hindering the particle coarsening [16,17]. Increasing the reaction temperature, without agitation, does not further coarsen the particles. The agglomerate growth was not significantly affected by increasing the temperature over 200 °C.

Figure 5. FE-SEM micrographs of $Ca_3Cr_2Si_3O_7$ powders produced under hydrothermal conditions at 220 °C for 24 h using a 5.0 M KOH solution (12.5 mL) and 3.0 mole Si^{4+} precursor mixed content; the experiments were conducted at a stirring speed (**a**) 50 rpm and (**b**) 0 rpm. (**c**) Micrograph of the powders prepared at 240 °C without autoclave stirring under the same experimental conditions.

In addition, detailed crystalline structure features of the popcorn agglomerates were revealed by TEM observations. The analyses were conducted on residual powders prepared for 72 h at low (200 °C) and high (240 °C) temperatures; the TEM micrographs are shown in Figure 6. These images revealed that the bulk morphology of the popcorn-shaped agglomerates is irrespective of the treatment temperature. However, the size of the constituent anhedral particles was slightly increased from 8.5 ± 1.7 nm (200 °C) up to a mean size of 11.5 ± 2.0 nm by increasing the temperature to 240 °C. Details associated with the crystallinity and the self-assembly architecture of the fine anhedral $Ca_3Cr_2Si_3O_{12}$ particles were revealed by the HR-TEM observations (Figure 6b,d). Interestingly, these observations suggest that the anhedral particles produced at 200 °C exhibit a distorted atomic ordering. Nevertheless, some particles at the surface revealed that the assembly that took place in the basal plane with index (400), see Figure 6b. These Miller indexes correspond to an interplanar distance of 0.305 nm for the cubic garnet structure. On the contrary, at 240 °C, a remarkable atomic stacking occurred on the anhedral so that these particles exhibited a high crystallinity with the atomic ordering along a preferential direction indicated by the Miller index (123). This crystallographic plane was indexed with a calculated lattice fringe spacing of 0.317 nm, as portrayed in Figure 6d. Based on these interpretations, we argue that the reactivity of OH^- ions in the hydrothermal media caused a marked variation in the growth and the spontaneous self-assembling of the anhedral particles. Therefore, the dissolution–crystallisation mechanism reaction kinetics proceeded slowly at 200 °C, causing disruptions in the atomic stacking. The faster kinetics achieved at 240 °C caused a rapid solute supersaturation in the alkaline solvent, leading to correspondingly rapid embryo crystallisation and spontaneous epitaxial growth in the preferential crystallographic direction; this process is analogous to that recently determined for other silicate inorganic pigments [16,17]. This process, it is argued, controls the 3D hierarchical assembly, resulting in the formation of the Victorian green popcorn particles.

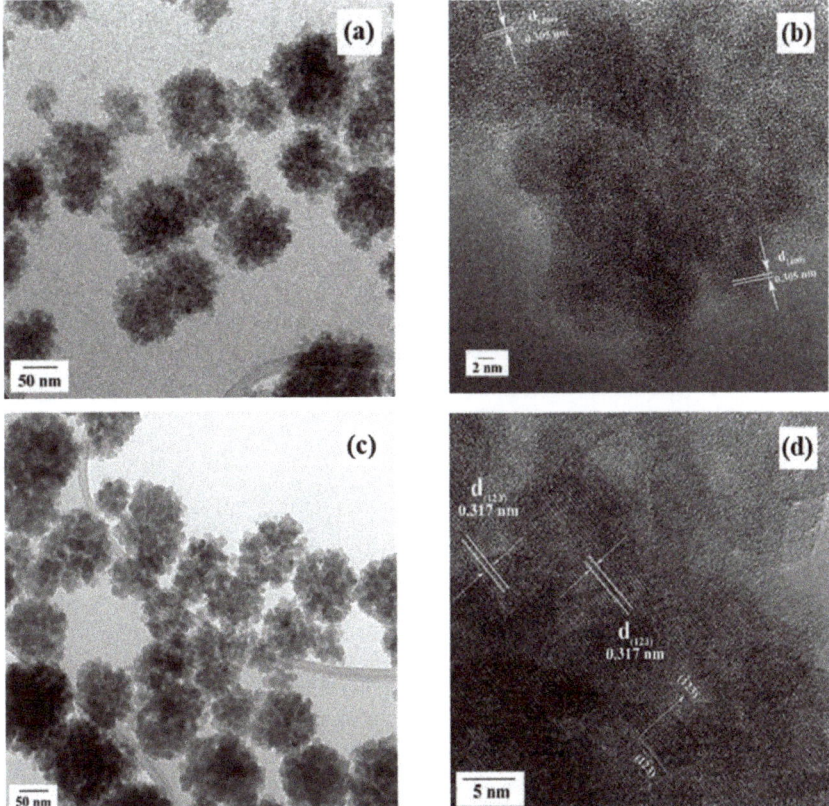

Figure 6. TEM (**a,c**) and HR-TEM (**b,d**) images of the popcorn shaped $Ca_3Cr_2Si_2O_{12}$ agglomerates hydrothermally prepared for 72 h without autoclave stirring and 3.0 mole Si^{4+} content using 12.5 mL of 5.0 M KOH as solvent. The one-pot hydrothermal treatment was carried at 200 (**a,b**) and 240 °C (**c,d**), respectively.

3.5. Diffuse Reflectance and Chromatic Properties of $Ca_3Cr_2Si_3O_{12}$ Victorian Green Powders

Near infrared to UV-vis diffuse reflectance analysis was carried out on the $Ca_3Cr_2Si_3O_{12}$ powders prepared at 200 and 240 °C and various reaction intervals without agitation. These samples produced with a stoichiometric Si^{4+} molar content (3.0 mole) exhibited differences in the particle size, green tonality, and structural crystallinity. The NIR reflectance data shown in Figure 7 were collected over wavelengths of 600–2500 nm using pelletised disks (10 mm diameter and 1 mm thickness). Generally, all the Victorian green powders prepared at 200 and 240 °C for both short and long reaction intervals exhibited an absorption peak at 833.33 nm. Furthermore, a marked increase in the reflectance took place in the NIR spectrum from 1000 and 2500 nm for all the samples, but all the pigments prepared at 200 °C (Figure 7a) had a slight reduction of 6% between 1300 and 1800 nm compared to those prepared at 240 °C. Interestingly, the pigment powders synthesised at 240 °C over a reaction interval of 24 h only had a tiny reflectance decay of approximately 2% below the maximum reflectance (98%) determined for powders produced over 72 h; see the inset in Figure 7b. These results, which were taken together with the crystalline structural differences revealed by HR-TEM, suggest that the uvarovite pigment NIR is maximised by improvement in the atomic stacking [22–24]. This is suggested to occur due to uvarovite dissolution–recrystallisation, which is promoted at reaction intervals over 24 h under hydrothermal conditions at 240 °C. It is worth mentioning that these results are especially relevant in highlighting a potential application for the prepared Victorian green pigments. In terms of energy, 52% of the sunlight reaches the Earth's atmosphere falls within the spectrum of the NIR region (700–2500 nm). The incidence of this radiation on the surface of dark-coloured objects causes them to heat up. The reflectance behaviour of the uvarovite $Ca_3Cr_2Si_3O_{12}$ powders, as suggested by the NIR reflectance analyses of Figure 7, indicates that a potential application is as a "cold" pigment because it absorbs little NIR radiation. Such surface coatings could have a remarkable impact on energy-saving applications where solar radiation causes unwanted heating [24]. In general, bright colour TiO_2 pigment has an 80% NIR radiation, resulting in a low heat up. This pigment is used to reduce the total solar energy absorbed. Furthermore, the $Ca_3Cr_2Si_3O_{12}$ powders have similar NIR reflectance properties to those submicron-sized blue pigments $YIn_{0.8}Mn_{0.2}O_3$ [22] and $YIn_{0.9}Mn_{0.1}O_3$-ZnO [24], which have a 90% near infrared reflectance property. Similarly, yellow $BiVO_4$ pigments exhibit a reflectance above 80%, and the reflectance was maximised due to the $BiVO_4$ polymorph formation provoked by the treatment temperature [25].

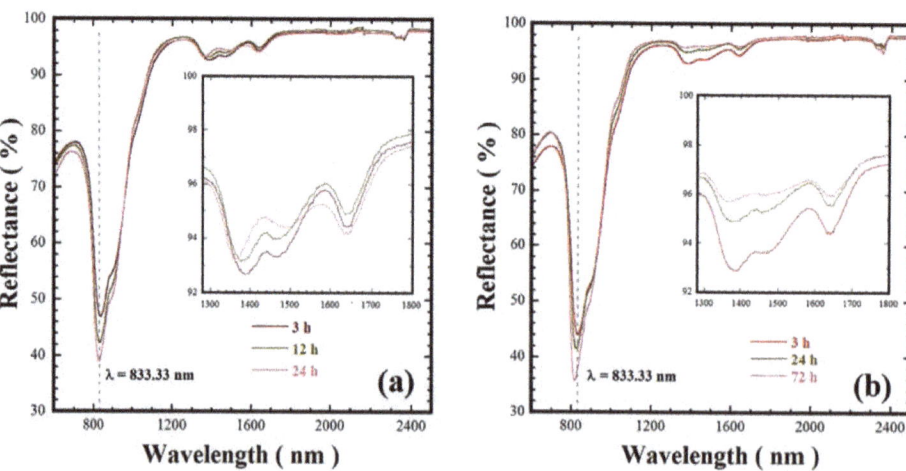

Figure 7. Near infrared reflectance spectra of the Victorian green pigments hydrothermally prepared in a 5.0 M KOH solution (12.5 mL) with 3.0 mole of Si^{4+} and no autoclave agitation. The powder pigments were synthesised at various reaction intervals and temperatures of (**a**) 200 and (**b**) 240 °C.

Victorian green powders colour characterisation was conducted using the CIE-L*a*b* colour space. The chroma value was calculated using the mathematical expression $C_{ab}^* = \sqrt{(a^*)^2 + (b^*)^2}$ (details reported elsewhere [25]). Table 2 summarises the chromatic coordinate L*a*b* and the chroma values determined for various pigments prepared under different conditions. The RGB colour coordinates obtained by transforming the L*a*b* values, and the pigment colour tonality associated with its RGB coordinates, are also given in Table 2. Generally, the results revealed that the powder colour is consistent with the standard CIELab coordinates of the Victorian green pigment. Although there were no significant variations in the chromatic values of single-phase uvarovite pigments, lighter bright green tonalities were associated with powders having chroma "C_{ab}^*" values within the range of 12.98–17.75. The pigments with chroma values ranging from 18.94 to 22.44 have a slightly darker green tonality. Based on these results, we surmise that the green tonality tuning is further enhanced by the microstructural parameters of the $Ca_3Cr_2Si_3O_{12}$ powders, particularly the popcorn-shaped agglomerate size and the refinement of the crystalline structure. This inference agrees with the experimental results obtained with the one-pot hydrothermal processing, because at temperatures over 220 °C in the alkaline medium (5.0 M KOH), well crystalline anhedral particles are produced, which are responsible for the variation on the $Ca_3Cr_2Si_3O_{12}$ green pigment hue.

The Si^{4+} deficiency did not affect the self-assembly process of the anhedral particles. However, based on the chemical equilibrium (Equation (2)), the hydrolysed metal cations reacted with the OH^-, giving rise to the formation of complex hydroxide species ($Ca(OH)_z^{n+}$ and $Cr(OH)_m^{n+}$). This phenomenon caused a local reduction in the OH^- concentration, which slowed down the coarsening of the hierarchical 3D popcorn-shaped agglomerates, and this reaction took place using low Si^{4+} molar contents (2.2 and 2.6 mole). Hence, the particle size effect is the phenomenon responsible for trigger the bright green hue according to the natural light scattering mechanism. This mechanism is further enhanced in the nanometric-sized 3D popcorn-shaped agglomerates consisting of fine anhedral-shaped crystals (14.4–28.2 nm, see Table 1). Additionally, the dark green tonalities caused by the light interaction occur due to an increase in the crystallinity of the anhedral particles, together with the well-formed faceted surfaces (Figure 7a,b). Furthermore, the compact popcorn-shaped 3D arrangement could physically enhance the dynamic light reflection, giving the $Ca_3Cr_2Si_3O_{12}$ powders the properties of cool pigment, consequently reducing the light absorbance [4,11]. Hence, based on the optical measurements, we argue that the nano-sized hierarchical 3D popcorn-shaped $Ca_3Cr_2Si_3O_{12}$ agglomerate powders have potential application as cold pigments. Other applications include printing ink preparation, acrylic paints, and decorative purposes.

Table 2. Summaries of the particle sizes, CIE-L*a*b* values, chroma, RGB parameters, and colour of $Ca_3Cr_2Si_3O_{12}$ pigments prepared under stirred and static conditions; at different temperatures, reaction intervals, Si^{4+} deficiency with respect to the nominal concentration; and alkaline KOH media with different concentrations.

Sample ID	KOH Solution [M]	Temperature (°C)	Time (h)	Stirring Speed (RPM)	Si^{4+} Nominal (mole)	Average Agglomerate Size (nm)	CIELab Coordinates			RGB Colour Coordinates			Chroma C_{ab}^*	Colour Hue
							L*	a*	b*	R	G	B		
Gel CCS	-	-	-	-	-	-	64.31	−16.74	1.16	123	165	153	16.78	
CCS8	0.0	220	24	50	3.0	-	32.80	−43.49	14.88	0	93	51	45.97	
CCS7	0.01	220	24	50	3.0	-	41.75	−35.60	5.75	0	113	88	36.07	
CCS1	5.0	220	24	50	3.0	-	58.97	−17.78	−3.63	102	151	148	18.16	
CCS2	5.0	220	24	50	3.0	87.0 ± 17.0	58.03	−22.13	−3.66	88	151	145	22.44	
CCS4	2.5	220	24	50	3.0	-	67.04	−16.89	5.41	133	172	153	17.75	
CCS13	5.0	220	24	0	3.0	148.0 ± 3.0	59.98	−22.21	−4.24	93	156	151	22.62	
CCS21	5.0	240	24	0	3.0	156.0 ± 3.0	64.16	−18.75	−3.06	114	165	160	19.00	
CCS23	5.0	240	72	0	3.0	99.0 ± 20.0	62.99	−18.68	−3.10	111	162	157	18.94	
CCS24	5.0	200	72	0	3.0	66.0 ± 23.0	60.57	−20.04	−3.78	100	157	152	20.40	
CCS14	5.0	220	12	0	3.0	112.0 ± 15.0	66.52	−17.22	−1.78	125	171	164	17.31	
CCS31	5.0	220	12	0	2.6	114.0 ± 17.0	60.79	−19.48	−4.13	102	157	153	19.92	
CCS33	5.0	220	12	0	2.2	90.0 ± 28.0	72.44	−12.95	−0.79	151	185	179	12.98	

4. Conclusions

The synthesis of Victorian green pigment was successfully prepared in a highly concentrated KOH solution for the first time by the one-pot hydrothermal method. The nanosized aggregates crystallisation proceeded via a single-step chemical reaction that was achieved by the dissolution–crystallisation mechanism. The new processing approach is highly efficient, because the synthesis occurred even for a short reaction interval (3 h) at 200 °C without stirring. These conditions are adequate to synthesise nano-sized anhedral irregular particles; simultaneously, these particles underwent a spontaneous self-assembly that produce new nanometric 3D hierarchical popcorn-shaped agglomerates. Additionally, the colour tuning of the $Ca_3Cr_2Si_3O_{12}$ pigment was achievable with controlling the Si^{4+} deficiency (2.2–2.8 mole). The $Ca_3Cr_2Si_3O_{12}$ optical properties, colour hue, and NIR diffuse reflectance are affected by slight alterations in the particle size and crystallinity of the anhedral irregular particles forming the 3D hierarchical particle agglomerates. Hence, based on optical properties, the Victorian green powders prepared by the one-pot hydrothermal process have potential applications as a cold pigment source, and to prepare printing ink, acrylic paint, and other decorative purposes.

Supplementary Materials: The following are available online at https://www.mdpi.com/2079-4991/11/2/521/s1, Figure S1: XRD diffraction results of treatments conducted at different temperature without stirring, Figure S2: The Rietveld refinement plots were calculated for the cubic $Ca_3Cr_2Si_3O_{12}$ and various secondary phases, Figure S3: FE-SEM micrographs of Victorian green pigments prepared with different Si^{4+} mole contents, Figure S4: DTA of $Ca_3Cr_2Si_3O_{12}$ prepared at various experimental conditions, Table S1: Summary of the chemical composition of the single-phase $Ca_3Cr_2Si_3O_{12}$, Tables S2–S7: CIF files of the crystalline phases determined in the XRD patterns of the powders prepared in hydrothermal conditions.

Author Contributions: J.C.R.-A. conceptualize, designed and organize the research work; L.A.A.-C. and J.L.R.-G. performed the experiments; Z.M.-V. carried out the chemical and crystalline structural characterization and the data analysis; K.Y. provide the infrastructure to conduct the hydrothermal experiments under agitation; A.T. and J.R. preformed the colour and reflectance characterisation. G.S. and J.C.R.-A. analysed the results and wrote the manuscript. All authors have read and agreed to the published version of the manuscript.

Funding: A part of this research was supported by federal research budget (C-3000) of the Centre for Research and Advanced Studies of the NPI.

Acknowledgments: J.C.R.-A.: Z.M.-V. and J.L.R.-G. are indebted to CONACYT-SNI. Many thanks are also given to T. Matsuzaki and S. Yanagimoto at the Center of Advanced Marine Core Research, Kochi University, Japan, for their assistance on the FE-Scanning Electron Microscopy observations; and D. Vazquez-Obregon for his help in the thermal analysis characterisation of the powders conducted at Technological Institute of Saltillo. Special acknowledge is offer to Jonathan Willis-Richards for his particular technical suggestions regarding the content and English correction of the present research work.

Conflicts of Interest: The authors declare that they have no known competing financial interests or personal relationships that could have influenced the work reported in this paper.

References

1. Parthasarathy, G.; Balaram, V.; Srinivasan, R. Characterization of green garnets from an archean calc-silicate rock, Bandihalli, Karnataka, India: Evidence for a continuous solid solution between Uvarovite and Grandite. *J. Asian. Earth. Sci.* **1999**, *17*, 345–352. [CrossRef]
2. Valenzano, L.; Pascale, F.; Ferrero, M.; Dovesi, R. Ab Initio quantum-mechanical prediction of the IR and Raman spectra of $Ca_3Cr_2Si_3O_{12}$ Uvarovite Garnet. *Int. J. Quantum Chem.* **2010**, *110*, 416–421. [CrossRef]
3. Izawa, M.R.M.; Cloutis, E.A.; Rhind, T.; Mertzman, S.A.; Poitras, J.; Applin, D.M.; Mann, P. Spectral reflectance (0.35–2.5 μm) properties of garnets: Implications for remote sensing detection and characterisation. *Icarus* **2018**, *300*, 392–410. [CrossRef]
4. Andrut, M.; Wildner, M.; Beran, A. The crystal chemistry of birefringent natural Uvarovites. Part IV. OH defect incorporation mechanisms in non-cubic garnets derived from polarised IR spectroscopy. *Eur. J. Mineral.* **2002**, *14*, 1019–1026. [CrossRef]
5. Gréaux, S.; Yamada, A. Density variations of Cr-rich garnets in the upper mantle inferred from the elasticity of Uvarovite garnet. *C. R. Geosci.* **2019**, *351*, 95–103. [CrossRef]

6. Novak, G.A.; Gibbs, G.V. The crystal chemistry of the silicate garnet. *Am. Mineral.* **1971**, *56*, 791–825.
7. Carda, J.; Monros, G.; Esteve, V.; Amigo, J.M. Cation distribution by powder X-ray diffraction in Uvarovite-Grossularite garnets solid solutions synthesised by the sol-gel method. *J. Solid. State. Chem.* **1994**, *108*, 24–28. [CrossRef]
8. Antao, S.M.; Salvador, J.J. Crystal chemistry of birefringent Uvarovite solid solutions. *Minerals* **2019**, *9*, 395. [CrossRef]
9. Verger, L.; Dargaud, O.; Chassé, M.; Trcera, N.; Rousse, G.; Cormier, L. Synthesis, properties and uses of chromium-based pigments from the manufacture de sèvres. *J. Cult. Herit.* **2018**, *30*, 26–33. [CrossRef]
10. Hummel, F.A. Synthesis of Uvarovite. *Am. Miner.* **1950**, *35*, 324–325.
11. Isaacs, T. Synthesis of Uvarovite. *Nature* **1963**, *4887*, 1291. [CrossRef]
12. De Villiers, J.P.R.; Muan, A. Liquidus-solidus phase relations in the system $CaO-CrO-Cr_2O_3-SiO_2$. *J. Am. Ceram. Soc.* **1992**, *75*, 1333–1341. [CrossRef]
13. Carda, J.; Monrós, G.; Escribano, P.; Alarcon, J. Synthesis of Uvarovite. *J. Am. Ceram. Soc.* **1989**, *72*, 160–162. [CrossRef]
14. Llusar, M.; Monrós, G.; Tena, M.Á.; Vicent, J.B.; Badenes, J.A. Effect of synthesis methods and aging on the synthesis of Uvarovite garnet by ceramic and sol-gel processes. *Br. Ceram. Trans. J.* **1999**, *98*, 113–121. [CrossRef]
15. Carda, J.; Monrós, G.; Tena, M.A.; Escribano, P.; Rincón, J.M. Composition microheterogeneity of Uvarovite-Grossularite garnets. *J. Am. Ceram. Soc.* **1994**, *77*, 160–162. [CrossRef]
16. Corona-Martínez, D.A.; Rendón-Angeles, J.C.; Gonzalez, L.A.; Matamoros-Veloza, Z.; Yanagisawa, K.; Tamayo, A.; Alonso, J.R. Controllable synthesis of $BaCuSi_2O_6$ fine particles via a one-pot hydrothermal reaction with enhanced violet colour hue. *Adv. Powder Technol.* **2019**, *30*, 1473–1483. [CrossRef]
17. Rendón-Angeles, J.C.; Quiñones-Gurrola, J.R.; López-Cuevas, J.; Gonzalez, L.A.; Matamoros-Veloza, Z.; Perez-Ramos, E.; Yanagisawa, K.; Tamayo, A.; Alonso, J.R. Rapid one-pot hydrothermal reaction for preparing $BaCu_2Si_2O_7$ fine particles with controlled blue colour tonality. *Ceram. Int.* **2021**, in press. [CrossRef]
18. Antano, S.M. Crystal chemistry of birefringent hydrogrossular. *Phys. Chem. Minerals.* **2015**, *42*, 455–474. [CrossRef]
19. O'neill, B.; Bass, J.D.; Rossman, G.R.J. Electron Spectros. Elastic properties of hydrogrossular garnet and implications for water in the upper mantle. *J. Geophys. Res.* **1993**, *98*, 20031–20037. [CrossRef]
20. Zhu, Y.; Seong, G.; Noguchi, T.; Yoko, A.; Tomai, T.; Takami, S.; Adschiri, T. Highly Cr-Substituted CeO_2 Nanoparticles Synthesized Using a Non-equilibrium Supercritical Hydrothermal Process: High Oxygen Storage Capacity Materials Designed for a Low-Temperature Bitumen Upgrading Process. *ACS Appl. Energy Mater.* **2020**, *3*, 4305–4319. [CrossRef]
21. Litwinowicz, A.A.; Takami, S.; Asahina, S.; Hao, X.; Yoko, A.; Seong, G.; Tomai, T.; Adschiri, T. Formation dynamics of mesocrystals composed of organically modified CeO_2 nanoparticles: Analogy to a particle formation model. *Cryst. Eng. Comm.* **2019**, *21*, 3836–3843. [CrossRef]
22. Li, J.; Subramanian, M.A. Inorganic pigments with transition metal chromophores at trigonal bipyramidal coordination: $Y(In,Mn)O_3$ blues and beyond. *J. Solid State Chem.* **2019**, *272*, 9–20. [CrossRef]
23. Jose, S.; Jayaprakash, A.; Laha, S.; Natarajan, S.; Nishanth, K.G.; Reddy, M.L.P. $YIn_{0.9}Mn_{0.1}O_3$-ZnO nano-pigment exhibiting intense blue colour with impressive solar reflectance. *Dyes Pig.* **2016**, *124*, 120–129. [CrossRef]
24. Zhanga, Y.; Qib, H.; Liu, H.; Wang, S.; Yuan, L.; Hou, C. Thermal stable blue pigment with tuneable colour of $DyIn_{1-x}Mn_xO_3$ ($0 \leq x \leq 0.1$). *Dyes Pig.* **2018**, *156*, 192–198. [CrossRef]
25. Dolić, S.D.; Jovanović, D.J.; Štrbac, D.; Farc, L.Đ.; Dramićanin, M.D. Improved coloristic properties and high NIR reflectance of environment-friendly yellow pigments based on bismuth vanadate. *Ceram. Int.* **2018**, *44*, 22731–22737. [CrossRef]

Article

Large-Scale Synthesis Route of TiO$_2$ Nanomaterials with Controlled Morphologies Using Hydrothermal Method and TiO$_2$ Aggregates as Precursor

Wenpo Luo [1] and Abdelhafed Taleb [1,2,*]

[1] Institut de Recherche de Chimie Paris, PSL Research University Chimie ParisTech—CNRS, 75005 Paris, France; Wenpo.Luo@chimieparistech.psl.eu
[2] Sorbonne Université, 75231 Paris, France
* Correspondence: abdelhafed.taleb@sorbonne-universite.fr; Tel.: +33-1-85-78-41-97

Abstract: TiO$_2$ of controlled morphologies have been successfully prepared hydrothermally using TiO$_2$ aggregates of different sizes. Different techniques were used to characterize the prepared TiO$_2$ powder such as XRD, XPS, FEGSEM, EDS, and HRTEM. It was illustrated that the prepared TiO$_2$ powders are of high crystallinity with different morphologies such as nanobelt, nanourchin, and nanotube depending on the synthesis conditions of temperature, time, and additives. The mechanism behind the formation of prepared morphologies is proposed involving nanosheet intermediate formation. Furthermore, it was found that the nanoparticle properties were governed by those of TiO$_2$ nanoparticles aggregate used as a precursor. For example, the size of prepared nanobelts was proven to be influenced by the aggregates size used as a precursor for the synthesis.

Keywords: TiO$_2$ nanoparticles; aggregates; morphologies

Citation: Luo, W.; Taleb, A. Large-Scale Synthesis Route of TiO$_2$ Nanomaterials with Controlled Morphologies Using Hydrothermal Method and TiO$_2$ Aggregates as Precursor. *Nanomaterials* **2021**, *11*, 365. https://doi.org/10.3390/nano11020365

Academic Editors: Goran Drazic and Jimyeong Seong

Received: 30 November 2020
Accepted: 28 January 2021
Published: 1 February 2021

Publisher's Note: MDPI stays neutral with regard to jurisdictional claims in published maps and institutional affiliations.

Copyright: © 2021 by the authors. Licensee MDPI, Basel, Switzerland. This article is an open access article distributed under the terms and conditions of the Creative Commons Attribution (CC BY) license (https://creativecommons.org/licenses/by/4.0/).

1. Introduction

Recently, tremendous efforts have been devoted to developing innovative strategies to synthesize nanomaterials with the desired morphologies and properties. Particularly the one-dimensional (1D) structure of TiO$_2$ nanomaterials exhibits interesting properties compared to other TiO$_2$ nanoparticles: it has lower carrier recombination rate and higher charge carrier mobility, thanks to the grain boundaries and junctions absence. In fact, the electron diffusion takes place through the junctions between nanoparticles, inducing slower charge transfer by several orders of magnitude [1]. In addition, it favors light scattering in the photoanode, which increases the light harvesting [2]. Among the studied morphologies and materials, semi conducting nanostructured materials such as nanowires, nanobelts, and nanotube have received particular attention, due to their use as photoanaode for potential applications in different areas such as photovoltaic [2], photo catalysis [3], gas sensing [4], and water photo-splitting [5].

Tuning the size and the morphology of materials is becoming a challenging goal in materials science. Over the past few years, various synthesis methods and protocols have been developed to control the semi-conducting nanomaterials morphology, including vapor–liquid–solid (VLS) [6], solution–liquid–solid (SLS) [7], template-based synthetic approaches [8,9], arc discharge [10], laser ablation [11], chemical vapor deposition [12], microwave [13,14], and sol–gel [15]. Among these synthesis methods, which mostly brought contamination to the synthesis products, the hydrothermal technique has been proven to be a simple and straightforward method using noncomplex apparatus, scalable for large production, with high chemical purity, allowing a large rang of nanomaterial sizes and morphologies [16,17]. Furthermore, the morphology of prepared TiO$_2$ nanomaterials by using hydrothermal method was demonstrated to depend on the concentration of alkaline solution, the synthesis temperature and time, the material precursor used [17,18], additives, Pressure, pH, and the reaction medium [19–25].

Additionally, the hydrothermal method allows the control of the nanoparticles aggregation [26]. The most reported strategy to control the morphology of oxide nanomaterials is based on using organic surfactant, which adsorbs on a selected crystallographic plan of growing nucleus, leading to a change of its orientation and growth rate. This results in controlling the morphology of the obtained nanomaterial at the final growth stage [27,28]. Additionally, strategies based on aggregation/coalescence of nanomaterials were reported and demonstrated to be efficient in controlling the morphology of the final synthesized powder [29]. The exfoliation step was also reported to be a crucial step in the formation mechanism of prepared morphologies [29]. Most of the studies are based on nanomaterials aggregation/coalescence processes, and to the best of our knowledge, very few are based on exfoliation/aggregation/coalescence processes to explain synthesized morphologies. In the case of TiO_2 nanomaterials, there is still a misunderstanding of the mechanism behind the formation of reported morphologies and particularly nanotube, nanobelt, and nanourchin. Some authors claimed that the $Na_2Ti_3O_7$ nanosheets exfoliation step is the crucial step in the mechanism formation of different morphologies, whereas other authors stated that it is the dissolution of TiO_2 nanoparticles into TiO_6 octahedra, followed by $Na_2Ti_3O_7$ nucleation and growth, forming a nanosheet in a later stage [29]. Furthermore, it is well accepted that different polymorphs of TiO_2 nanomaterials are formed by different arrangements of TiO_6 octahedra. In fact, the growth of anatase tetragonal polymorph proceeds through face sharing arrangements of TiO_6 octahedras, whereas the rutile tetragonal phase growth takes place through edge-sharing arrangements. Furthermore, the Brookite phase is obtained by TiO_6 octahedra assembly, sharing their edge and corner; whereas in Ti_2O (B) (bronze) phase, Ti^{4+} ion form two distinct geometries with oxygen: octahedron in one case and a square pyramidal in the other. In addition, to homogeneous size and morphology, prepared TiO_2 nanomaterials using hydrothermal method exhibit several characteristics such as high crystallinity, an accurate control of different crystallinity phases from anatase to rutile depending on the synthesis and annealing temperatures, and high specific surface [30]. It is well accepted that the anatase polymorph possesses a higher band gap energy (3.3 eV) than that of the rutile polymorph (3 eV).

In the present work, different morphologies of TiO_2 have been successfully prepared hydrothermally using TiO_2 aggregates made of TiO_2 nanoparticles as a precursor. The mechanism behind the morphology control of prepared nanomaterials was discussed. It was found that the prepared TiO_2 nanomaterials properties were governed by those of TiO_2 nanoparticles aggregate. By controlling TiO_2 nanoparticles and aggregate sizes, it has been demonstrated that it is possible to control the TiO_2 nanobelt sizes.

2. Materials and Methods

2.1. Synthesis of TiO_2 Nanoparticles

For the synthesis of TiO_2 nanoparticles, titanium (IV) oxysulfate hydrate ($TiOSO_4$, Sigma Aldrich, St. Louis, MO, USA) precursor was used. Furthermore, the synthesis of TiO_2 aggregates has been performed using a hydrothermal synthesis technique. The $TiOSO_4$ precursor solution was prepared by dissolving 6.4 g of $TiOSO_4$ (2.5 M) in 16 mL of distilled water under constant stirring of 750 r/min and temperature of 45 °C for 2 h to get a clear solution. Then the solution of $TiOSO_4$ was transferred into a Teflon-lined stainless-steel autoclave of 25 mL capacity. The heating rate was of 2.5 °C/min, and during the synthesis, the temperature was maintained at different temperatures of 100, 200 and 220 °C for 6 h depending on the aggregate size required. After this synthesis in autoclave, a white TiO_2 powder was obtained and was washed six times in distilled water and two times in ethanol. Then the powder was dried overnight in the oven and annealed in air at temperature of 500 °C for 30 min with the heat rate of 5 °C/min. For nanourchin, nanotube, and nanobelt synthesis, 0.5 g powder of TiO_2 aggregate was introduced in a Teflon-lined autoclave of 25 mL capacity. Then, the autoclave was filled with 10 M NaOH solution up to 80% of the autoclave capacity. During the synthesis, the temperature was maintained at different temperatures of 100, 150, and 220 °C with the heating rate of 2.5 °C/min and the synthesis time of 360, 180, and 15 min, depending on the required morphology. Afterwards, synthesis nanobelt particles are subjected to the washing and annealing protocols to obtain at the end of these processes: sodium titanate. The latter product was washed many times with diluted HCl solution to attain a pH value of 1. After that, the suspension was washed with distilled water several times to reach a pH value of 7. Finally, the obtained powder was dried overnight in the oven, and annealed in air at temperature of 500 °C for 30 min, with the heat rate of 5 °C/min.

All the chemicals are of analytical grade and used without further purification. The water used in all the experiments was purified by Milli Q System (Millipore, electric resistivity 18.2 MΩ.cm).

2.2. The Characterizations of TiO_2 Films

The morphological investigations of the prepared films were achieved with a high-resolution Ultra 55 Zeiss FEG scanning electron microscope (FEGSEM) operating at an acceleration voltage of 10 kV and the high-resolution transmission electron microscope HRTEM using JEOL 2100 Plus microscope.

The crystalline structure of TiO_2 was determined by an X-ray diffractometer (Siemens D5000 XRD unit) in 2θ range from 20° to 80° by $0.07°/s^{-1}$ increasing steps operating at 40 KV accelerating voltage and 40 mA current using Cu Kα radiation source with λ = 1.5406 Å.

The chemical compositions of all the samples were determined by the FEGSEM using a Princeton Gamme-Tech PGT, USA, spirit energy dispersive spectrometry EDS system, and by X-ray photoelectron spectroscopy XPS realized with X-ray photoelectron spectroscopy (XPS), and for the measurements we used a Thermo K Alpha analyzer system equipped with an AL Kα X-ray source (hυ = 1486.6 eV; spot size 400 μm).

3. Results and Discussion

Various powders were prepared using the alkali hydrothermal synthesis method and varying synthesis temperatures and reaction times. To prepare these powders, TiO_2 aggregates of spherical shape and different sizes were prepared and used as precursors. The FEG-SEM characterization of precursor powders are shown in Figure 1, and it can be observed that the sizes of spherical aggregates are ranging from 50 to 200 nm.

The XRD method was used to characterize the crystalline phase of TiO_2 aggregate precursors, and the obtained results are depicted in Figure 2. Several well-resolved peaks were observed and are all assigned to TiO_2 anatase phase (JCPDS No. 21-1272), which is proof of the high purity of the prepared precursor powders. Additionally, Scherer analysis was used to calculate the average crystallite sizes at the half-maximum width of the intense peak corresponding to (101) crystallographic plane, and were found to be 9.8, 24.7, and 30.4 nm, for the synthesis temperatures of 100, 200, and 220 °C, respectively.

White powders were obtained using TiO_2 aggregate precursors whatever the preparation conditions, and their corresponding morphologies are depicted in Figure 3. As it can be observed, at the synthesis temperature of 100 °C, the morphology of the prepared powder is nanourchin-like with a stretched sheet-like network (Figure 3a), whereas at a temperature of 150 °C, the morphology is still nanourchin-like but with a more rolled nanosheet-like network (Figure 3b). From these experiments, it is clear that the temperature increase favors the nanosheet scrolling. This could be explained by the fact that the crystallization enhanced by the temperature increase tends to induce the microstructure to change into rolled nanosheet structure. In fact, to reduce the surface energy of rolled structure, nanosheets reduce the defects and the distortion energy [31]. At a higher temperature of about 200 °C, the FEGSEM characterization of prepared white powder is depicted in Figure 3c,d. It can be observed that TiO_2 powder is of nanobelt-like and nanotube morphologies, with monodisperse size. The insert of Figure 3d shows a sticking of several distinguishable nanobelts along their axis direction, forming bundles of nanobelts as a building unit. It can also be observed that their thickness is homogeneous and it is of about 10 nm, their diameter is ranging from 50 to 100 nm with length of around 10 mm. In addition, the nanobelt surface is smooth at the magnification scale, and no contamination was observed. As indicated in Figure 3d, some curved nanobelts were observed, which gives an indication about their high elasticity. From the described experiments, it is clear that the synthesis temperature is an important parameter in the morphology control of TiO_2 nanomaterials.

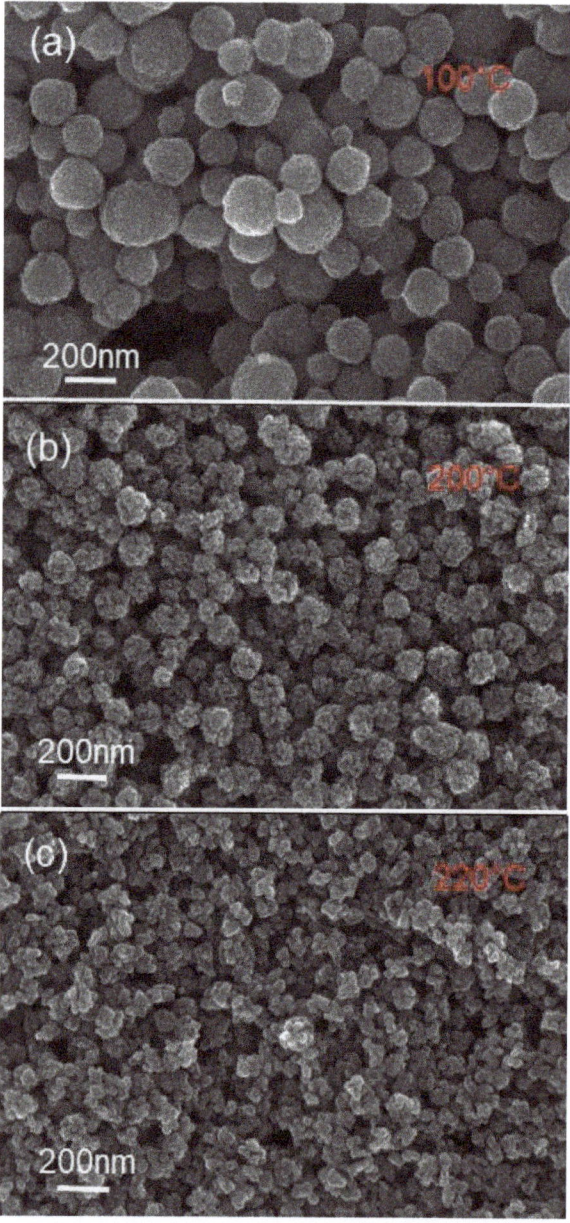

Figure 1. FEGSEM images of TiO_2 aggregates obtained at different synthesis temperatures: (**a**) 100, (**b**) 200, and (**c**) 220 °C.

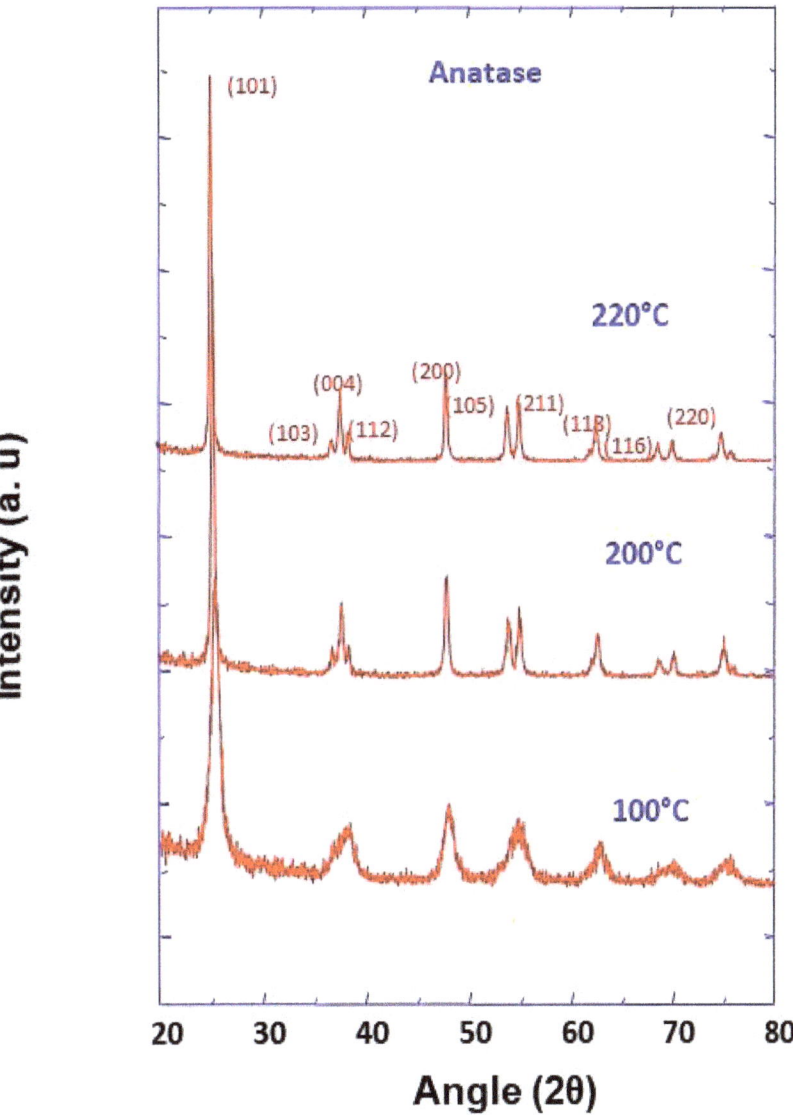

Figure 2. XRD pattern of TiO_2 nanoparticle aggregates prepared at different synthesis temperatures as indicated.

The crystalline structure and phase of prepared TiO_2 nanobelt, nanotube, and nanourchin-like powders were studied by the X-ray diffraction method. The obtained XRD patterns are presented in Figure 4, and they show well-resolved peaks in the case of nanourchin and nanotube mophologies attributed to (-511) and (020) crystallographic planes of pure $TiO_2(B)$ phase (JCPDS No. 35-0088) (Figure 4a–c). In the case of TiO_2 with nanobelt morphology, the observed XRD peaks indicates that the prepared powder is a mixture of anatase (JCPDS 21-1272) and brookite (JCPDS 29-1360) phases (Figure 4d).

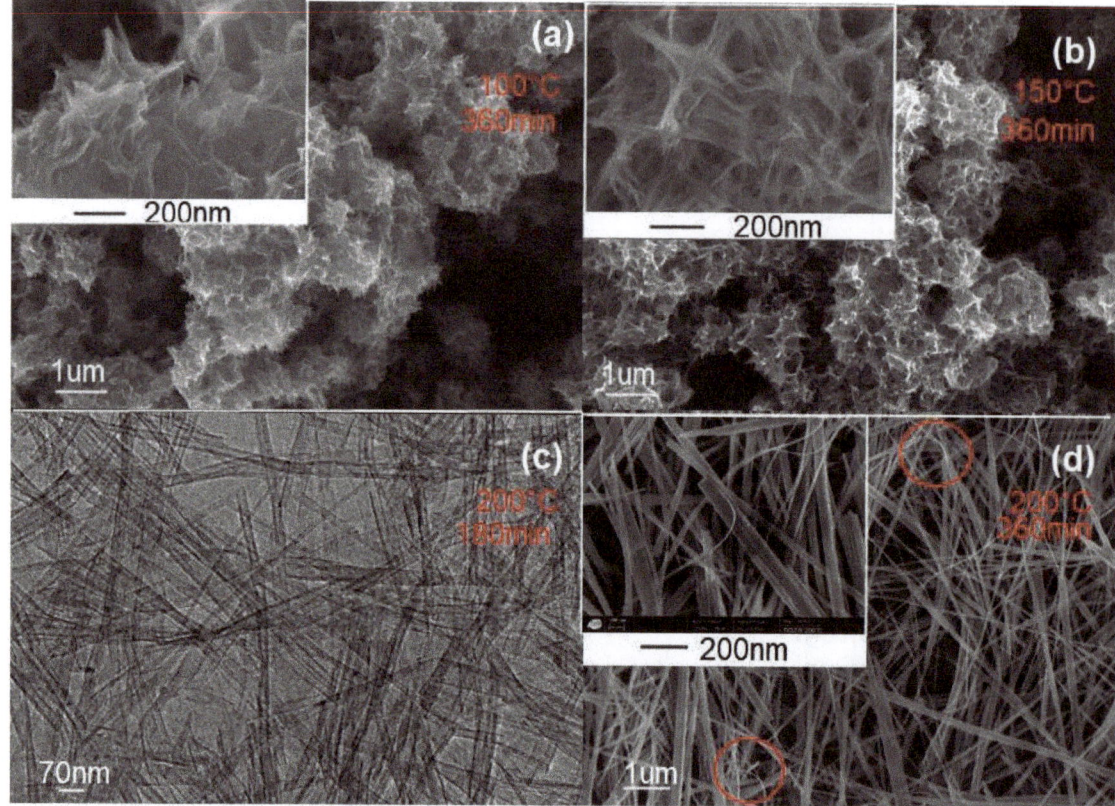

Figure 3. FEGSEM images of TiO$_2$ nanoparticles with different morphologies obtained at different synthesis times and temperatures: (**a**) Nanourchin prepared at conditions of 100 °C and 360 min, (**b**) Nanourchin prepared at conditions of 150 °C and 360 min, (**c**) TEM image of Nanotube prepared at conditions of 200 °C and 180 min, and (**d**) Nanobelts prepared at conditions of 200 °C and 360 min.

Additionally, among all the peaks, the most intense is the one corresponding to (121) crystallographic plane of brookite. Further details of crystallinity are provided by HRTEM depicted in Figure 5, clearly well resolved lattice planes are shown, and the insert electron diffraction shows well resolved spots (Figure 5b). These spots are the signature that the individual nanobelt is a single crystal. The interplanar distance of about 0.88 nm measured from HRTEM image is assigned to (100) crystallographic plane of brookite, indicating that the growth takes place along the (100) crystallographic plane, which is in good agreement with the result from XRD experiments in terms of brookite formation.

Furthermore, the chemical composition of the powder was provided by XPS analysis, and the obtained spectra are depicted in Figure 6. The XPS survey spectrum in Figure 6a of TiO$_2$ aggregates precursor shows intense peaks corresponding to O1s and Ti2p core levels, and the very weak intensity of the peak corresponding to Na1s. However, the XPS survey spectrum corresponding to TiO$_2$ nanobelt-like and nanourchin-like powders (Figure 6b) shows intense and well resolved peak, corresponding to the core level of Na1s, which is a signature of the formation of sodium titanate (Na$_2$Ti$_3$O$_7$), in addition to those of O1s and Ti2p. It was reported that Na$_2$Ti$_3$O$_7$ is constituted by corrugated strips of edge-sharing TiO$_6$ octahedra [29]. The width of each strips is about three-octahedra, and they are connected through their corner to form stepped layers. Within the sticking layers, sodium cations are located at the positions between the layers.

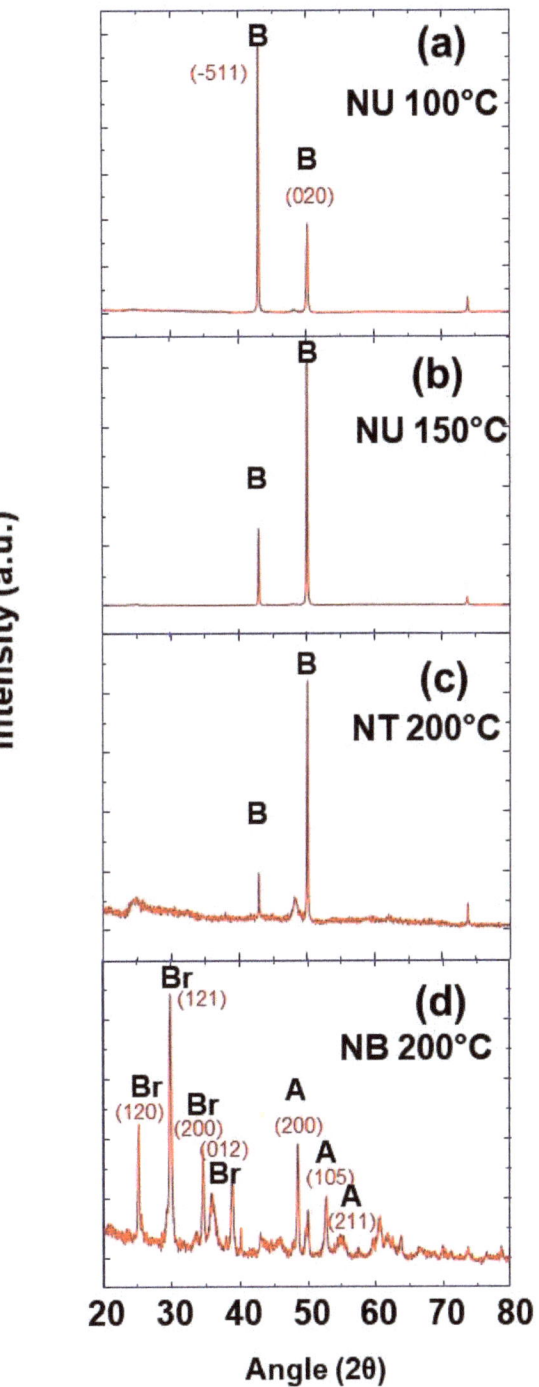

Figure 4. XRD pattern of TiO$_2$ nanoparticles with different morphologies prepared at different synthesis temperatures as indicated, (**a**) nanoursin 100 °C, (**b**) nanoursin 150 °C, (**c**) nanotube 200 °C and (**d**) nanobelt 200 °C (Br: Brookite; A: Anatase; B: TiO$_2$-B).

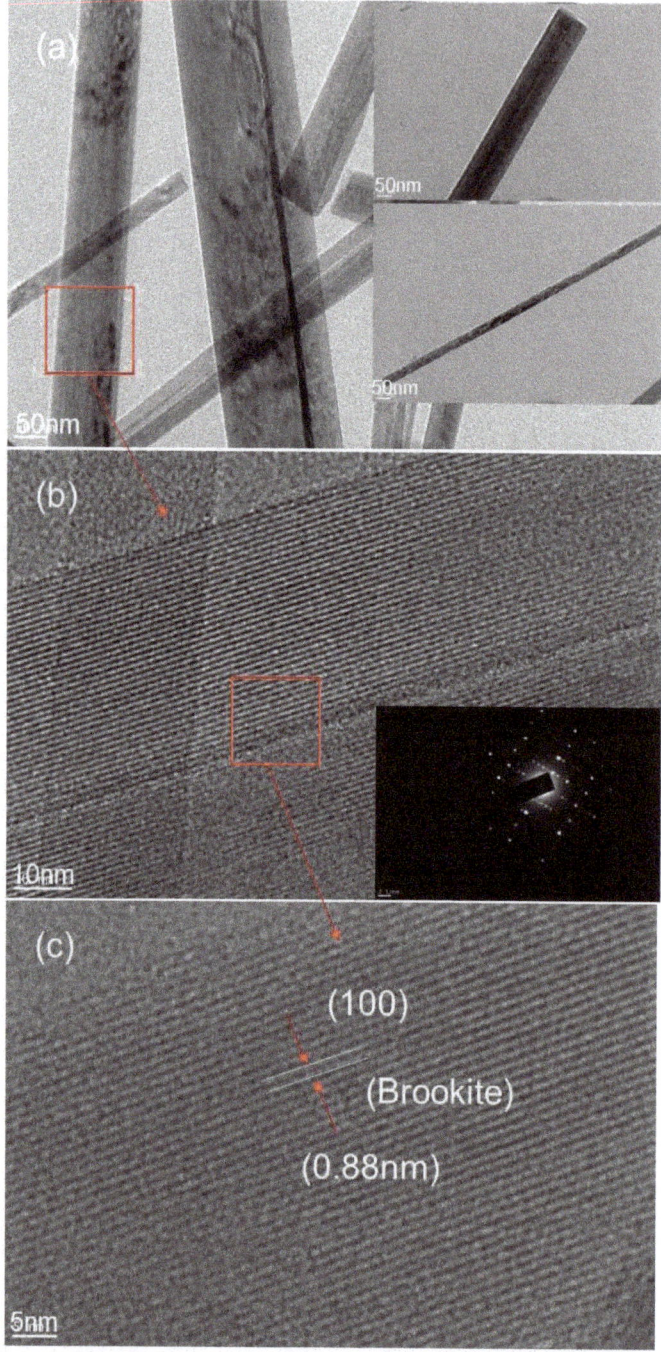

Figure 5. (a) TEM images of TiO$_2$ nanobelt obtained at synthesis temperature of 200 °C a synthesis time of 6 h at different magnifications; (b) the corresponding HRTEM showing inter atomic crystallographic planes and the insert show the corresponding electron diffraction; (c) another magnification of TiO$_2$ nanobelt.

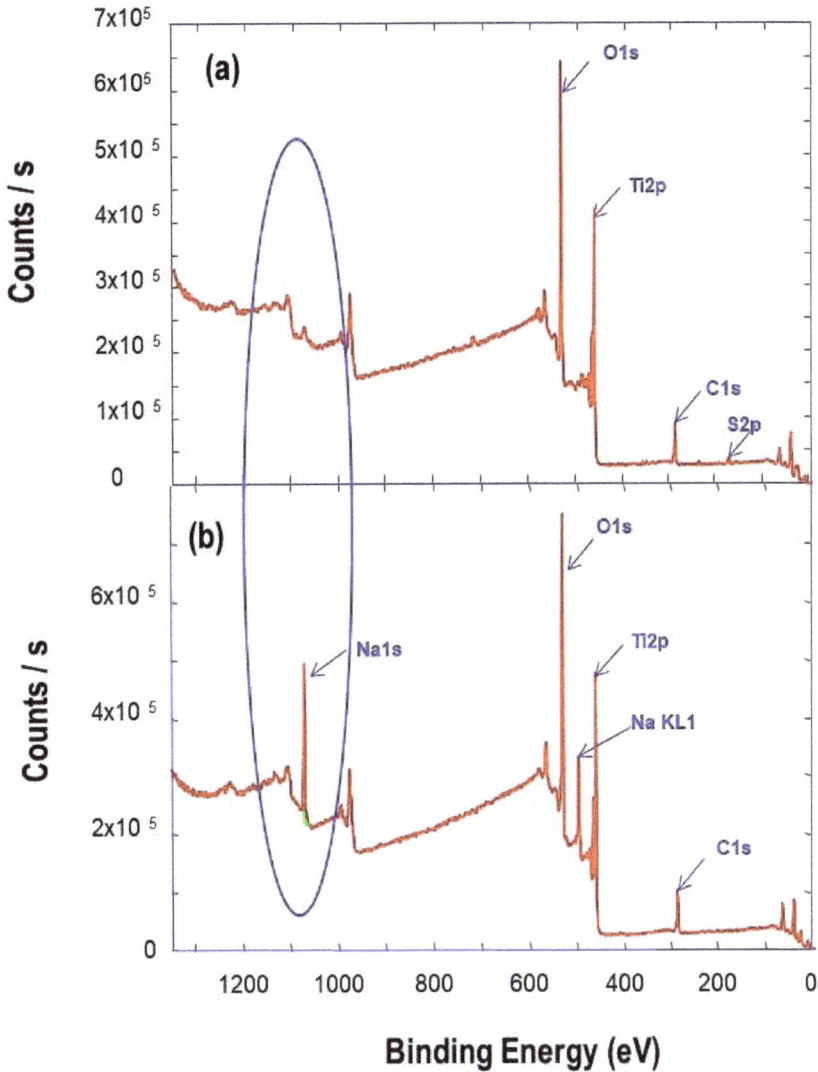

Figure 6. XPS survey spectra of prepared TiO_2 (**a**) aggregate precursor (**b**) nanobelts after synthesis.

In Figure 7a, it is important to note that nanourchin-like nanoparticles show more enrolled nanosheet with more dense structure, as a consequence of the annealing process. The EDS analyses have been performed to determine the chemical composition of TiO_2 nanoparticles, after just synthesis, or after washing and annealing processes. In Figure 7 the obtained EDS spectra are depicted; it should be noted that, on the EDS spectrum of TiO_2 nanoparticles, after synthesis shows the presence of Na peak Figure 7b, whereas it is absent in the spectrum after the washing and annealing processes in Figure 7c. In fact, during the washing processes of $Na_2Ti_3O_7$ by HCl, Na^+ ions were exchanged by H^+ ions. These results are a clear evidence of the important role played by Na^+ ions in the formation of TiO_2 nanobelts, nanotube, and nanourchin morphologies.

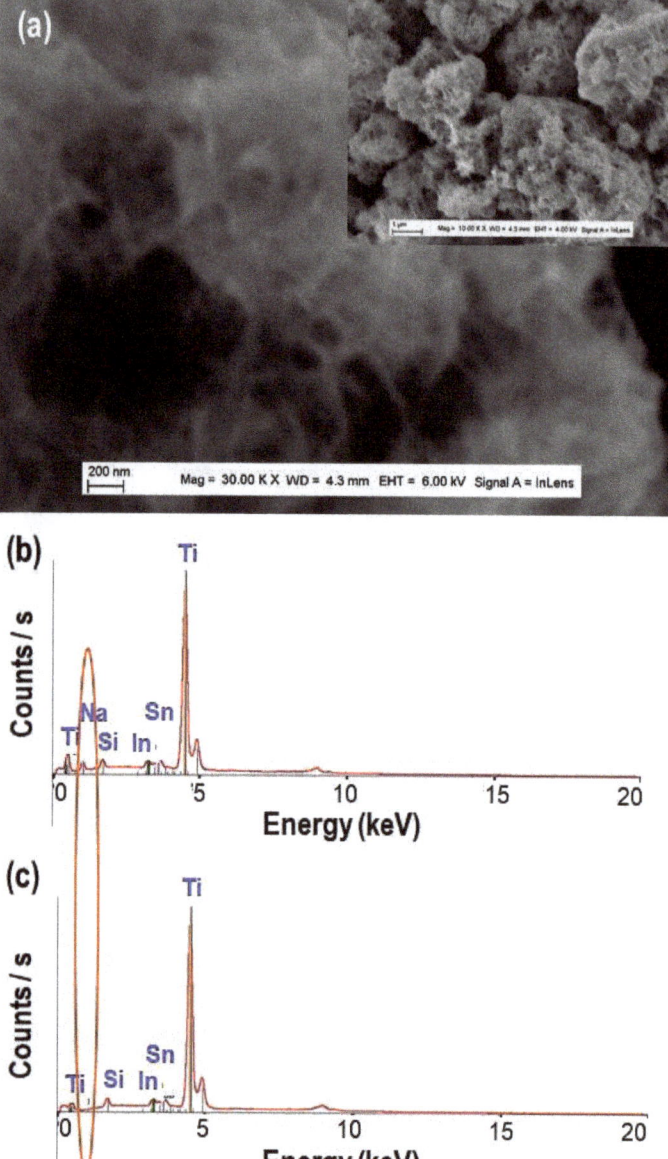

Figure 7. (**a**) FEGSEM images of TiO$_2$ nanourchin obtained at synthesis temperatures of 150 °C, after washing and annealing; (**b**,**c**) the corresponding EDS spectrum obtained just after synthesis and after washing and annealing processes, respectively.

The details of TiO$_2$ nanobelt and nanotube formation mechanisms are further investigated by using high resolution transmission electron microscopy (HRTEM). The influence of hydrothermal reaction time on the morphology of prepared TiO$_2$ nanomaterials is studied at 15, 180, and 360 min. At short reaction time of about 15 min, the morphology of prepared powder is mainly stretched nanosheet-like, with some minor rolled sheet. Closer analysis of prepared powder (Figure 8) shows different stages of the same formation mechanism.

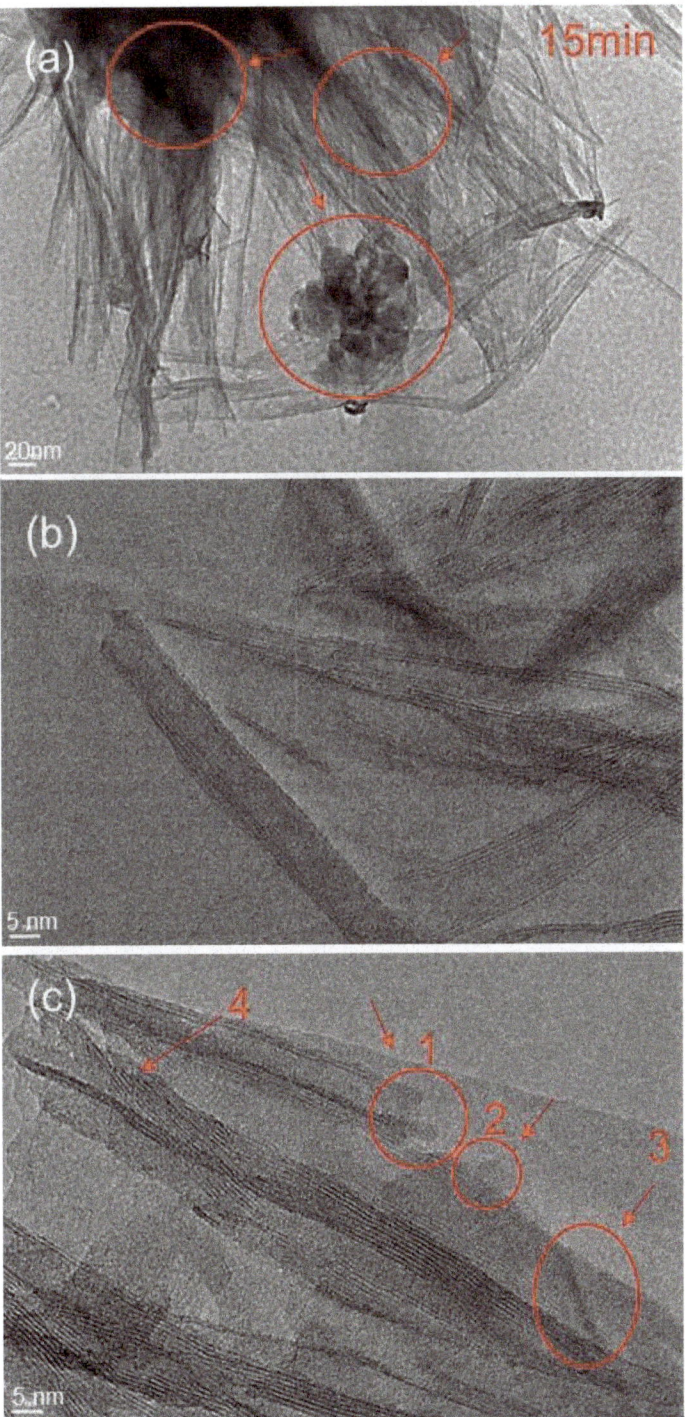

Figure 8. HRTEM images with different magnifications (**a**–**c**) of TiO$_2$ nanotube prepared at synthesis temperature of 200 °C and synthesis time of 15 min.

In fact, the observed nucleation stage can be considered as an integrated growth process of nanobelt structure, from aggregates made of nanoparticles of about 20 nm diameter to nanobelt of several micrometers in length. Similar evolution was observed by other authors [4,29]. Thus, we may assume that the morphologies shown in Figures 8 and 9 represent different stages of the nanobelt growth process.

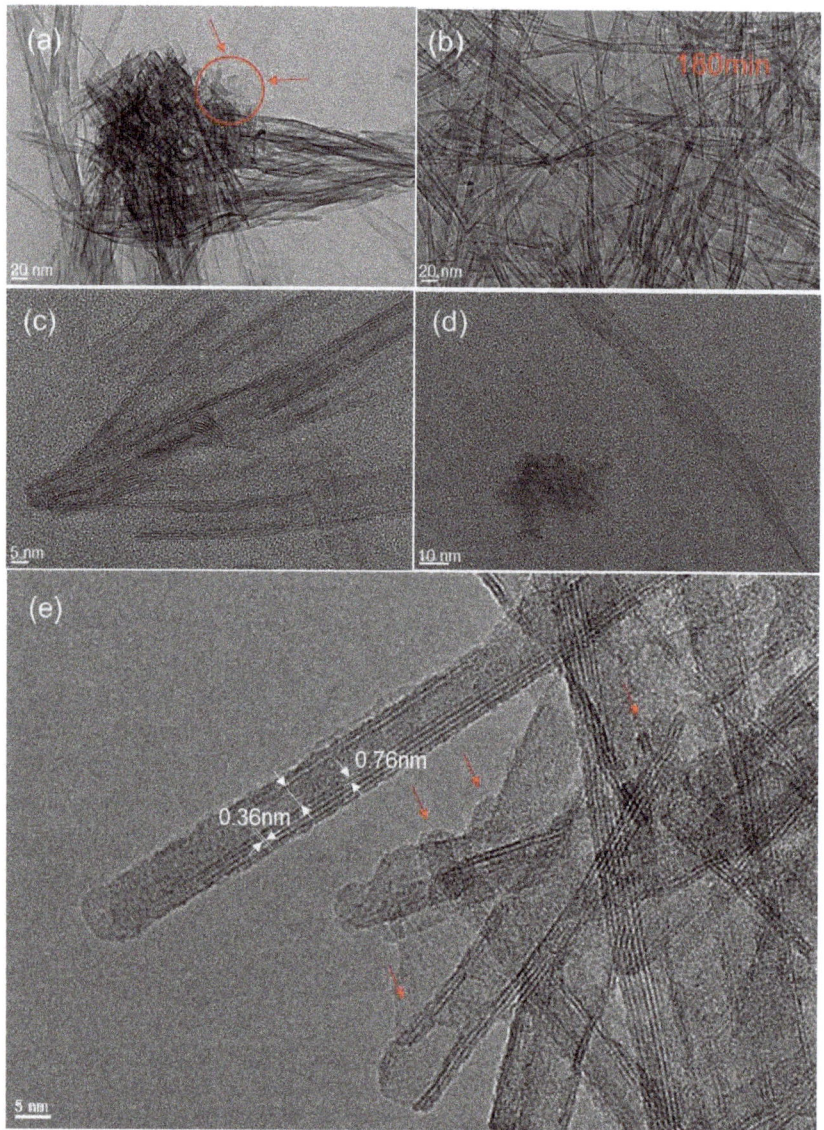

Figure 9. HRTEM images with different magnifications (**a–e**) of TiO$_2$ nanotube prepared at synthesis temperature of 200 °C and synthesis time of 180 min.

It can be observed that at the earlier stage (reaction time of 15 min) of the nanobelt growth process, coalesced nanoparticles coexist with nanosheet like particles, indicated by zones in Figure 8a,c. Nanoparticles were located at the nanotube edges (region 1 and 2 in Figure 8c), and beside this simple attachment, an alignment of coalesced nanoparticles takes place (region 3 in Figure 8c). In addition, the nanosheet shows both stretched and rolled structures. The indicated region 4 in Figure 8c shows the starting process of nanosheet rolling. However, all these steps are a consequence of different nanobelt growth stages, which will evolve in a later stage to a nanotube structure observed in Figure 9 and nanobelt structure shown in Figure 10. However, at closer inspection of the nanosheet structure at an earlier stage, with a synthesis time of 15 min, we find that it presents an assembly of nanoparticles, whose sizes range from 5 to 20 nm, as indicated in selected region of Figure 8a,c. This proves that these nanoparticles and aggregates are the primary building units for the nanosheet formation process. Furthermore, it is well accepted in the literature that the key point for the formation of nanobelt-like structure is the formation of sodium titanate nanobelt intermediate, in which the sodium ion (Na^+) is inserted into space between TiO_6 octahedra layers, balancing their negative charges [4,29]. From the present experiments, it can be inferred that the aggregate of TiO_2 nanoparticles split up into nanosheets as a consequence of Na^+ insertion and their rolling in a second stage to form nanotube in an intermediate stage. Typical TEM and HRTEM patterns of TiO_2 nanotube are depicted in Figure 9, with similar structure of nanotube obtained using TiO_2 nanoparticles in terms of asymmetrical walls. It can be seen that the nanotube exhibits four layers on one side and two layers on the other (Figure 9e), which indicates that the nanotubes are formed by the scrolling of several layers of nanosheet, as previously observed by other authors. The interplane on both sides is of 0.36 nm, which corresponds to the (010) crystallographic plane, and is the characteristic of monoclinic H_2Ti_3O7. It was reported for the same materials that the nanotube growth takes place along the (010) direction. Additionally, the interlayer distance between rolled nanosheets is about 0.76 nm closer to different reported values [29].

From XPS and EDX analysis in Figures 6 and 7, it is clearly demonstrated that the sodium ions are incorporated in the TiO_2 nanobelt, nanotube, and nanourchin, which suggest that it plays a role in their formation mechanisms. These observations indicate that nanobelts are formed by an orderly sticking of nanosheet and their coalescence in later stage; whereas nanourchins are formed by random assembly of the nanosheets.

The size dependence of the TiO_2 nanobelt on the size of TiO_2 aggregate precursor was demonstrated. Different sizes of TiO_2 aggregate precursors were used to prepare TiO_2 nanobelt, and the obtained results are depicted on Figure 10. It can be observed that the nanobelt length tends to increase with the increasing of the TiO_2 aggregate precursor size. Additionally, the TiO_2 nanobelt width increases from 50 to 200 nm (Figure 10), when the TiO_2 aggregate precursor size increases from 50 to 200 nm (Figure 1). This confirms that TiO_2 nanoparticles play a role in the formation of different observed morphologies. In fact, if we assume that the formation of observed morphologies goes through the TiO_2 dissolution and precipitation, the TiO_2 nanoparticles size will not have any effects on the final nanoparticle size. Additionally, the observation of TiO_2 nanoparticles during the nanotube formation supports the mechanism through which sodium ions (Na^+) induce exfoliation of TiO_2 aggregates by insertion into the space between TiO_6 octahedra layers and their coalescence to form nanosheets at later stage. Furthermore, the present results provide additional arguments to support some reported works in the literature and contradict others [30,32], in which it was claimed that during the hydrothermal synthesis process, TiO_2 is dissolved through Ti–O–Ti bonds breaking and formation of sodium titanate nanosheet [29], which is converted to hydrogen titanate during the washing step and at a later stage to TiO_2 nanobelt after the annealing process.

It can be seen from the XRD results that the nanobelt powder, at different synthesis stages (Figure 11), shows a changing of crystalline structure. The TiO_2 aggregates precursor is of anatase phase, with tetragonal structure, in which TiO_6 octahedra are sharing their face and get stacked in a one-dimensional zigzag chain. During the synthesis of $Na_2Ti_3O_7$ nanobelts, a crystalline transition takes place, and TiO_2 anatase phase is transformed into an orthorhombic structure. In fact, the formation of sodium titanate nanobelt intermediate is obtained through the insertion of sodium ion (Na^+) into the space between TiO_6 octahedra layers, inducing the distortion of the initial structure.

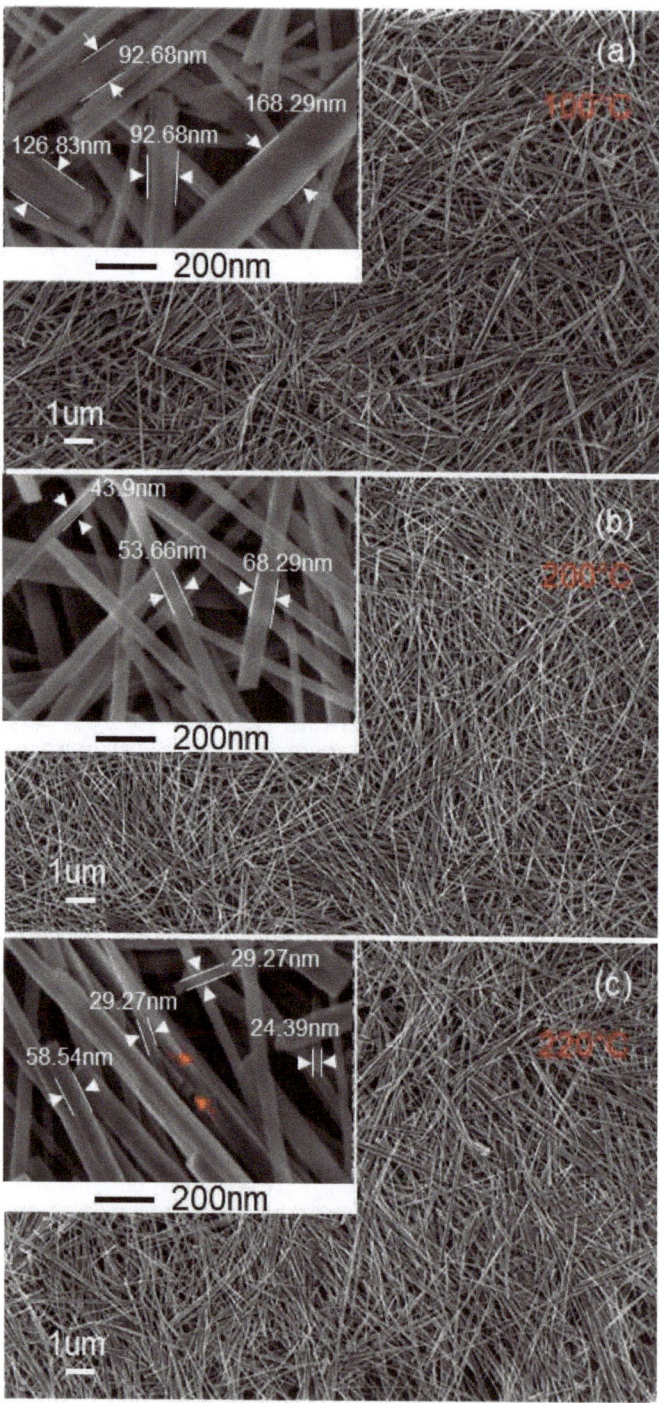

Figure 10. FEGSEM images of TiO$_2$ nanobelts prepared at synthesis temperature of 200 °C and using TiO$_2$ aggregate precursors of different sizes prepared at temperatures of (**a**) 100 °C, (**b**) 200 °C, and (**c**) 220 °C, respectively.

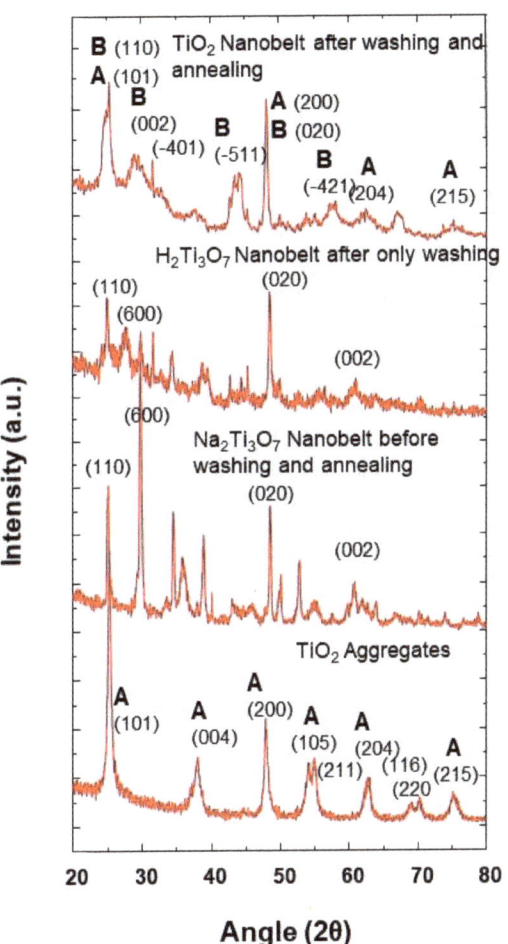

Figure 11. XRD pattern of TiO$_2$ nanobelt at different synthesis stages as indicated and at the synthesis temperature of 100 °C (A: Anatase; B: TiO2-B).

From these XRD results obtained at the synthesis temperature of 100 °C, it can be inferred that the anatase TiO$_2$ aggregate structure changes are a consequence of Na$^+$ insertion and a strong repulsion between Na$^+$ ions, which induces a distortion of the anatase crystalline structure. Similar behavior is observed with the insertion of Na+ ion in the case of Na ion batteries charging/discharging cycles [33]. However, after the washing step with hydrochloric acid solution, the H$_2$Ti$_3$O$_7$ nanobelts are obtained as a consequence of proton exchange processes of sodium trititanate. From Figure 11, it can be seen that this exchanging of steps and the resulting orthorhombic structure of H$_2$Ti$_3$O$_7$ (JCPDS Card No. 47-0124) are accompanied by some XRD peak modifications, in terms of the intensity enhancement of some peaks, and their decrease for some others [34,35]. These modifications indicate the distortion of the initial structure after ion exchanges. Additionally, after the annealing process and the removal of protons, a mixture of anatase (JCPDS 21-1272) and TiO$_2$-B (JCPDS 35-0088) phases is obtained at the synthesis temperature of 100 °C. The obtained XRD pattern is similar to that obtained for the same mixture by Beuvier et al. [36]. A phase transition was observed when the morphology changed from nanotube to nanobelt, but with different compositions than those obtained at the synthesis temperature of 200 °C. It was reported by Zhang et al. that the TiO$_2$ nanoparticle size has a strong impact on the phase transformation during the growth of coalesced nanoparticles [37]. In addition, the temperature also plays an important role in the phase transformation of TiO_2 nanoparticles [26]. However, as when the temperature is changed the coalescence and/or growth of TiO_2 nanoparticles take place, both the temperature and the size contribute to the phase transformation and a formation of different phase mixtures depending on the used synthesis temperature 100 and 200 °C. Furthermore, as it can be observed from Figures 4 and 10, the peaks corresponding to the anatase phase are of lower intensity, which indicates that both of the latter synthesis temperatures produce a lower proportion of anatase, in agreement with different reported works in the literature [36]. During the synthesis process at a given temperature, the phase is also changed due to the insertion of different ions, and it is not necessary to dissolve and precipitate TiO_2 octahedra. Furthermore, from these results, it is worth noting that the synthesis temperature plays a crucial role in the phase control of prepared nanobelt powders.

4. Conclusions

Different morphologies of TiO_2 nanoparticles have been synthesized, in a large scale using hydrothermal synthesis technique and TiO_2 aggregate as a precursor. Both nanotube, nanourchin-like, and nanobelt-like nanoparticles were obtained at low temperatures and over short times. Furthermore, it is demonstrated that a morphology control of prepared TiO_2 powders could be achieved through the tuning of the synthesis temperature and time. The mechanisms formation of TiO_2 nanobelt-like, nanourchin-like, and nanotube nanoparticles are illustrated to involve TiO_2 nanoparticles coalescence and nanosheet intermediate, formed thanks to Na^+ ions exfoliation. Furthermore, it was found that the prepared TiO_2 nanomaterials properties were governed by those of TiO_2 nanoparticles aggregate. It has been demonstrated that it is possible to tune the nanobelt size by using different TiO_2 aggregates precursor sizes. Additionally, it was shown that the synthesis temperature enables the tuning of the phase's composition of the nanobelt powders. The investigation of prepared powders performance, as anode material for Li-ion batteries, is under progress in our group.

Author Contributions: A.T. conceived and designed the experiment and also wrote the paper; W.L. performed experiments and also analyzed the corresponding data. All authors have read and agreed to the published version of the manuscript.

Funding: This research was funded by the European Union's Horizon 2020 research and innovation program under the Marie Sklodowska-Curie grant agreement No. 734276.

Informed Consent Statement: Not applicable.

Acknowledgments: The authors would like to thank Pierre Dubot from Université Paris Est Créteil, CNRS, ICMPE, UMR 7182, 2-8 rue Dunant F-94320 Thiais, France; for the XPS experiments. The authors would like also to thank the Chinese Scholar Council for supporting W.L. with a scholarship.

Conflicts of Interest: The authors declare no conflict of interest. The funders had no role in the design of the study; in the collection, analyses, or interpretation of data; in the writing of the manuscript; or in the decision to publish the results.

References

1. Fisher, A.C.; Peter, L.M.; Ponomarev, E.A.; Walker, A.B.; Wijiayantha, K.G.U. Intensity Dependence of the Back Reaction and Transport of Electrons in Dye-Sensitized Nanocrystalline TiO_2 Solar Cells. *J. Phys. Chem. B.* **2000**, *104*, 949–958. [CrossRef]
2. Tan, B.; Wu, Y. Dye-sensitized solar cells based on anatase TiO_2 nanoparticle/nanowire composites. *J. Phys. Chem. B.* **2006**, *110*, 15932–15938. [CrossRef] [PubMed]
3. Schneider, J.; Matsuoka, M.; Takeuchi, M.; Zhang, J.; Horiuchi, Y.; Anpo, M.; Bahnemann, D.W. Understanding TiO_2 Photocatalysis: Mechanisms and Materials. *Chem. Rev.* **2014**, *114*, 9919–9986. [CrossRef] [PubMed]
4. Zhao, Z.; Tian, J.; Sang, Y.; Cabot, A.; Liu, H. Structure, Synthesis, and Applications of TiO_2 Nanobelts. *Adv. Mater.* **2015**, *27*, 2557–2582. [CrossRef] [PubMed]
5. Sun, B.; Shi, T.; Peng, Z.; Sheng, W.; Jiang, T.; Liao, G. Controlled fabrication of Sn/TiO_2 nanorods for photoelectrochemical water splitting. *Nanoscale Res. Lett.* **2013**, *8*, 1–8. [CrossRef]

6. Trentler, T.J.; Hickman, K.M.; Goel, S.C.; Viano, A.M.; Gibbons, P.C.; Buhro, W.E. Solution-liquid-solid growth of crystalline III-V semiconductors: An analogy to vapor-liquid-solid growth. *Sci.* **1995**, *270*, 1791. [CrossRef]
7. Wang, F.; Dong, A.; Buhro, W.E. Solution–Liquid–Solid Synthesis, Properties, and Applications of One-Dimensional Colloidal Semiconductor Nanorods and Nanowires. *Chem. Rev.* **2016**, *116*, 10888–10933. [CrossRef]
8. Martin, C.R. Nanomaterials: A membrane-based synthetic approach. *Sci.* **1994**, *266*, 1961–1966. [CrossRef]
9. Xu, G.R.; Wang, J.N.; Li, C.J. Template direct preparation of TiO_2 nanomaterials with tunable morphologies and their photocatalytic activity research. *Appl. Surf. Sci.* **2013**, *279*, 103–108. [CrossRef]
10. Choi, Y.C.; Kim, W.S.; Park, Y.S.; Lee, S.M.; Bae, D.J.; Lee, Y.H.; Kim, J.M. Catalytic Growth of β-Ga_2O_3 Nanowires by Arc Discharge. *Adv. Mater.* **2000**, *12*, 746–750. [CrossRef]
11. Morales, A.M.; Lieber, C.M. A laser ablation method for the synthesis of crystalline semiconductor nanowires. *Science* **1998**, *279*, 208. [CrossRef] [PubMed]
12. Duan, X.; Lieber, C.M. General synthesis of compound semiconductor nanowires. *Adv. Mater.* **2000**, *12*, 298–302. [CrossRef]
13. Li, L.; Qin, X.; Wang, G.; Qi, L.; Du, G.; Hu, Z. Synthesis of anatase TiO_2 nanowire by modifying TiO_2 nanoparticles using the microwave heating method. *Appl. Surf. Sci.* **2011**, *257*, 8006–8012. [CrossRef]
14. Wang, H.E.; Zheng, L.X.; Liu, C.P.; Liu, Y.K.; Luan, C.Y.; Cheng, H.; Bello, I. Rapidmicrowave synthesis of porous TiO_2 spheres and their applications in dye sensitized solar cells. *J. Phys. Chem. C* **2011**, *115*, 10419–10425. [CrossRef]
15. Crippa, M.; Callone, E.; D'Arienzo, M.; Müller, K.; Polizzi, S.; Wahba, L.; Scotti, R. TiO_2 nanocrystal grafted on macroporous silica: A novel hybrid organic-inorganic sol-gel approach for the synthesis of highly photoactive composite material. *Appl. Catal. B.* **2011**, *104*, 282–290. [CrossRef]
16. Poudel, B.; Wang, W.Z.; Dames, C.; Huang, J.Y.; Kunwar, S.; Wang, D.Z.; Ren, Z.F. Formation of crystallized titania nanotubes and their transformation into nanowires. *Nanotechnology* **2005**, *16*, 1935–1940. [CrossRef]
17. Asiah, M.N.; Mamat, M.H.; Khusaimi, Z.; Abdullah, S.; Rusop, M.; Qurashi, A. Surfactant-free seed-mediated large-scale synthesis of mesoporous TiO_2 nanowires. *Ceram. Int.* **2015**, *41*, 4266.
18. Cui, L.; Hui, K.N.; Hui, K.S.; Lee, S.K.; Zhou, W.; Wan, Z.P.; Thuc, C.N.H. Facile microwave-assisted hydrothermal synthesis of TiO_2 nanotubes. *Mater. Lett.* **2012**, *75*, 175–178. [CrossRef]
19. Byrappa, K.; Adschiri, T. Hydrothermal technology for nanotechnology. *Prog. Cryst. Growth Charact. Mater.* **2007**, *53*, 117–166. [CrossRef]
20. Yu, J.; Su, Y.; Cheng, B.; Zhou, M. Effects of pH on the microstructures and photocatalytic activity of mesoporous nanocrystalline titania powders prepared via hydrothermal method. *J. Mol. Catal. A Chem.* **2006**, *258*, 104–112. [CrossRef]
21. Mamaghani, A.H.; Haghighat, F.; Lee, C.S. Role of titanium dioxide (TiO_2) structural design/morphology in photocatalytic air purification. *Appl. Catal. B Environ.* **2020**, *269*, 118735. [CrossRef]
22. Hoang, S.; Guo, S.; Hahn, N.T.; Bard, A.J.; Mullins, C.B. Visible light driven photoelectrochemical water oxidation on nitrogen modified TiO_2 nanowires. *Nano Lett.* **2012**, *12*, 26–32. [CrossRef] [PubMed]
23. Pan, X.; Zhao, Y.; Liu, S.; Korzeniewski, C.L.; Wang, S.; Fan, Z. Comparing graphene-TiO_2 nanowire and graphene-TiO_2 nanoparticle composite photocatalysts. *ACS Appl. Mater. Interface* **2012**, *4*, 2944–3950. [CrossRef] [PubMed]
24. Chen, J.Z.; Ko, W.Y.; Yen, Y.C.; Chen, P.H.; Lin, K.J. Hydrothermally processed TiO_2 nanowire electrodes with antireflective and electrochromic properties. *ACS Nano.* **2012**, *6*, 6633–6639. [CrossRef] [PubMed]
25. Kasuga, T.; Hiramatsu, M.; Hoson, A.; Sekino, T.; Niihara, K. Formation of titanium oxide nanotube. *Langmuir* **1998**, *14*, 3160–3163.
26. Taleb, A.; Mesguich, F.; Hérissan, A.; Colbeau-Justin, C.; Yanpeng, X.; Dubot, P. Optimized TiO_2 nanoparticle packing for DSSC photovoltaic applications. *Sol. Energy Mater. Sol. Cells.* **2016**, *148*, 52–59. [CrossRef]
27. Wang, Y.D.; Zhang, S.; Ma, C.L.; Li, H.D. Synthesis and room temperature photoluminescence of ZnO/CTAB ordered layered nanocomposite with flake-like architecture. *J. Lumin.* **2007**, *26*, 661. [CrossRef]
28. Wang, H.X.; Li, X.X.; Tang, L. Effect of surfactants on the morphology and properties of TiO_2. *Appl. Phys. A.* **2020**, *126*, 448. [CrossRef]
29. Wu, D.; Liu, J.; Zhao, X.; Li, A.; Chen, Y.; Ming, N. Sequence of events for the formation of titanate nanotubes, nanowire and nanobelts. *Chem. Mater.* **2006**, *18*, 547–553. [CrossRef]
30. Mamaghani, A.H.; Haghighat, F.; Lee, C.S. Photocatalytic oxidation technology for indoor environment air purification: The state-of-the-art. *Appl. Catal. B Environ.* **2017**, *203*, 247–269. [CrossRef]
31. Ding, L.; Chen, J.; Dong, B.; Xi, Y.; Shi, L.; Liu, W.; Cao, L. Organic macromolecule assisted synthesis of ultralong carbon&TiO_2 nanotubes for high performance lithium-ion batteries. *Electrochem. Acta* **2016**, *200*, 97–105.
32. Diebold, U.; Ruzycki, N.; Herman, G.S.; Sellonii, A. One Step towards Bridging the Materials Gap: Surface Studies of TiO_2 Anatase. *Catal. Today* **2003**, *85*, 93–100. [CrossRef]
33. Massaro, A.; Munoz-Garc, A.B.; Maddalena, P.; Bella, F.; Meligrana, G.; Gerbaldi, C.; Pavone, M. First-principles study of Na insertion at TiO_2 anatase surfaces: New hints for Na-ion battery design. *Nanoscale Adv.* **2020**, *2*, 274. [CrossRef]
34. Yao, B.D.; Chan, Y.F.; Zhang, X.Y.; Zhang, W.F.; Yang, Z.Y.; Wang, N. Formation mechanism of TiO_2 nanotubes. *Appl. Phys. Lett.* **2003**, *82*, 281. [CrossRef]
35. Wang, W.; Varghese, O.K.; Paulose, M.; Grimes, C.A.; Wang, Q.; Dickey, E.C. A study on the growth and structure of titania nanotubes. *J. Mater. Res.* **2004**, *19*, 417. [CrossRef]

36. Beuvier, T.; Richard-Plouet, M.; Mancini-Le Granvalet, M.; Brousse, T.; Crosnier, O.; Brohan, L. TiO$_2$(B) Nanoribbons as negative electrode material for lithium ion batteries with high rate performance. *Inorg. Chem.* **2010**, *49*, 8457–8464. [CrossRef]
37. Zhang, H.; Banfield, J. Understanding polymetric phase transformation behavior during growth of nanocrystalline aggregates Insghts from TiO$_2$. *J. Phys. Chem. B* **2000**, *104*, 3481–3487. [CrossRef]

Review

Behavior of Silicon Carbide Materials under Dry to Hydrothermal Conditions

Nicolas Biscay [1], Lucile Henry [1], Tadafumi Adschiri [2], Masahiro Yoshimura [3] and Cyril Aymonier [1,2,*]

[1] CNRS, University of Bordeaux, Bordeaux INP, ICMCB, UMR 5026, 33600 Pessac, France; nicolas.biscay@icmcb.cnrs.fr (N.B.); lucile.henry@icmcb.cnrs.fr (L.H.)
[2] WPI-Advanced Institute for Materials Research (WPI-AIMR), Tohoku University, 2-1-1 Katahira, Aoba-ku, Sendai 980-8577, Japan; tadafumi.ajiri.b1@tohoku.ac.jp
[3] Department of Materials Science and Engineering, National Cheng Kung University, No. 1, University Road, Tainan 70101, Taiwan; masahiroyoshimura75@gmail.com
* Correspondence: cyril.aymonier@icmcb.cnrs.fr

Abstract: Silicon carbide materials are excellent candidates for high-performance applications due to their outstanding thermomechanical properties and their strong corrosion resistance. SiC materials can be processed in various forms, from nanomaterials to continuous fibers. Common applications of SiC materials include the aerospace and nuclear fields, where the material is used in severely oxidative environments. Therefore, it is important to understand the kinetics of SiC oxidation and the parameters influencing them. The first part of this review focuses on the oxidation of SiC in dry air according to the Deal and Grove model showing that the oxidation behavior of SiC depends on the temperature and the time of oxidation. The oxidation rate can also be accelerated with the presence of H_2O in the system due to its diffusion through the oxide scales. Therefore, wet oxidation is studied in the second part. The third part details the effect of hydrothermal media on the SiC materials that has been explained by different models, namely Yoshimura (1986), Hirayama (1989) and Allongue (1992). The last part of this review focuses on the hydrothermal corrosion of SiC materials from an application point of view and determine whether it is beneficial (manufacturing of materials) or detrimental (use of SiC in latest nuclear reactors).

Keywords: silicon carbide; wet oxidation; supercritical fluids; supercritical water oxidation; hydrothermal corrosion; nanocarbon films

1. Introduction

Ceramics have been used as dielectric, magnetic and optical materials. The oxide ceramics are chemically stable at high temperature and have good refractory properties but poor thermal-shock resistance. This is not the case for the non-oxide ceramics. Non-oxide ceramics show a high thermal conductivity, which leads to excellent thermal-shock resistance. Non-oxide ceramics are composed essentially of borides, nitrides and carbides, of which silicon carbide (SiC) is the most widely used. SiC was originally discovered in 1891 by Acheson under the name of "carborundum" [1]. SiC materials have low density, and they exhibit a high degree of hardness and toughness due to an important degree of crosslinking of covalent bond. These properties justify their use for not only aerospace and automotive parts, but also in nuclear applications. In 1975, Yajima et al. elaborated a process for producing SiC materials in a fiber shape by pyrolysis of organosilicon polymers [2]. This process allows for continuous fiber production, generating fibers with a small diameter and with good flexibility to be used for designing composite materials. Carbon-based materials reinforced with SiC fibers have higher mechanical properties: these then constitute thermostructural composites for high-performance applications [3].

At high temperatures, silicon carbide undergoes passive and active oxidation, which contribute to its degradation.

Passive oxidation is responsible for both the formation of a silica layer on the top of the surface and for the active oxidation for the release of volatile oxides. The material is not able to withstand high mechanical properties as the oxidation is occurs, and dramatic failures can result when exposed to stress. Moreover, in the aerospace field, water and corrosive gases are released by the propulsion system. This, along with high temperatures, is expected to enhance the degradation of SiC materials. However, SiC materials need to operate properly for a defined range of temperatures and various gas compositions. In that way, the whole oxidation process needs to be characterized, and all the influential parameters need to be well understood.

What kind of oxidation behavior will the material exhibit when exposed to dry atmosphere? Which parameters can influence the oxidation kinetics? For example, what is the influence of water on the oxidation behavior? What are the effects of hot water and pressurized atmospheres? How do high temperature and high pressure water modify the surface properties?

The aim of this paper is to review, firstly, the dry oxidation of SiC materials, as the literature has already provided a comprehensive background of this phenomena along with accurate kinetic models.

In the first part, the oxidation of SiC under dry conditions, and the Deal and Grove model for the passive oxidation of silicon, are explained. The parameters which can influence the oxidation behavior of SiC are studied. The nature of silica scale and oxidant species are discussed, as well as the influence of crystal faces and impurities, to lead to the conclusion of the rate-determining step of the SiC oxidation.

In the second part, the effect of water vapor onto SiC is studied and a mixed oxidation regime which is in competition with the passive oxidation regime, is expressed. Then, the parameters which influence the oxidation of SiC are reviewed.

The third part focuses on the importance of understanding the ability of water to accelerate SiC degradation. Hydrothermal conditions are disastrous for SiC materials and lead to chemical corrosion through three possible reactions: wet (air) oxidation, supercritical water oxidation and hydrolysis. Then, three models for the interaction of SiC with water are proposed, and their validity is assessed by the microstructural study of the corroded surface. Finally, another SiC corrosion mechanism is discussed which occurs under hydrothermal conditions—tribochemical corrosion.

The last part deals with supercritical water medium and its interaction with SiC materials. This interaction can be either profitable or detrimental depending on the desired application.

2. Dry Oxidation of Silicon Carbide Materials

In Figure 1, a scheme is provided, detailing the layout of the following part.

As a silicon-based ceramic, silicon carbide is unstable in air. At high temperatures and under a dry atmosphere, SiC materials undergo passive (1) or active oxidation (2).

$$SiC\ (s) + 3/2\ O_2\ (g) \rightarrow SiO_2\ (s) + CO\ (g) \tag{1}$$

$$SiC\ (s) + O_2\ (s) \rightarrow SiO\ (g) + CO\ (g) \tag{2}$$

The SiO_2 layer formed according to Equation (1) at the surface has a low permeability to oxygen, so it can act as a protective barrier to prevent further oxidation of the bulk material. This protective effect tends to be limited at high temperatures, as the layer can interact with and react with SiC [4–6]:

$$SiC\ (s) + 2\ SiO_2\ (s) \rightarrow 3\ SiO\ (g) + CO\ (g) \tag{3}$$

During passive oxidation, the silica film grows, and an increase of mass is observed. However, during active oxidation of SiC, the oxygen reaches the bulk material through cracks or due to the failure of the protective layer, and a mass reduction is observed.

The model for oxidation of SiC shows a relationship of passive oxidation occurring generally at low temperature and high partial pressure of O_2. The contrary is seen for the active oxidation. Only the passive oxidation is explored in the following sections.

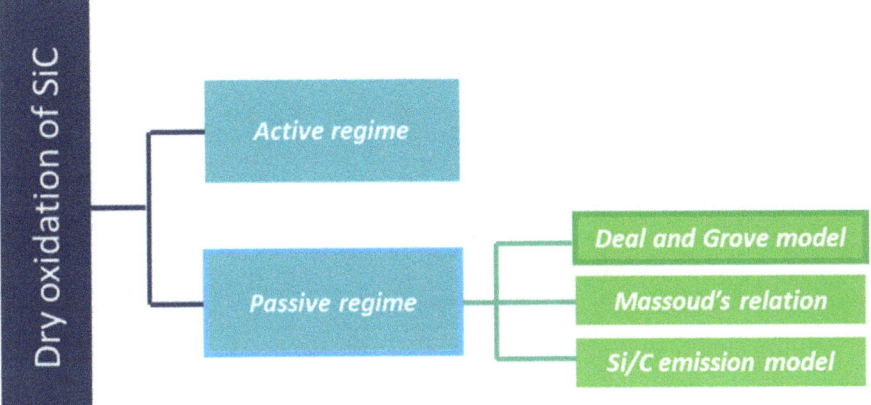

Figure 1. Layout of the dry oxidation of SiC part.

2.1. The Passive Oxidation Regime

During passive oxidation, the mobile species diffuse through the lattice via cracks or pores. Then, these species react with silicon at the SiO_2/Si interface or SiO_2/SiC interface, leading to the growth of the oxide scale [4]:

Jacobson concluded that five mechanisms were involved in the oxidation process of SiC [7]:

1. Transport of molecular oxygen gas to the oxide surface,
2. Diffusion of oxygen through the oxide film (Figure 2),
3. Reaction at the oxide/ceramic interface (Figure 3).
4. Transport of product gases (CO, CO_2) (Figure 4)
5. Transport of product gases away from the surface.

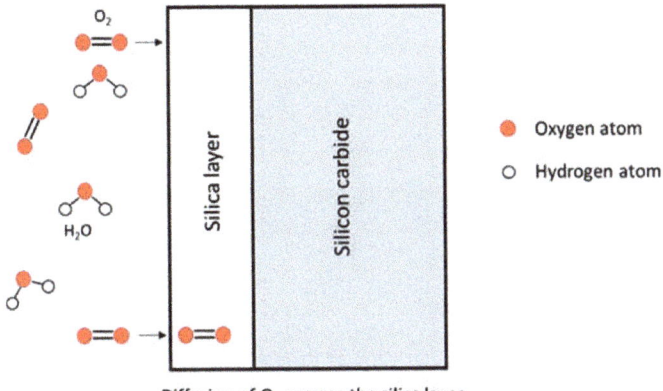

Figure 2. Scheme of diffusion of oxygen.

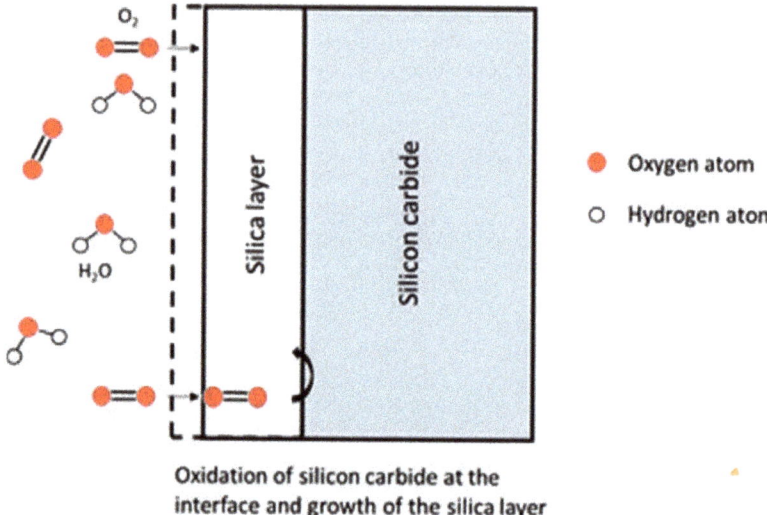

Figure 3. Scheme of the chemical reaction at the interface.

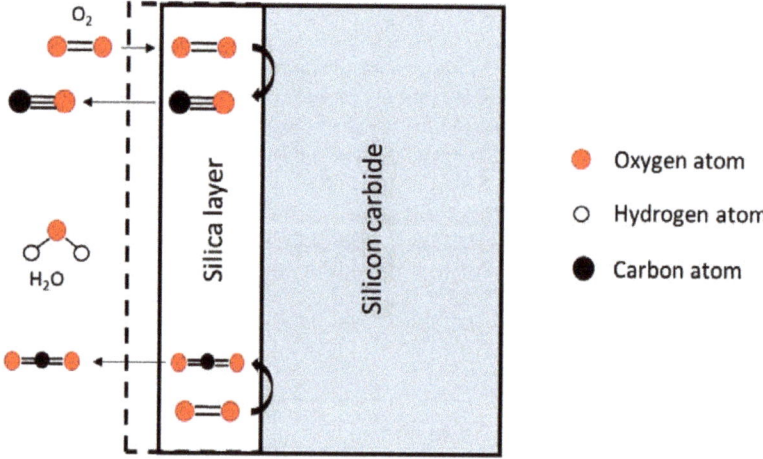

Figure 4. Scheme of the chemical reaction at the interface.

Between 500 and 800 °C, the oxidation of SiC fibers generates voids and releases gaseous compounds which create a porous silica scale at the SiO_2/SiC interface. These two facts are probably responsible for the loss of its mechanical properties [8].

Between 1200 and 1400 °C, the silica layer forms rapidly and seals off the surface porosity. The layer tends to delay the inward diffusion of oxygen.

The kinetics of the silica growth depend on the thickness of the oxide layer, which directly relates to the time of oxidation. For thick layers, or a long period of oxidation, the diffusion of oxygen limits the growth of silica, and the kinetics follow a parabolic law. For example, Zheng et al. showed that the parabolic regime of single crystal SiC occurred between 1200 and 1500 °C under pressures from 10^{-3} up to 1 Bar [9]. For thin layers, or a short period of oxidation, the reaction at the interface is the limiting step, and the kinetics follows a linear law. For silica growth which does not follow neither the parabolic nor the linear law, Deal and Grove formulated a linear parabolic law.

Recently, Park et al. used XPS to precisely characterize the chemical nature of the silica layer [10] and discovered that it consists of several oxidation states corresponding to SiO, Si_2O_3 and SiO_2 chemical environments.

2.2. Kinetic Models for Si and SiC Oxidation

2.2.1. The Deal and Grove Model

In 1965, Deal and Grove developed the kinetic law for the thermal oxidation of silicon (Figure 5) [11] and established a general equation:

$$Ax_0 = B(t+\tau) \quad (4)$$

where t is referring to the oxidation time, and the quantity, τ, corresponds to a shift in time).

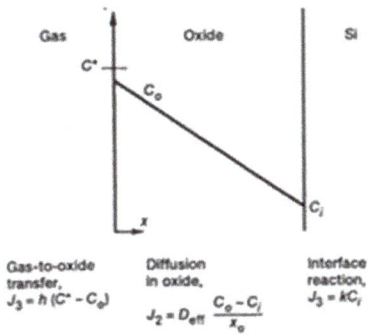

Figure 5. Model for the oxidation of silicon at the two boundaries of the oxide layer. Reprinted with permission from Reference [11]. Copyright 1965 AIP.

In this model, both a diffusion process and a model of oxidation are related in Figure 6. The reactions occur at the two boundaries of the oxide layer (x_0).

The diffusion of oxidant species through the oxide layer is expressed according to Fick's first law. Moreover, the model considers the gas phase transport to the oxide and Henry's law for the interface reaction. The subsequent relations are as follows:

$$B = 2D_{eff}C^*/N \quad (5)$$

$$C^* = KP \quad (6)$$

$$A = 2D_{eff}(1/k + 1/h) \quad (7)$$

where B is the parabolic rate constant in units of (oxide thickness)^2/time, D_{eff} is the effective coefficient diffusion of the oxidant species (oxide thickness)^2/time, C^* is the equilibrium concentration of the oxidant in the oxide, N is the number of oxidant species into a volume unit of oxide layer, K is the Henry's law constant and P the partial pressure of the oxidant. The A parameter is linked to k (reaction rate at the interface) and to the coefficient h (flux of oxidant species entering the oxide).

The Equation (6) can be written as follows:

$$x_0 = \frac{A}{2}\left[\left(1 + (t+\tau)/\left(A^2/4B\right)\right)^{1/2} - 1\right] \quad (8)$$

At long oxidation times ($t + \tau \gg A^2/4B$), a thick oxide layer is created:

$$x_0^2 = B(t + \tau) \tag{9}$$

This yields the parabolic law of oxidation growth. As B is proportional to D_{eff}, the oxide growth is limited by the diffusion of O_2 through the oxide.

At shorter times ($t + \tau \ll A^2/4B$), the Equation (8) can be written as follows:

$$x_0 = \frac{B}{A}(t + \tau) \tag{10}$$

where B/A is proportional to the chemical–surface reaction rate constant. Therefore, a linear regime is obtained for the thin oxide formation because the limiting step is controlled by the interface reaction [12].

Thus, Harris et al. were able to correlate this relationship to experimental data obtained over a temperature range of 700–1300 °C, with a partial pressure of 0.1 to 1.0 bar, and for an oxide thickness between 30 and 2000 nm, for both oxygen and water oxidant species.

This model illustrates the passive oxidation of silicon, but multiple studies demonstrated that this oxidation kinetics model fits well with the data obtained for silicon carbide materials as well.

2.2.2. Massoud Empirical Relation

Despite fitting well with most experimental results, the Deal–Grove model is not adapted to the early stages of oxidation (nanometer scale). Indeed, for thickness lower than 1 nm, the oxidation rates are very high and cannot be fitted by linear-parabolic kinetics predicted by the Deal–Grove model. Therefore, Massoud et al. [13] experimentally studied the kinetics of oxidation for very low thicknesses (50 nm). Their hypothesis was that the high oxidation rate can be modeled by adding an exponential term that decays with increasing thickness in the Deal–Grove model.

The modified model is then given by Equation (11):

$$\frac{dX}{dt} = \frac{B}{2X_{ox} + A} + C_2 e^{-\left(\frac{X_{ox}}{L_2}\right)} \tag{11}$$

This model fits well with data obtained for low thicknesses of oxide, but it is important to emphasize that it is an empirical solution that does not precisely take into account the physical mechanisms occurring during the early stages of oxidation [14].

Until very recently, no model could unify the Deal–Grove model with a model that takes into account both the early stages of oxidation and the physical mechanisms associated with them.

2.2.3. Si and C Emission Model

Goto et al. [15] developed a model based on the emission of Si and C atoms during the oxidation. It is a model based on the Si atoms emission model that had been previously developed [16]. The main difference between this model with the Deal–Grove model is that additional mechanisms for oxide growth surface are considered. Contrary to the D–G model that only considers the formation of oxide at the Si-oxide interface, this model considers the emission of Si atoms that can form oxide by two different ways:

- If the oxide layer is thin enough, the Si atoms can diffuse through it and instantly react with the oxidant atmosphere,
- The Si atoms can also encounter oxidant molecules in the oxide layer itself, and react with it.

The growth rate is then obtained by summing the 3 contributions to oxide formation:

$$N_0 \frac{dX}{dt} = kC_O^1(1 - v_{Si}) + \int_0^X \kappa(C_O)^2 C_{Si} dx + \eta \left(C_O^S\right)^2 C_{Si}^S \tag{12}$$

With ν the emission ratio, κ the oxidation rate of Si inside SiO_2, η is the oxidation rate of Si on the oxide surface, and superscript S is related to the position of the atom in the oxide layer.

Goto et al. thus modified the Si atom emission model to apply it to SiC materials. A schematic view of the model is given in Figure 6.

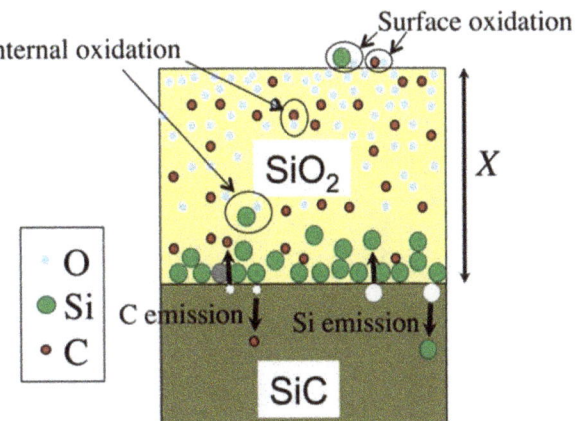

Figure 6. Schematic view of the Si and C emission model. Reprinted with permission from [17].

Using the work of Kageshima et al. the interfacial reaction rate is defined by the following:

$$k = k_0 (1 - \frac{C_{Si}^I}{C_{Si}^0})(1 - \frac{C_C^I}{C_C^0}) \quad (13)$$

where C_x^I and C_x^0 are the concentration of the species at the interface and the solubility limit of the species, respectively.

By modifying the diffusion equations of the Si emission model, diffusions for the reactive species are given as follows:

$$\frac{\partial C_{Si}}{\partial t} = \frac{\partial}{\partial x}\left(D_{Si}\frac{\partial C_{Si}}{\partial x}\right) - R_1 - R_2 \quad (14)$$

with

$$R_1 = \eta C_O^S C_{Si}^S \text{ and } R_2 = \kappa_1 C_{Si} C_O + \kappa_2 C_{Si} (C_O)^2 \quad (15)$$

$$\frac{\partial C_C}{\partial t} = \frac{\partial}{\partial x}\left(D_C\frac{\partial C_C}{\partial x}\right) - R_1' - R_2' \quad (16)$$

with

$$R_1' = \eta' C_O^S C_C^S \text{ and } R_2' = \kappa_1' C_C C_O + \kappa_2' C_C (C_O)^2 \quad (17)$$

$$\frac{\partial C_O}{\partial t} = \frac{\partial}{\partial x}\left(D_O\frac{\partial C_O}{\partial x}\right) - R_1 - R_2 - R_1' - R_2' - R_3 \quad (18)$$

with

$$R_3 = h\left(C_O^S - C_O^0\right) \quad (19)$$

By numerically solving these equations and using experimental obtained values, it is possible to use Equation (14) to determine the oxide growth rate.

As an example, results are given on Figure 7 for a fixed temperature and different partial pressures of oxygen for the C-face and the Si-face [18].

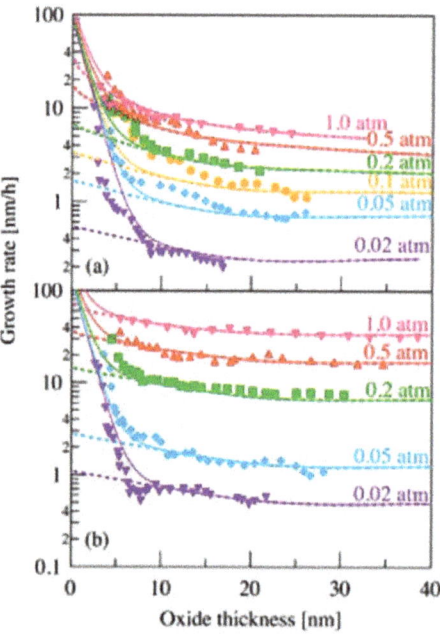

Figure 7. Comparison between experimental results and Si/C emission model for 1100 °C oxidation (**a**) Si-face and (**b**) C-face. Reprinted with permission from Reference [18]. Copyright 2019 Elsevier.

The results given here show notable similarity between the model and the experimental results. This validates the model even for very small thicknesses of silicon oxide.

In the next section, the parameters influencing the oxidation of SiC materials are discussed. It is important to point out that the discussion is based on studies using the Deal–Grove model, as it is still the most used model to describe SiC oxidation. Recent studies on the parameters influencing SiC oxidation and using the Si and C emission model include several references [19–24].

2.3. Parameters Which Can Influence the Oxidation of SiC Materials

Numerous authors calculated the activation energy E_A of the parabolic oxidation of SiC thanks to Arrhénius Equation (9):

$$X^n = Cte^{-\frac{E_A}{RT}} \qquad (20)$$

where C is a constant, X is the oxide thickness, t is the oxidation time and T is the temperature. The activation energy corresponds to the minimum energy required for a chemical reaction to occur. The data for the activation energy for linear oxidation and for parabolic oxidation are shown below, in Tables 1 and 2:

* Please note that linear parabolic refers to the case when both linear and parabolic regimes are observed.

Table 1. Activation energies for linear oxidation of SiC in dry atmosphere.

Types of SiC	T (°C)	Kinetics	Activation Energy for Linear Oxidation (kJ/mol)
Powder SiC (green) short heating and cooling cycles	1100	Linear law	No data [25]
Si slices (111)	900–1200	Linear parabolic	193 [11]
Single crystals SiC (thin oxide face)	970–1245	Linear parabolic	355 [12]
RF-Sputtered thin α-SiC films (C-face)	950–1100	Linear law	155–200 [26]
Single crystal Si	800–1100	Linear parabolic	155 [27]
Single crystal SiC (fast oxidation face)	800–1100	Linear parabolic	159 [27]
Single crystal SiC (slow oxidation face)	800–1100	Linear parabolic	330 [27]
CVD-SiC thick films (fast oxidation face)	800–1100	Linear parabolic	170 [27]
CVD-SiC thick films (slow oxidation face)	800–1100	Linear parabolic	334 [27]

Table 2. Activation energies for parabolic oxidation of SiC in dry atmosphere.

Types of SiC	T (°C)	Kinetics	Activation Energy for Linear Oxidation (kJ/mol)
Single crystals SiC (thick oxide face)	970–1245	Linear parabolic	196 [12]
HfB2 + 20 v/o SiC composite	1350–1550	Parabolic law	452 [6]
Hot-pressed SiC	1200–1400	Parabolic law	481 [28]
Hot-pressed SiC	1200–1500	Parabolic law	134–389 [29]
Sintered α-SiC	1200–1500	Parabolic law	155–498 [29]
Single-crystals Si	1200–1400	Linear parabolic	120 [30]
Single-crystal SiC (green) (fast-grow face)	1200–1400	Linear parabolic	121–297 [30]
Single-crystal SiC (green) (slow-grow face)	1200–1400	Linear parabolic	339 [30]
Controlled nucleation thermally deposited SiC	1200–1400	Linear parabolic	142–293 [30]
Sintered α-SiC	1200–1400	Linear parabolic	217–289 [30]
Hot-pressed SiC	1200–1400	Linear parabolic	221 [30]
CVD-SiC	1550–1675	Linear parabolic	345 (amorphous silica) and 387 (cristobalite) [31]
Single crystal SiC (green) (C face)	1200–1350	Parabolic law	120 [9]
Single crystal SiC (green) (C face)	1350–1500	Parabolic law	260 [9]
Single crystal SiC (green) (Si face)	1350–1500	Parabolic law	223–298 [9]
CVD-SiC	1200–1400	Linear parabolic	142 [32]

Table 2. Cont.

Single crystal Si	800–1100	Linear parabolic	96 [27]
Single crystal SiC (fast oxidation face)	800–1100	Linear parabolic	99 [27]
Single crystal SiC (slow oxidation face)	800–1100	Linear parabolic	292 [27]
CVD-SiC thick films (fast oxidation face)	800–1100	Linear parabolic	94 [27]
CVD-SiC thick films (slow oxidation face)	800–1100	Linear parabolic	285 [27]
CVD-SiC	1200–1500	Linear parabolic	118 [33]
CVD-SiC	1397–1737	Linear parabolic	210 [34]
Types of SiC	**T (°C)**	**Kinetics**	**Activation energy for parabolic oxidation (kJ/mol)**
Powder SiC (black)	1000–1200	Parabolic law	209 [35]
Powder SiC (green) Oxidation time <30 min	1000–1200	Parabolic law	117 [35]
Powder SiC (green) Oxidation time >60 min	1000–1200	Parabolic law	263 [35]
Powder SiC (green) short time oxidation	1100–1300	Parabolic law	209 [25]
Single-crystals SiC (green)	1200–1500	Parabolic law	276 [36]
High purity SiC	900–1200	Parabolic law	85 (amorphous silica) and 65 (cristobalite) [37]
High purity SiC	1380–1556	Parabolic law	190 [38]
Si slices (111)	900–1200	Linear parabolic	119 [11]
Si slices (111)	1000–1200	Parabolic law	125 [39]
Powder SiC	1200–1500	Parabolic law	632 [40]
Polycrystalline CVD SiC	1477–1627	Linear parabolic	1130 [41]
Self-bonded SiC (50/50 α/β)	1000–1300	Parabolic law	No data [42]

2.3.1. Interpretation of the Activation Energy Values

The first data on the oxidation of SiC powders are all in the same order of magnitude, lying between 85 and 209 kJ/mol [11,35–38]. According to the authors, four remarks can be made:

1. Their data differ depending on the fitting of experimental values and the nature of SiC samples,
2. Their data differ due to the presence of impurities from either the sample or the apparatus, or the gas phases present. Thus, the oxidation rate is determined by the nature and concentration of impurities as well as other physicochemical parameters,
3. The oxidation period seems to affect the oxidation kinetics:
    - ➤ For short oxidation times, a thin amorphous oxide film is created, and the kinetics of the oxide growth follow a linear regime, which implies that this mechanism is surface-controlled,
    - ➤ For long oxidation times, the oxidation rates decrease as the oxide layer grows. The kinetics follow a parabolic regime, meaning that the mechanism proceeds by gas diffusion through the oxide layer [25,28],
    - ➤ An initial period of sixty minutes is observed between 1100 and 1300 °C for which the parabolic law fits the data well. Then, at short oxidation times,

the silica growth follows a linear regime with an increase of oxidation rates, meaning that the silica scale loses its protective effect [25,35].

4. Finally, the rate determining step of the oxidation is thought to be either the inward diffusion of oxygen or the outward diffusion of CO (i.e., product gases).

First, Deal and Grove established a kinetic model for oxidation of silicon under wet and dry atmospheres [11]. They defined a parabolic constant, which expresses a diffusion-controlled mechanism, and a linear constant, which expresses a surface-controlled mechanism. As the parabolic activation energy of the silicon oxidation is close to the value of oxygen permeation through fused silica (113 kJ/mol)—given by Norton [43] from the literature data—Motzfeldt concluded that oxidation rates of silicon and silicon carbides were similar [44]. It can be concluded that, in both cases, the diffusion of oxygen controls the oxidation kinetics. However, the initial period of SiC oxidation was not implemented into the Deal and Grove model.

Secondly, Jorgensen et al. proposed that the oxidation rate decrease could be due to the crystallization of the scale over long oxidation times [37]. Amorphous silica is produced by the oxidation reaction and can be transformed into cristobalite above 1200 °C, that slows down the diffusion of oxygen slows down.

Finally, the next section deals with the effect of impurities on SiC oxidation and the oxidation time dependence of silica growth.

2.3.2. Nature of the Silica Layer

Jorgensen et al. [37,45] claimed that, for low temperatures, the silica layer was mainly consisted of amorphous silica, but at higher temperatures and/or after long periods of time, crystallization occurred. Costello et al. confirmed that the activation energy increases with the temperature and/or with the oxidation time. The low activation energy values (134 and 155 kJ/mol) could be attributed to the diffusion of molecular oxygen through an amorphous scale, whereas crystallization of silica could occur at high temperature. This is likely why the highest values (398 and 498 kJ/mol) are recorded [29]. Thus, the transport of oxygen through crystalline scales is thought to be slower than through amorphous [31], as the oxidation rates decreased by a factor of thirty when the scale crystallization was completed [33]. A representation of the phase transitions of amorphous silica layer with time during the SiC oxidation are related on the scheme below (Figure 8):

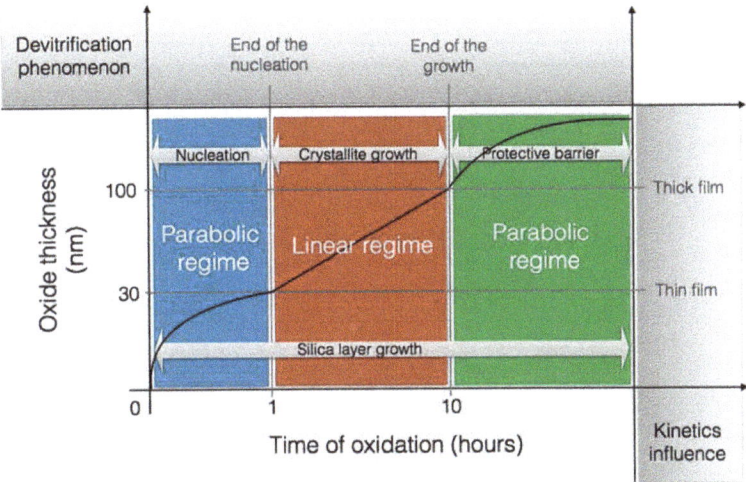

Figure 8. Representation of the kinetics of silica growth and the transition phases of amorphous silica versus time of oxidation in the temperature range of 1000–1300 °C.

The purpose of this scheme is to show that the devitrification of silica is in competition with the growth of the oxide layer during the nucleation period and the crystallite growth. However, when the scale is fully crystalline, silica growth is the only mechanism driving the kinetics.

As it can be seen, there is an initial period which seems to correspond with the nucleation of the cristobalite crystals within the amorphous scale. Deal and Grove did not consider this initial period in their model because it was not possible to measure such low thicknesses. However, they suggested the existence of another oxidation regime when the oxide thickness was below 30 nm. At very short oxidation times, silica growth follows a parabolic regime controlled by oxygen diffusion. Then it follows a linear regime controlled by a surface reaction. Finally, it returns to a parabolic regime at long oxidation times. It seems that two processes compete, one increasing the protective property of the oxide scale and the other one degrading it.

At the end of the crystal nucleation, the protective property of the scale is not maintained anymore, as the cristobalite growth generates numerous defects through which speed up oxygen diffusion [46]. Thus, in this second stage, the oxidation mechanism follows linear kinetics both because the silica layer does not limit the diffusion of oxygen anymore and because the crystallization rate is linear under dry atmospheres [47].

For the first stage, devitrification of silica has already occurred, but the layer is still protective. The hypothesis could be that the impurities, which act as crystallization starting point, induce a local decrease of the oxide viscosity. The consequence is a decrease of interfacial stress, which limits the crystallite growth and, therefore, retards the devitrification process. This hypothesis is supported by the work of Wei and Halloran who demonstrated that the devitrification of silica into cristobalite can be avoided by adding mullite grains (i.e., impurities) into vitreous silica [48]. In this first stage, the scale is still protective as the oxygen diffusion proceeds through the amorphous scale; however, it proceeds via cracks and pores as the crystal grows. At this point, defects allow a fast diffusion of oxygen through the scale, and the kinetics follow a linear regime.

Finally, a second hypothesis can be proposed: the impurities can affect the oxidation behavior of polycrystalline material by forming either high or low protective films, depending on the impurity's concentration. For low concentrations, the silica growth will show an initial oxidation period, and at the end of the second stage, a crystalline layer is formed with low permeability. However, for high concentrations, a rapid initial growth rate is observed, which decreases with the crystal sizes [47]. In this case, the crystal growth is limited, and a partially crystalline layer is obtained. The consequence is that the silica scale shows a high level of microporosity, allowing fast diffusion of oxygen.

This second hypothesis is supported by the work of Costello et al. who demonstrated that a high level of impurities and nucleation sites in SiC materials led to a greater susceptibility to crystallization of the scale and complicated the oxidation behavior [30]. It was found that the presence of cations, impurities and additives (such as aluminum or carbon atoms) led to the formation of a viscous layer with high permeability to oxygen, which was responsible for the increase of the oxidation rates. The crystallization of the scale can lead either to low oxidation rates—when a continuous layer of spherulitic crystals (cristobalite) is formed—or to high oxidation rates—when these crystals are randomly dispersed in the amorphous matrix and locally increase grain boundaries [46].

2.3.3. Crystal Faces Effects

In the silicon carbide structure, one carbon atom is linked to 4 atoms of silicon, forming CSi_4 at their vertices. Double layers of atom are formed exhibiting one carbon face $(000\bar{1}/\bar{1}\bar{1}\bar{1})$ and one silicon face $(0001)/(111)$ referenced as C-face and Si-face (Figure 9) [27].

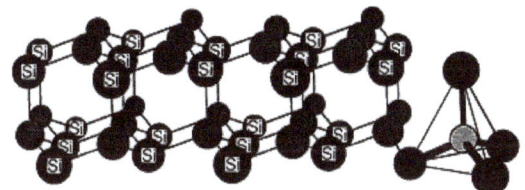

(0 0 0 $\bar{1}$)/($\bar{1}$ $\bar{1}$ $\bar{1}$) C-terminated fast oxidation face of SiC

(0 0 0 1)/(111) Si-terminated slow oxidation face of SiC

Figure 9. Schematic showing different terminations of the (000$\bar{1}$)/($\bar{1}\bar{1}\bar{1}$) and (0001)/(111) faces of SiC. Stacking in this case is for cubic SiC. A representative CSi4 tetrahedron is also shown. Reprinted with permission from Reference [27]. Copyright 1996 The American Ceramic Society/Wiley.

The network is predominantly covalent and exhibits two general crystalline forms: the cubic β-SiC and the α-SiC. These networks have different crystallographic polytypes depending on the stacking of the tetrahedral bilayers. The main polytypes are the cubic 3C (ABC) with a Zincblende crystal structure and the hexagonal (4H for ABCB or 6H for ABCACB), with a Wurtzite crystal structure (Figure 10).

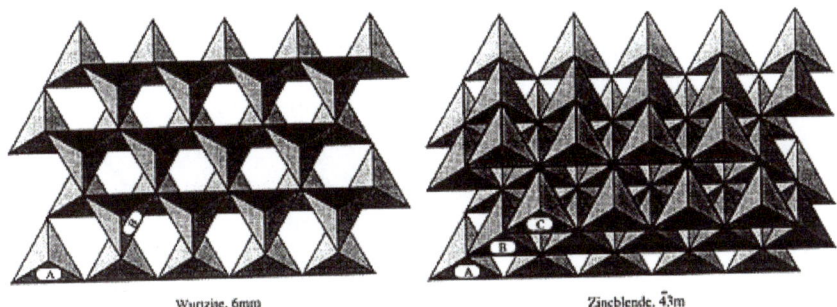

Wurtzite, 6mm   Zincblende, $\bar{4}$3m

Figure 10. Tetrahedral representation of close-packed planes in cubic and hexagonal SiC. Top schematic shows Zincblende (cubic) structure with CSi4 tetrahedra stacking in ABC periodicity; bottom schematic shows Wurtzite (hexagonal) structure with tetrahedral stacking in AB periodicity. Reprinted with permission from Reference [27]. Copyright 1996 The American Ceramic Society/Wiley.

Furthermore, it is interesting to note that single crystals of SiC are often hexagonal (α) and that CVD-SiC samples generally crystallize in the cubic (β) crystalline. The different crystal structures and the different atomic natures of the C- and Si-faces lead to different oxidation behaviors which is not the case for silicon crystals [49].

The oxidation behavior of the (110, 111, 311, 511 and 100) faces of silicon single crystals

Lewis and Irene obtained different oxidation rates for the thermal oxidation of single crystal Si depending on the crystal faces (110, 111, 311, 511 and 100). It appears that, for thin oxide film, the growth of silica depends on the density of atoms, which is specific to each orientation. Although the oxidation rates increase with the silicon atom density, the development of intrinsic oxide stress becomes dominant for thick oxide films. This affects the transport of oxidant species to the interface and leads to a decrease of oxidation rates below 1100 °C [49]. The influence of the C- and Si-face on the oxidation rates is discussed below.

The oxidation behavior of the slow and fast oxidation faces.

Harris first demonstrated that the oxidation of the two crystal faces of SiC platelets followed different kinetic laws in accordance with the Deal and Grove model [12]. Between

1000 and 1300 °C, the C-face showed faster oxidation than the Si-face. The oxidation of the C-face follows parabolic law and leads to the formation of a thick oxide layer whereas the oxidation of the Si-face follows linear kinetics and a thin oxide layer is created, as seen in Figure 11 below.

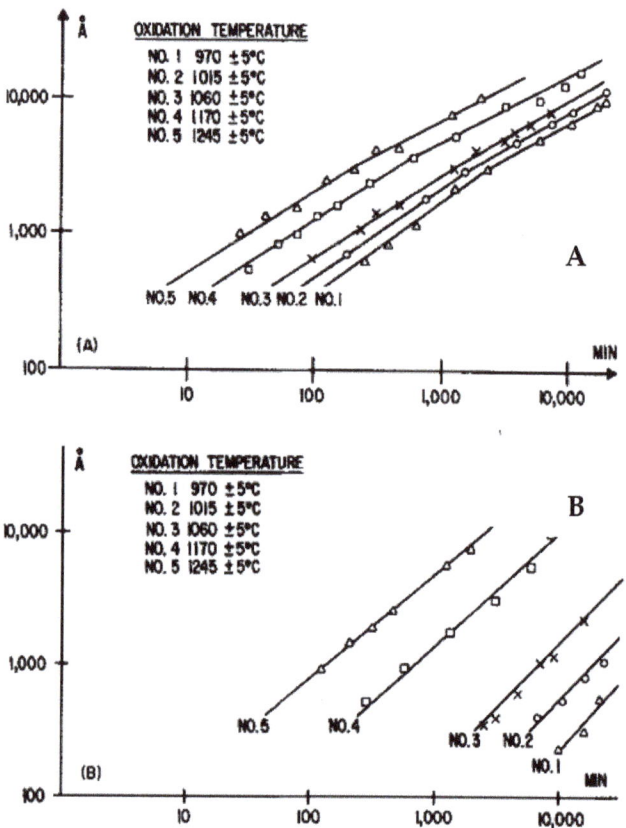

Figure 11. Oxide thickness vs. oxidation time on (**A**) thick oxide side and (**B**) thin oxide side. Reprinted with permission from Reference [12]. Copyright 1974 The American Ceramic Society/Wiley.

Many authors confirmed that the oxidation behavior of the fast oxidation face (C-face) of CVD-SiC and single-crystal SiC is similar to that of single-crystal silicon over the temperature range of 1200–1400 °C [9,12,25,27,30,50]. The slow oxidation faces (Si-face) of CVD-SiC and single-crystal SiC also exhibit similar oxidation behavior and activation energy. Firstly, the activation energy is identical for silicon and C-faces of silicon carbide materials, and the crystalline structure of SiC does not seem to have any on its value. Secondly, a change in the chemical composition of the oxide scale is noted [27] which gives evidence for different diffusional processes. An inner layer of unknown composition was found at the SiC/SiO_2 interface of the Si-slow oxidation face. This layer had higher refractive index and may have had lower permeability, which could explain the change in activation energy. Later, XPS analysis revealed a C-rich region in the oxide scale in the form of silicon oxycarbides that had been formed by the reaction of silica with the carbonaceous species released via Equations (1–3) [51]. The presence of C-rich regions was revealed, as well, by EELS, and these C-rich regions took the form of carbon clusters with a thickness of 10–15 Å [52].

Harris demonstrated that the oxidation rate of the C-face was increasing when temperatures and oxidation time periods were decreasing.

The oxidation behavior of the C-face depending on temperature:

The same observations were made by Zheng et al. [9]: two different activation energies for the oxidation of the C-face were recorded between 1200 and 1500 °C, but only one was recorded for the Si-face of single crystal SiC. The explanation could be that two oxidation kinetics are competing—one dominant at high temperatures and for long oxidation periods, and the other one being dominant at low temperatures. In fact, if the temperature and the oxidation time periods increase, the linear oxidation rate of the material will become dominant over the parabolic law, and the oxidation rate will increase [12]. This implies that the growth of the oxide is surface-reaction controlled and not oxide-diffusion controlled.

Another explanation is a change in the diffusion mechanism of the oxidant species. Costello et al. submitted the idea that if lattice diffusion occurs [30], high activation energy (400 kJ/mol) will be recorded in comparison to a classic oxygen permeation, which is associated to a low activation energy (of 120 kJ/mol).

Zengh et al. agreed and demonstrated that the activation energy was influenced by the chemical change of the diffusing species. Below 1350 °C, a low activation energy value is obtained, and the major diffusing species are molecular oxygen. However, above 1350 °C, a high activation energy occurs, which seems to be the consequence of ionic diffusion [9]. Nevertheless, Ogbuji et al. recorded only one activation energy during parabolic oxidation over the range 1200–1500 °C, which was equal to the value obtained by Deal and Grove [33]. The experiments were performed under highly dry oxygen with clean samples and apparatus, so the conclusion was that only the permeation of oxygen through the scale was limiting the oxidation of SiC, up to 1500 °C. A review on the growth of silica during the oxidation of SiC details these observations [53].

2.3.4. Oxidation Rate-Determining Step

Initially, Pultz and Singhal [28], recorded high activation energies and assumed that the oxygen transport was not the rate controlling step of the oxidation. Singhal concluded that it might be the desorption of volatile carbonaceous products released at the SiC/SiO_2 interface. However, Antill et al. [42] revealed that the pressure of CO_2 had no influence on the kinetics over the range 0.2–1 bar, so the pressure was not controlling the reaction. Moreover, he assumed that the diffusion of carbonaceous species through the silica layer did not affect the layer's protective property, as silicon and silicon carbide demonstrated similar reactivity between 1200 and 1300 °C.

Nevertheless, the discovery of some C-rich clusters at the SiC/SiO_2 interface may suggest that the trapping of carbonaceous species (CO or CO_2) could limit the diffusion out into the gas phase, and, thus, limit the whole oxidation process. However, Zheng et al. established the profile concentration of C18O molecules and demonstrated that the concentration was constant through the silica scale [9]. This confirms fast carbon transport out of the silica scale. The carbon diffusion was also not dependent on the oxygen partial pressure, whereas the oxidation rates of SiC increased with the oxygen pressure. Thus, the transport of carbonaceous species could not be the rate-controlling step of the oxidation. As suggested by Narushima et al., it is possible that the diffusion of oxygen ions into the silica network is why high activation energy values are observed [31] Gavrikov et al. investigated the defect generation and passivation of the Si-face of SiC dry oxidation [54]. An abrupt model which describes the transition between crystalline SiC and amorphous SiO_2 was used to perform calculations of SiC oxidation reactions. Thus, the mechanism kinetic was designed for a rigid silica scale (i.e., at temperatures above 1300 °C and without water hydration), as described below (Figure 12):

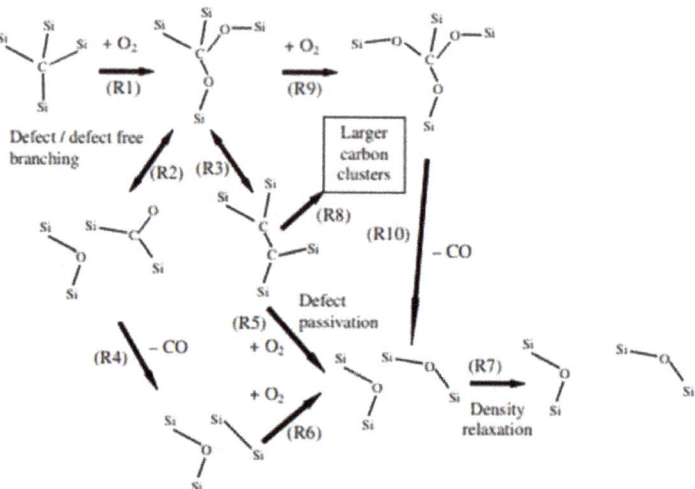

Figure 12. The scheme of the kinetic mechanism of the (0001) SiC surface oxidation process. Reprinted with permission from Reference [54]. Copyright 2008 AIP Publishing.

The mechanism described above can be explained within four steps:

1. The penetration of oxygen into the silica scale:

    First, the transport of oxygen at the Si-face is considered (R1, on Figure 12). The calculated activation energy is 250 kJ/mol for molecular diffusion of oxygen through the oxide layer.

    However, the oxygen diffusion can proceed through lattice oxygen atoms with higher calculated activation energies. The value is about 390 kJ/mol, and therefore the lattice diffusion of oxygen is not considered to contribute to the SiC oxidation.

2. Reaction of oxygen with SiC:

    The oxidized bonds Si-O-C are created at the SiC/SiO_2 interface and three chemical reaction paths are observed:

    The first one consists of the formation of a carbonyl defect: $Si_2 = CO$ (R2, on Figure 12), which is followed by CO desorption (R4, on the Figure 12). DFT calculations predict a high activation energy of 350 kJ/mol, but this path proceeds faster at high temperatures. Then, further oxidation creates new silica units (R6, on the Figure 12) with an activation energy of 190 kJ/mol.

    The second path consists of the formation of C_2 dimer intermediates: $Si_2 = C = C = Si_2$ (R3, on Figure 12), which evolve into larger aggregates (R8, on the Figure 12). This creates carbon defects near the interface.

    The last path consists of further oxidation of the Si-O-C bonds (R9, on the Figure 12) followed by CO desorption (R10, on the Figure 12). DFT calculations show that three oxidized bonds are necessary for CO desorption, and that the activation energy is 190 kJ/mol.

3. Passivation reactions of carbon defect by O_2:

    Calculations show that the carbon defect could react with an oxygen molecule in singlet state to give new silica entities: $Si_2 = C = C = Si_2 + O_2 \rightarrow Si_2 = C = C = O + O = Si_2$ (R5, on Figure 12). The activation energy for this is 190 kJ/mol.

    For temperatures above 900 °C, the calculations shown that the reaction rate of carbon with an oxygen in a triplet state is higher. An activation energy of 60 kJ/mol is calculated for the dissociation of the oxygen molecule: $Si_2 = C = C = Si_2 + O_2 \rightarrow Si_2 = C = C(Si)-O-Si + O$.

4. Density relaxation process:

The new silica units undergo structure relaxation (R7, on Figure 12) in order to decrease the bond density at the interface and thus lower the stress energy. The activation energy is 190 kJ/mol.

The interest of this model is to underline the complexity of the oxidation mechanism of SiC. As it is described, ten reactions are involved in the oxidation process, but only one can control the silica growth kinetics. Gavrikov et al. used an abrupt model for the SiC/SiO_2 interface, in which a high activation energy was predicted for the oxygen penetration. Thus, they assumed that oxygen diffusion (R1) was the rate-determining step of the oxidation process, whereas CO desorption was competing with carbon defect generation.

2.4. Conclusions

The oxidation of SiC materials can be either passive or active, leading to the formation of a silica layer in the passive case, and leading to oxide gaseous compounds in the active case. From a general point of view, the oxidation of these materials leads to the degradation of SiC and to the loss of its mechanical properties. For high-performance applications, it is necessary to better understand this phenomenon, which is why kinetics model for the passivation of silicon are studied example, Deal and Grove elected a linear parabolic regime for the oxidation of silicon under dry and wet atmospheres.

Different parameters can influence the oxidation behavior of SiC materials, and, thus the kinetics:

- At short oxidation times, a gas diffusion mechanism is dominant (parabolic regime) whereas at long times, a surface-reaction mechanism is dominant (linear regime),
- For the gas diffusion mechanism, the temperature plays an important role: at low temperatures, the oxygen diffusion is molecular, whereas above 1350 °C, the diffusing species is ionic oxygen. Therefore, a C-rich inner oxide layer is created on the Si-faces,
- The oxidation behavior becomes complicated when the crystallization of amorphous silica takes part in the oxidation process. This reduces the oxidant transport and leads to the decrease of oxidation rates.
- Finally, the presence of impurities is not negligible and may have an impact in all these studies. A high degree of impurities will enhance both the crystallization of the scale and the formation of defects, which allows for faster oxygen diffusion.

From these points, researchers started to investigate the oxidation mechanism of SiC to explain the differences in activation energy values.

A large amount of research has been performed on the isothermal, passive oxidation of silicon-based ceramics. Jacobson [7] highlighted that the combination of various secondary effects (on the outer circle below) and the complexity of the combustion environments makes the oxidation process difficult to understand. This directed fundamental studies to focus on the center of the circle—pure materials and pure oxygen environments (Figure 13).

As shown, the SiC oxidation is influenced by both numerous operating parameters and by intrinsic material properties. In the following part, the effect of water molecules on the oxidation of SiC materials is reviewed.

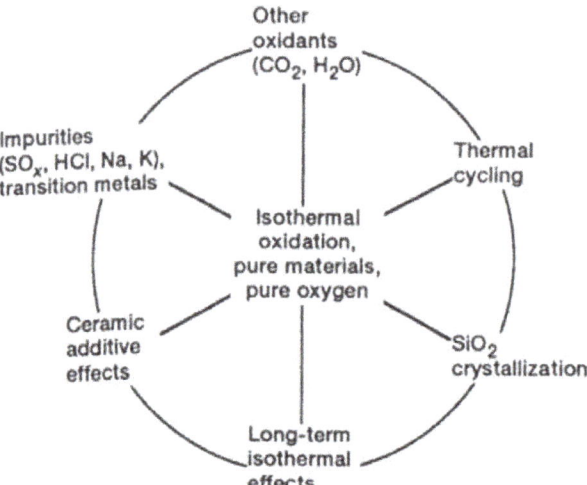

Figure 13. Complications to isothermal oxidation due to additive-containing ceramics and combustion environments. Reprinted with permission from Reference [7]. Copyright 1993 The American Ceramic Society/Wiley.

3. Wet Oxidation of Silicon Carbide Materials

Figure 14 gives a scheme describing the layout of the next section.

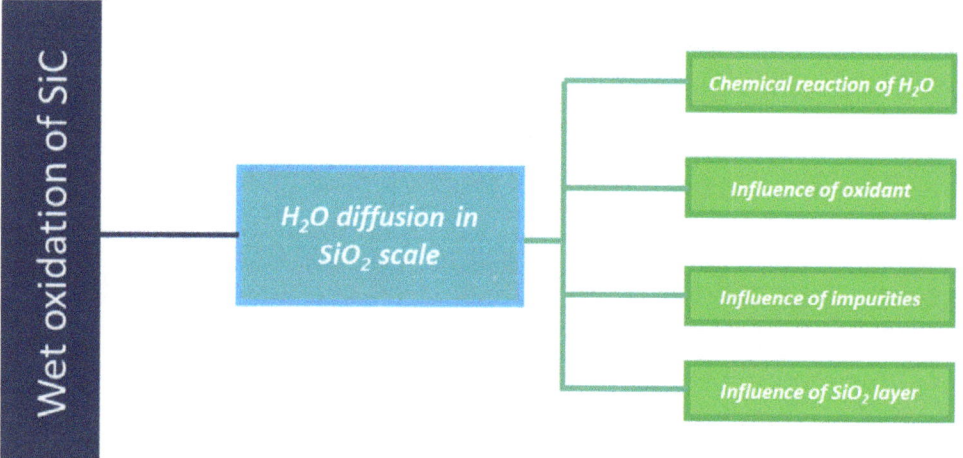

Figure 14. Layout of the wet oxidation part of the review.

Nakatogawa [35], Suzuki [55] and later Jorgensen et al. [38,46] studied the effects of water vapor on the oxidation of silicon carbide powders. Jorgensen et al. found out that the presence of water vapor in O_2 accelerated the oxidation rate of SiC and the nucleation rate of cristobalite.

The interaction of water with the SiC material produces a silica layer which can be described by the following hydrothermal oxidation reaction 21:

$$\text{SiC (s)} + 3 \text{ H}_2\text{O (g)} \rightarrow \text{SiO}_2 \text{ (s)} + \text{CO(g)} + 3\text{H}_2 \text{ (g)} \tag{21}$$

However, some authors have proposed that Equation (22) competes with (21) [56]:

$$\text{SiC (s)} + 4\text{H}_2\text{O (g)} \rightarrow \text{SiO}_2 \text{ (s)} + \text{CO}_2 \text{ (g)} + 4\text{H}_2 \text{ (g)} \tag{22}$$

Chaudhry used Auger Electron Spectroscopy to assess the chemical composition of silica films grown on SiC materials under dry and wet oxygen. The content of carbon atoms within the film was higher for wet oxidation (14 at%) than for dry oxidation (2 at%). Following the (19) equation, CO_2 molecules are produced at the interface. As the diameter of CO_2 (3.0 Å) is bigger than the ring width of SiO_2 (2.5 Å) [57], the molecule is easily trapped within the growing scale.

Furthermore, the high content of carbon can result from the high reaction rates of water with SiC, which then releases a large amount of carbonaceous species at the SiC/SiO_2 interface. Due to this, the silica layer catches more volatile species. The higher permeation of water, as opposed to the permeation of oxygen, through the oxide scale explains the high oxidation rates [46,58]. A new oxidation regime is described in the next section in order to explain the fast kinetics of SiC wet oxidation.

3.1. The Two Competitive Oxidation Regimes

Two regimes of oxidation compete depending on the operating parameters.

The passive oxidation regime is described by Deal and Grove using linear parabolic kinetics for the wet and dry oxidation of silicon [11]. From this model, the calculated permeation of oxygen through silica was one thousand times less than that of water. In other words, a water content in oxygen gas of less than 25 ppm will affect both the surface kinetics during linear oxidation and the diffusional kinetics during the parabolic oxidation [59].

The mixed oxidation regime is described by Rosner et al. and Opila using Tedmon's treatment of paralinear kinetics for Cr and Fe-Cr alloys [59]. In this model, both diffusive and gas/oxide interface processes occur simultaneously, and the oxide scale grows to a limiting thickness.

During the oxidation of CVD-SiC in a 50% H_2O/O_2 gas mixture, between 1200 and 1400 °C, the two oxidation reactions are (23) and (24) [42,60–64]:

$$\text{SiC (s)} + 3\text{H}_2\text{O (g)} \rightarrow \text{SiO}_2 \text{ (s)} + 3\text{H}_2 \text{ (g)} + \text{CO (g)} \tag{23}$$

$$\text{SiC (s)} + 2\text{H}_2\text{O (g)} \rightarrow \text{Si(OH)}_4 \text{ (g)} \tag{24}$$

The rate of reaction (22) is described by the parabolic rate constant for oxide formation k_p, whereas that of reaction (24) is described by the linear rate constant for oxide volatilization k_1. The evolution of the oxide thickness x with the time t is described by the following relation:

$$\frac{dx}{dt} = \frac{k_p}{2x} - k_1 \tag{25}$$

Over long periods of time or high volatility rates, the rate for oxide formation equals the rate for oxide volatilization, and a steady state is reached. At that moment, SiC undergoes a linear recession given by the rate y_L which is proportional to the volatility rate of the oxide, k_1. The oxide thickness is reaching a maximum, x_L:

$$x_L = \frac{k_p}{2k_1} \tag{26}$$

This relation can be expressed in terms of weight change from the integrated form of Equation (25) and the figure below shows the evolution of dimensional change for SiC with the oxidation time (Figure 15):

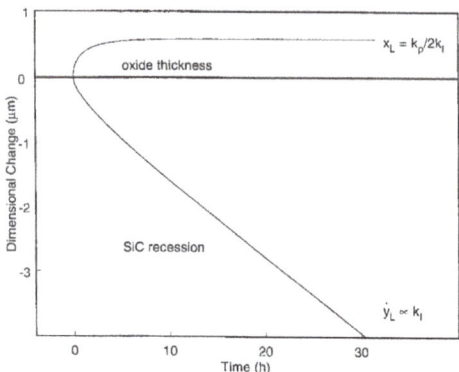

Figure 15. Dimensional change for SiC due to paralinear oxidation and volatilization. Oxide thickness and SiC recession curves calculated for k_p = 0.25 mil^2/h and k_1 = 0.21 mil/h (1 mil = 1 × 10^{-3} in. = 25.4 µm). Reprinted with permission from Reference [64]. Copyright 2003 The American Ceramic Society/Wiley.

The kinetics of Equation (22) are described by the parabolic rate constant for moderately thick scales, in which the rate constant k_p is dependent of the partial pressure of water vapor, $P_{H_2O}^n$ but not of gas velocity (24):

$$k_p \propto P_{H_2O}^n \tag{27}$$

where n is the power-law exponent. A n value of 1 is used, as other studies demonstrated that the oxidation mechanism was controlled by molecular water diffusion through the silica scale [11,65,66]. A reference temperature of 1316 °C was set to establish the oxidation rates under water vapor (25):

$$k_{p,\,1316°C} = 0.44 P_{H_2O}^1 \tag{28}$$

One can see that the oxidation rate increases with the partial pressure of water. From reaction (22) it leads to an increase of gaseous compounds at the SiC/SiO$_2$ interface responsible for the high porosity of the oxide. This can be seen on the cross-section images of the oxide layer on a CVD-SiC substrate below (Figure 16):

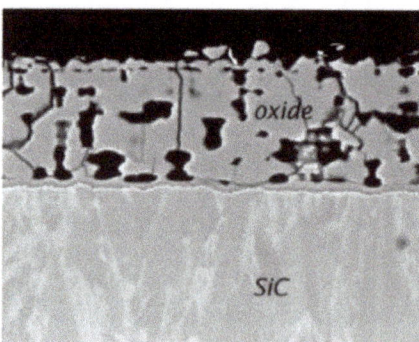

Figure 16. Representative oxide cross-section formed on SiO$_2$-formers for the CVD-SiC sample oxidized in a high pressure furnace at 1200 °C and under 10 bar of 15% H$_2$O, during 500 h, with a gas velocity of 0.05 cm/s. Reprinted with permission from Reference [64]. Copyright 2003 The American Ceramic Society/Wiley.

As the oxide layer decreases in density (i.e., increasing in porosity), the gas-phase transport is occurring through the pores. The increase of the distance from the surface to the SiC/SiO$_2$ interface does not follow the parabolic law anymore. Thus, the solid-state transport of water is considered to be rate-determining through the dense part of the silica scale.

To summarize, Opila et al. demonstrated that below 1100 °C, the wet oxidation of SiC follows linear–parabolic kinetics and, above 1100 °C, the oxidation follows simple parabolic kinetics. In the first case, the oxidation is surface-controlled whereas, in the other case, it is diffusion-controlled [64].

Finally, Tortorelli et al. observed the oxide scale after the wet oxidation of pure silicon and CVD-SiC materials [67]. In contradiction with Opila, all the oxide scales were crystalline after a 1200 °C exposure under wet atmosphere [66]. For these samples, a two-layer oxide scale is observed for which the layer at the interface is dense and amorphous, whereas the upper one is thick and crystalline for silicon samples, as shown below on Figure 17.

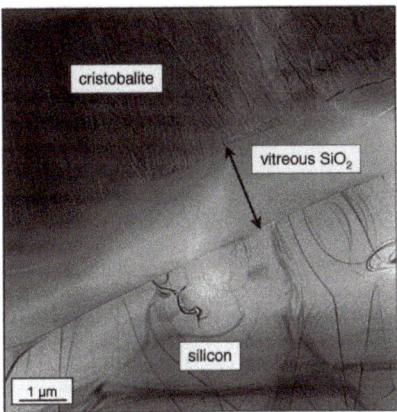

Figure 17. Cross-sectional TEM image of the thin underlying dense layer on pure silicon exposed for 500 h in air + 15 vol% H$_2$O at 1200 °C and 10 bar. Reprinted with permission from Reference [67]. Copyright 2003 The American Ceramic Society/Wiley.

For CVD-SiC samples, carbonaceous species are released by the oxidation under water pressure (1.5 bar). At low speed flow, the volatile species are trapped in the layer, leading to pore formation, whereas at high speed flow, the volatilization of the scale is dominant and no layer remains (Figure 18):

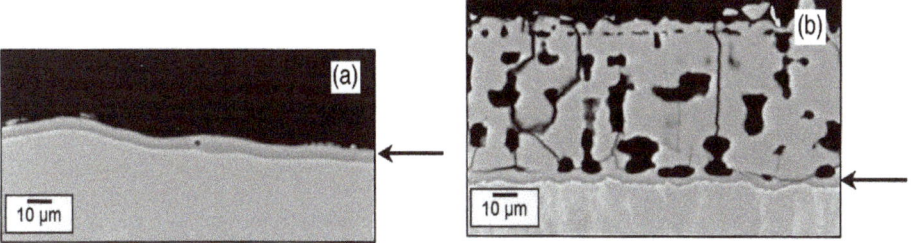

Figure 18. CVD-SiC seal coats after exposures in 1.5 atm H$_2$O at the high gas velocity (~30 m/s) of (**a**) Solar Turbines combustor at ~1200 °C for 5016 h and (**b**) low gas velocity (~3 cm/min) of a laboratory furnace (ORNL rig) at 1200 °C after 500 h. Reprinted with permission from Reference [67]. Copyright 2003 The American Ceramic Society/Wiley.

As was observed by Opila [63], the thickness of the thin film was constant with time whereas the crystalline oxide layer was continually increasing. Therefore, the volatilization of the cristobalite layer is considered not to contribute to the SiC recession. In addition, the high rate of the cristobalite growth could be predicted neither by the paralinear model nor by the presence of impurities. However, the fact that the dense layer does not thicken with time could be the expression of another paralinear regime. In this new model, the thickness of the dense layer becomes constant as the rate of its formation is equal to the rate of its transformation into a porous scale [66]. In the model of Haycock, two density conversion factors are used for relating the thickness of reacted material to the thickness of the dense and the porous scales. Agreement of the parabolic rate constant with the experimental data shows that the paralinear model is valid at high water pressure and low gas-flow velocities.

To conclude, the paralinear model is used depending on the experimental conditions as temperature, pressure, gas velocities and oxidation time. Moreover, Tortorelli et al. demonstrate that the transformation from a dense to a porous layer leads to the formation of a non-protective scale. As a consequence, this formation has a great influence on kinetics. The study of the activation energy, nature of the silica scale, and the diffusion of species leads to better understand of the wet oxidation mechanism of SiC.

3.2. Main Characteristics of Wet Oxidation of SiC Materials

Numerous authors calculated the activation energy of the oxidation of SiC, for which the values lie between 46 and 428 kJ/mol. However, there are two exceptions of 531/656 and 527 kJ/mol, which were reported by Palmour et al. and Singhal et al., respectively [58,68,69]. The data for linear and parabolic oxidation are reported in the two tables below (Tables 3 and 4):

Table 3. Activation energies for linear oxidation of SiC in wet atmosphere.

Types of SiC	Oxidant Species	T (°C)	Kinetics	Activation Energy for Linear Oxidation (kJ/mol)
Si slices (111)	Wet O_2	900–1200	Linear parabolic	190 [11]
Powder SiC	Wet air	1200–1400	Linear law	146 [57]
Powder SiC	Wet O_2, Ar, N_2	1500	Linear law	No data [70]
Single-crystal α-SiC (Si + C faces)	84% vol H_2O	850–1050	Linear parabolic	109 [51]
Single-crystal α-SiC (C face)	84% vol H_2O	850–1050	Linear parabolic	200 [51]
Single-crystal β-SiC (C face)	98 °C water + O_2	1000–1200	Linear parabolic	251 [69]
Single-crystal β-SiC (C face)	98 °C water + Ar	1000–1200	Linear parabolic	280 [69]
CVD-SiC	84% vol H_2O	1000–1250	Linear parabolic	309 [71]
RF-Sputtered thin α-SiC films	84% vol H_2O	950–1100	Linear law	205–218 [25]
CVD-SiC	10% H_2O in O_2	1550–1650	Linear parabolic	428 [72]
Powder-SiC	50%H_2O/50%O_2	1200–1400	Parabolic law	No data [61]
Sintered α-SiC	12.3% H_2O, 2.1% O_2, 11.0% CO_2, 71.8% N_2 50% H_2O/50% O_2	1316	Paralinear	No data [63]
CVD-SiC		1100–1400	Paralinear	No data [65]
Sintered α-SiC	50%H_2O/50%O_2	1100–1400	Paralinear	No data [65]
Fused quartz	50%H_2O/50%O_2	1100–1400	Paralinear	No data [65]

Table 4. Activation energies for parabolic oxidation of SiC in wet atmosphere.

Types of SiC	Oxidants	T (°C)	Kinetics	Activation Energy for Parabolic Oxidation (kJ/mol)
High purity SiC	H_2O/Ar	1218–1514	Parabolic law	102 [46]
Si slices (111)	Wet O_2	900–1200	Linear parabolic	68 [11]
Si slices (111)	90 °C water + O2	1000–1200	Parabolic law	85 [40]
Si slices (111)	Steam	1000–1200	Parabolic law	102 [40]
SiC (50/50 of α/β)	Wet O_2	1000–1300	Parabolic law	No data [43]
Hot pressed SiC	3% H_2O in O_2	1200–1400	Parabolic law	527 [58]
Single crystals Si (100)	H_2O/O_2 (1 to 2000 ppm)	780–980	Linear parabolic	No data [62]
Single crystals Si (100)	H_2O/N_2 (1 to 2000 ppm)	780–980	Linear parabolic	No data [62]
Single-crystal β-SiC (C face)	98 °C water + O_2	1000–1200	Linear parabolic	531 [69]
Single-crystal β-SiC (C face)	98 °C water + Ar	1000–1200	Linear parabolic	656 [69]
CVD-SiC	84% vol H_2O	1000–1250	Linear parabolic	209 [71]
Pressureless-sintered α-SiC	H_2O in Air (10 to 40% vol)	1300	Parabolic law	No data [73]
CVD-SiC	10% H_2O in O_2	1550–1650	Linear parabolic	397 [74]
CVD-SiC in fused quartz tubes	10% H_2O in O_2	1200–1400	Linear parabolic	41 [32]
CVD-SiC in high purity Al_2O_3 tubes	10% H_2O in O_2	1200–1400	Linear parabolic	249 [32]
CVD-SiC	H_2O/O_2 (10 to 90% vol)	1200–1400	Parabolic	28-156 [66]
CVD-SiC	H_2O/Ar	1200–1400	Parabolic	No data [66]
Si, Sintered α-SiC, CVD-SiC	Air + 15% vol H_2O	1200	Paralinear (adapted from the model of Haycock)	No data [67]

3.2.1. Activation Energy of SiC Wet Oxidation

According to the Deal and Grove model, the parabolic rate expresses a diffusion-controlled mechanism whereas the linear rate expresses a surface-controlled mechanism.

For the linear rate, most of the values are between 109 and 218 kJ/mol, but the authors do not give any explanation for the oxidation mechanism [26,50,57]. High values (around 300 kJ/mol) [71] suggested that the breaking of Si-C bonds (290 kJ/mol) is the rate-controlling step of the linear growth of silica.

Most of the parabolic values are between 100 and 300 kJ/mol, which is comparable with the data obtained for the dry oxidation of SiC. In accordance with Jorgensen et al., Deal and Grove and Singhal et al. [11,46,62] the activation energy of SiC was affected when water vapor was added to oxygen due to particular dissolution of water through the silica layer.

However, low parabolic activation energy of 41 and 68 kJ/mol were recorded by Opila and Deal and Grove [11,32]. The first explanation for such low values is that the experiments were performed under highly controlled-atmosphere with low levels of impurities, and the second is that Opila showed agreement with the Deal and Grove model. However,

Opila et al. suggested that only water vapor diffusion could be the rate-determining step of the parabolic growth of silica.

In opposition, Singhal et al. obtained high activation energy (526 kJ/mol), which was close to the activation energy for dry oxygen (481 kJ/mol), and concluded that the impurities could lead to the devitrification of the oxide scale. According to Ainger, water has the capacity to enhance the crystallization of cristobalite [40], leading to cracks and defect formation into the amorphous layer. Therefore, the cracks would allow water, oxygen and impurities to diffuse faster, enhancing the oxidation rates.

3.2.2. Impurities Effect

First, Opila calculated the rate of oxidation under water vapor (10 wt% H_2O) according to Deal and Grove model of silicon oxidation [32]. The rate of oxidation was one order of magnitude higher when using sapphire tubes, and the activation energy increased from 41 to 249 kJ/mol. This could be a consequence of the transport of alkali ions to the interface induced by the combination of water vapor and impurities. The first hypothesis is that an amorphous sodium aluminosilicate layer is produced. This layer could demonstrate higher permeability to water which could raise the oxidation rate. The second hypothesis is that the presence of alkali ions (aluminum and sodium) allows the nucleation of the silica scale and enhances the transformation of cristobalite to tridymite [75]. Although there are limited data on the permeability of water through cristobalite, due to the conversion to the phase change of silica with temperature, the work of Jorgensen et al. [38], Antill et al. [43] and Lu et al. [25] showed that when tridymite and quartz formed, the oxide layer acted as a protective film to oxidant species. As a consequence, a tridymite layer should have a lower permeability to water than a cristobalite layer, and this shows that the second hypothesis does not seems to be valid.

Nevertheless, the α- and β-tridymite demonstrate the most open crystalline structure in comparison with quartz and cristobalite phases. Indeed, tridymite has more than 50% voids in the unit cell, according to the review paper of Lamkin et al. [76]. The conclusion is that the diffusion of oxygen and water species might be faster through tridymite than through the cristobalite network, and therefore the second hypothesis of Opila appears valid.

To conclude, one can see that the presence of impurities can modify either the permeability of the scale or the diffusion mechanism of oxidant species, leading to a change in the oxidation behavior of SiC materials. The following section deals with the determination of the primary oxidant species during wet oxidation of SiC.

3.2.3. Nature of the Oxidant Species

Jorgensen et al. carried out one experiment to determine the role of the silica nature in the oxidation behavior of SiC under dry and wet atmospheres [46]. First, the wet oxidation of SiC is performed to obtain a tridymite scale at the surface. Then, a second oxidation is performed under oxygen atmosphere. The two oxidation rates are similar; thus, it was concluded that the diffusing species were the same under the partial pressure of water and of oxygen.

In opposition, Opila [66] showed that the presence of oxygen with water vapor plays an important role in the SiC oxidation kinetics. Indeed, the activation energy is found to be inversely dependent on the partial pressure of oxygen [69]. Moreover, Irene and Ghez deduced that the oxidation rate of H_2O/O_2 mixtures was greater than the one calculated for simultaneous and isolated oxidation by water and oxygen as primary oxidants [60]. Thus, a kind of synergy is occurring when water is added to oxygen for SiC oxidation. In the next section, the high oxidation rates obtained with water could be related to the nature of the silica scale.

3.2.4. Nature of the Silica Layer

Opila et al. observed an amorphous silica scale after 100 h of oxidation [66] and concluded that water vapor had little effect on the crystallization rate of silica. The analysis of XRD pattern intensity of the oxide growth versus the content of water vapor confirmed this observation [74]. According to Figure 19, the water vapor content does not level up the relative intensity of cristobalite, thus, it does not enhance the crystallization of silica.

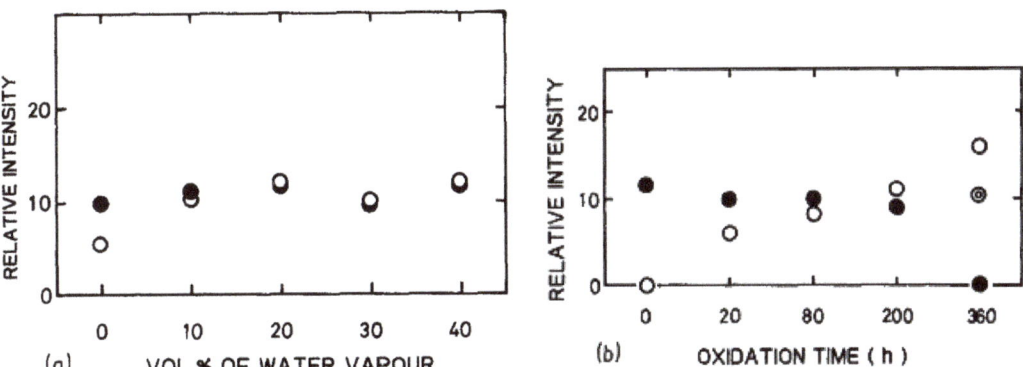

Figure 19. Relative intensities of X-ray diffraction peaks (**a**) against water vapor content (1300 °C, 100 h) and (**b**) against oxidation time (1300 °C, 20 vol% H_2O), (•) corundum, (o) cristobalite and (⊙) mullite. Reprinted with permission from Reference [74]. Copyright 1988 Springer Nature.

As corundum is present in the sintered materials, its peak intensity also appears in the XRD patterns. It was stated in the dry oxidation part that the oxide layer becomes more and more crystalline as the temperature and the oxidation time increase. For wet oxidation, Opila et al. concluded the same by showing that the crystallization is enhanced up to 1400 °C, at which the scale is fully crystalline (i.e., cristobalite) [64]. In addition, Maeda et al. demonstrated that the increase of the cristobalite content in the oxide layer with the oxidation time was faster when 20 vol% of H_2O was added to air [74]. Therefore, water vapor contributes to the devitrification of amorphous silica but does not increase its protective properties, as high oxidation rates are recorded under wet conditions [33,46,58].

In fact, Opila examined the oxidation kinetics of SiC in terms of Deal and Grove model under water/oxygen mixtures [66]. The oxidation rates increased with the water content, but the improvement was not as important as for the lowest contents—therefore the increase of oxidation rate is not linear with the increase of water content, as is shown on Figure 20.

For these experiments, the mix oxidation regime is not considered for temperatures above 1200 °C, as for long times and thick scales, the volatilization is negligible. So, the passive oxidation regime is applied, and, for high water contents, it is shown that the crystallization of the silica scale increases [40] which limits the water diffusion. The Arrhenius plots do not show temperature dependence, which seems to be correlated to a change in either the nature of the oxide scale or in the oxidation mechanism.

Finally, the crystallization rate of silica on silicon was found to follow parabolic kinetics in steam, whereas it obeys a linear regime under dry oxygen [48]. The contradiction is that water species engender higher oxidation rates in comparison with oxygen, but at the same time, the crystallization of the scale is catalyzed. It is known that the solubility of water in oxide is one thousand times higher than that one of oxygen, so high interaction of water with the oxide layer could be the reason for such oxidation behavior.

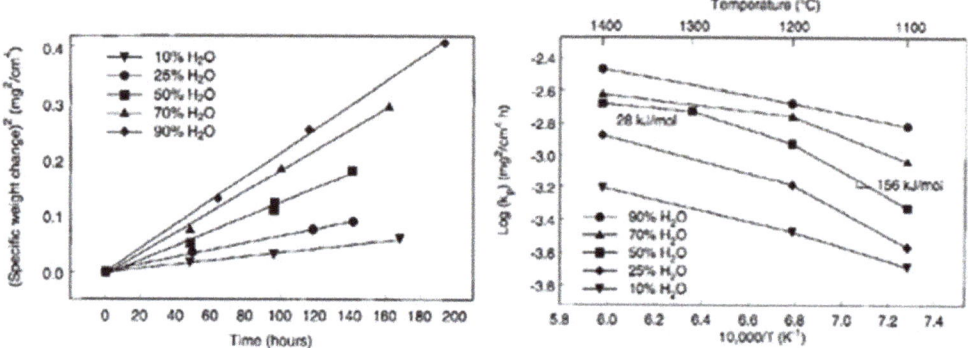

Figure 20. (**Left**)—Determination of parabolic rate constants for CVD SiC in H_2O/O_2 mixtures at a total pressure of 1 atm and a temperature of 1200 °C. (**Right**)—Temperature dependence of the parabolic oxidation rate constant for CVD SiC in the temperature range of 1100–1400 °C in various H_2O/O_2 mixtures. Reprinted with permission from Reference [66]. Copyright 1999 The American Ceramic Society/Wiley.

3.2.5. Particular Reaction of Water with the Oxide Layer

The high reactivity of water is noticed and a new mechanism of water diffusion could be linked to this reactivity [43]. Indeed, Cappelen attributed the high oxidation rates to the high permeation of H_2O molecules, by incorporation of (HO⁻) into SiO_2 [70]. Deal and Grove [11] obtained solubility values of water in silica, close to the one of Norton [44], which were one thousand times higher than the solubility of oxygen. The water in contact with silica could react through Equation (29):

$$Si\text{-}O + H_2O \rightarrow 2HO^- \qquad (29)$$

Based on the Wagner lattice defect model [77], the oxidant diffusion can occur via vacancies in the oxide lattice. If hydroxyl ions are produced, the parabolic rate constant will be proportional to the square root of pressure, as can be seen in the table below, with $H_2O_i^x$ for a water molecule at interstitial site, with HO_i' and $H_i^\cdot$, respectively for hydroxyl ions and protons at interstitial sites, with O_i'' for O^{2+} cations at interstitial sites and, finally, with $2h_i^\cdot$ for an interstitial site hole. Details of results are given on Table 5.

Table 5. Water partial pressure dependence for several water vapor defect species using standard Kroger–Vink notation (from Reference [64]).

Water Vapor Defect Species	Defect Formation Reaction	Mass Action Expression	Electro-Neutrality Expression	Water Vapor Partial Pressure Dependence	Power Low Exponent for Water
$H_2O_i^x$	$H_2O(g) = H_2O_i^x$	$K_1 = [H_2O_i^x]/P_{H2O}$	none	$[H_2O_i^x] \propto P_{H2O}^1$	1
no	$H_2O(g) = H_2 + \tfrac{1}{2}O_2$	$K_2 = [O_2]^{1/2} P_{H2}/P_{H2O}$	none	$[O_2] \propto \frac{P_{H2O}^{p2}}{P_{H2}^2}$	2
HO_i'	$H_2O(g) = HO_i' + H_i^\cdot$	$K_3 = [HO_i'][H_i^\cdot]/P_{H2O}$	$[HO_i'] = [H_i^\cdot]$	$[HO_i'] \propto P_{H2O}^{1/2}$	1/2
O_i''	$H_2O(g) = 2H_i^\cdot + O_i''$	$K_4 = [O_i''][H_i^\cdot]^2/P_{H2O}$	$[O_i''] = [H_i^\cdot]$	$[O_i''] \propto P_{H2O}^{1/3}$	1/3
O_i''	$H_2O(g) = 2h_i^\cdot + O_i'' + H_2$	$K_5 = [h_i^\cdot]^2[O_i'']P_{H2}/P_{H2O}$	$[O_i''] = [h_i^\cdot]$	$[O_i''] \propto \frac{P_{H2O}^{1/3}}{P_{H2}^{1/3}}$	1/3

The fact that n is not exactly equal to 1 could arise from the carrier gases used and from the nature (i.e., chemical composition, porosity) of the scale. For exponents close to 0.5, it seems that the diffusion mechanism involves molecular water and charged species.

The fact that the value of the power law exponent is not equal to 0.5 could come from two possibilities:
(1) The dissolution of water produces hydroxyl ions, but not all participate to the growth of the scale [43],
(2) The dissolution of water is not complete, and the diffusive species through silica are molecular water and hydroxyl ions [66].

Irene and Ghez investigated the second hypothesis [60] and underlined the particular role of water during wet oxidation of single-crystal Si. First, water is the primary oxidant species which allows the conversion of Si to SiO_2. Secondly, it has the capacity to loosen the SiO_2 network and thereby increase the diffusion of other oxidant species. The dissolution occurs by reaction to hydroxyl and the breaking of an oxygen bridge, as proposed by Moulson and Roberts [78] and confirmed by Wagstaff [48] (Figure 21):

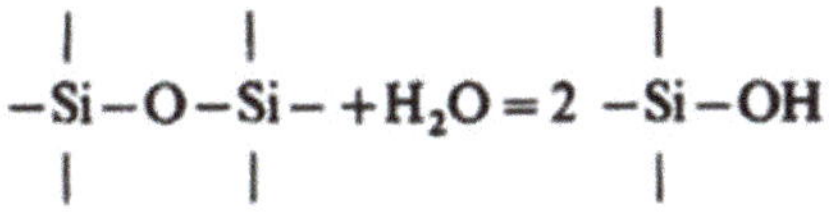

$$-Si-O-Si- + H_2O = 2\ -Si-OH$$

Figure 21. Dissolution of silica during wet oxidation. Reprinted with permission from Reference [68]. Copyright 1981 Acta Chemica Scandinavica, 1947–1999.

It was shown that the formation of silanol bonds is reversible, meaning that when the atmosphere is changing from wet to dry, the oxidation rates rapidly regress to that obtained under dry O_2 [68,71]. Hence, it is necessary to determine the rate-controlling step of the oxidation.

3.2.6. Oxidation Rate-Determining Step

Cappelen [70] showed that the quantity of CO (g) produced during wet oxidation (Equation (22)) is equal to the one produced during dry oxidation. Meanwhile, the oxidation rate under wet atmosphere is ten times higher than that of under dry conditions. As a result, the desorption of carbonaceous species at the SiC/SiO_2 interface is not the rate-controlling step of silica growth. Later, Narushima et al. [72] obtained a high activation energy value (around 200 kJ/mol) for the parabolic rate constant of wet oxidation. The same value was obtained for the dry oxidation of SiC when a crystalline layer of silica was obtained. This suggests that the rate-controlling step of both diffusion is identical, i.e., the diffusion of O^{2-} ions into the cristobalite film. In the case of wet oxidation, a high number of defects in the layer will be consistent with the O^{2-} fast diffusion through the layer and higher oxidation rates.

To conclude, the last phenomenon to take into account is the crystal orientation effect.

3.2.7. Crystal Orientation Effect

Lu et al. performed dry and wet oxidation on the C-face of SiC in order to create thin oxide film on SiC substrates [25]. It has been stated that the fast-oxidation face (C-face) of CVD-SiC and single-crystal SiC is similar to the one of single-crystal silicon over the temperature range of 1200–1400 °C [9,12,27,30,51]. Though, the oxidation rate of C- and Si-faces was determined to be slower than that of the single-crystal Si (100), depending on the temperature and time of oxidation. Indeed, the C-face rate of oxidation was about 2–4 times slower than the Si (001) material under dry oxygen and about 2–11 times slower under wet oxygen. For both conditions, by increasing the time of oxidation and the temperature, the oxidation rates of the C-face SiC and Si (001) tends to be equal. However, the crystallinity of the scale was not discussed and should be the reason why the oxidation rate of the two faces is not similar.

The most important remark is that similar linear activation energy (about 200 kJ/mol) is obtained for wet and dry oxidations of the C-face. Thus, the oxidation behavior of the

C-face of SiC thin films is the same under oxygen or water atmosphere, except that water allows higher reaction rate. To conclude, the rate-controlling step of the linear regime is the same for wet and dry oxidations of SiC materials.

3.3. Conclusions

From the parameters which can influence the wet oxidation behavior of SiC materials, the remarks made for the dry oxidation are still valid:

- At short oxidation time, a gas diffusion mechanism is dominant (parabolic regime) whereas at long times, a surface-reaction mechanism is dominant (linear regime),
- For the gas diffusion mechanism, temperature plays an important role: at low temperatures, the oxygen diffusion is molecular whereas above 1350 °C, the diffusing species are ionic oxygen,
- The oxidation behavior is complicated when the crystallization of amorphous silica takes part in the oxidation process. This reduces the oxidant transport and leads to the decrease of the oxidation rates,
- Finally, the presence of impurities is not negligible and could be involved in all these studies. On one hand, it enhances the crystallization of the scale which leads to an increase of defects and, on the other hand, it creates high-permeable and viscous oxides. Both mechanisms result in a faster oxygen diffusion.

Notably, the high reactivity of water has been highlighted, as it has the capacity to enhance the silica devitrification for which high crystallization rates are obtained. Water is able to improve the impurities' mobility through the network, which leads to the formation of highly permeable oxides.

Finally, the oxidation behavior of SiC is shown to be complex and ambiguous due to its dependence on two parameters: time and temperature. The dissolution and reaction of water with the silica scale seems to be the key to understand the whole oxidation process. When Si-OH groups form, the network relaxes; as a consequence, the SiO_2 viscosity, density, acoustic velocity and refractive index decrease, whereas the thermal expansion coefficient increases [79,80]. The introduction of water into oxide networks has been modeled by Doremus.

3.4. Dissolution-Reaction Model for Water through Oxide Scales

The dissolution models are necessary to understand the interaction mechanism of water with silicon-containing materials. First, Doremus established a model for water diffusion and reaction through glassy oxides with one hypothesis concerning the species mobility. For this model, a protective layer is considered to be formed during SiC wet oxidation [81,82] and a two-step mechanism is assumed when water is in contact with the oxide layer.

First, the solution of water molecule is possible by breaking Si-O-Si bridges (30):

$$H_2O + Si\text{-}O\text{-}Si = 2\ Si\text{-}OH \tag{30}$$

Then, Doremus assumed that the mobility of dissolved water is higher than that of one of the OH groups which belongs to the network. He suggested that dissolved water molecules jump from one cavity to another without any reaction with Si-O-Si bridges. In silica, the diffusion process proceeds via an interstitial mechanism [83]. However, another mechanism is possible in which water molecules react with Si-O-Si bridges on one site and are regenerated on another site. This is denoted as the interconversion–diffusion model [84].

When Equation (30) reaches equilibrium, the equilibrium constant, K_2, is defined by the following equation:

$$K_2 = \frac{S^2}{C} \tag{31}$$

where S is the concentration of hydroxyl groups and C the concentration of dissolved molecular water. This reaction is bimolecular and this expression assumes that hydroxyl groups can interact with their neighbor to regenerate a water molecule.

The diffusion of water through the oxide scale then needs to be expressed. Fick's law describes the diffusion of water as a movement of molecules along the x axis due to a gradient of concentration. Here, we consider Fick's second law (32):

$$\frac{\partial C(x,\,t)}{\partial t} = D\left(\frac{\partial^2 C(x,\,t)}{\partial x^2}\right) \tag{32}$$

where x and t represent the distance and time of diffusion and D the water diffusion coefficient (cm^2/s). The water diffusion coefficient can be calculated empirically through the Arrhénius relation, as diffusion is a thermally activated process (33):

$$D = D_0 T^n \exp\left(-\frac{E_A}{RT}\right) \tag{33}$$

where D_0 is the pre-exponential constant, R is the gas constant, E_A is the activation energy, T is the temperature and n is a temperature-dependence exponent. Usually, $n = 0$ for simplicity as the activation energy depends on the temperature, so the pre-exponential term has little influence on the diffusion coefficient of water [76].

Secondly, we need to consider the influence of the removal and generation of water molecules on the diffusion of water. Thus, a term ($\frac{\partial S(x,\,t)}{\partial t}$) is subtracted from Equation (32):

$$\frac{\partial C(x,\,t)}{\partial t} = D\left(\frac{\partial^2 C(x,\,t)}{\partial x^2}\right) - \frac{\partial S(x,\,t)}{\partial t} \tag{34}$$

where S is linked to C by a simple linear dependence in order to solve Equation (33):

$$K_1 = \frac{S}{C} \tag{35}$$

This relation assumes that OH groups are immobile which implies that the generation of water molecule is of first order.

Thus, Equation (34) is expressed as follows:

$$\frac{\partial C(x,\,t)}{\partial t} = \frac{D}{(1+K_1)}\left(\frac{\partial^2 C(x,\,t)}{\partial x^2}\right) \tag{36}$$

Now, it is clear that the diffusion coefficient of water D has changed and this model defines the effective diffusion coefficient, D_e, by the following expression (37):

$$D_e = \frac{D}{1+K_1} \tag{37}$$

where D_e is independent of the concentration of molecular water but depends on the initial concentration of hydroxyl group in the network:

→ When C >> S, the diffusion process is not influenced by the reaction, so $K_1 \approx 0$ and $D_e \approx D$
→ When S >> C, the effective diffusion coefficient decreases, as the reaction of molecular water occurs in the oxide layer, as described by the equilibrium Equation (31). Therefore, the effective diffusion coefficient takes the following form:

$$D_e = \frac{2SD}{K_2} \tag{38}$$

where D_e becomes dependent of the concentration of molecular water when the generation of water molecule is of second order.

In the literature, some studies reported the temperature and time-dependence of this model, as shown in the left graph on Figure 22.

For temperatures above 600 °C, it was found that the reaction (jump of water) is dominant over its diffusion. Therefore, the concentration of water is low, and hydroxyl groups are the main species. They diffuse in the oxide layer and the reaction is bimolecular. However, below 600 °C, the concentration of exchanged OH groups at the oxide surface is time dependent if the local equilibrium is not reached. Over a short time, OH groups are diffusing in the whole layer due to the presence of defects, and the reaction is bimolecular. However, over a long time, OH groups cannot diffuse far and tend to react with their neighbor, so the reaction is almost unimolecular. For bimolecular reactions, expressions (31) and (38) are applied and for unimolecular reactions, expressions (35) and (37) are valid.

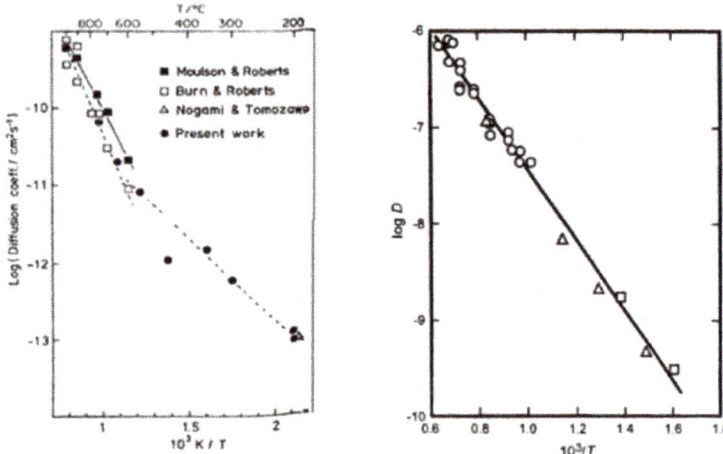

Figure 22. (**Left**) Arrhenius plot of the effective diffusion coefficient of water into silica glass. Reprinted with permission from Reference [85]. Copyright 1989 The American Ceramic Society/Wiley. (**Right**) Arrhenius plot of the diffusion coefficient of water into silica glass. Reprinted with permission from Reference [82]. Copyright 1999 Materials Research Society.

Furthermore, Doremus calculated D, the coefficient of water, from D_e, the effective coefficient of water, extracted from the data of Wakabayashi and Tomozawa, on the right graph seen in Figure 21 [86–88]. Equation (38) is used, and the linear regression shows agreement with the model. The activation energy for water permeation through silica is found to be equal to 70 kJ/mol. Another remark is that the D value is independent of time whereas the D_e value depends on the water solubility, diffusion time, water concentration or pressure, and type of diffusion experiment. However, it should not influence the activation energy.

Finally, Fortier and Giletti were able to correlate D_e with the measure of ionic porosity, Z, between 500 and 700 °C and under a water pressure of 100 MPa, calculated by the following [86]:

$$Z = 1 - \frac{V}{V_c} \tag{39}$$

where V is the total volume of atoms in the unit cell and V_c, the volume of the unit cell. The ionic porosity represents the openness of the oxide network and shows agreement with the activation energy E_A (kJ/mol). It can then be concluded that water is the diffusing species. The linear regression found for water permeation through eighteen glassy and seventeen crystalline oxides is:

$$E_A = 484 - 821Z \tag{40}$$

As a result, the higher the activation energy, the lower the ionic porosity, the lower the capacity of the oxide structure to open, and, therefore, the diffusion of water is slower.

To conclude on this model, the diffusion of water and oxygen species through amorphous and crystalline silica proceeds via two routes, according to Lamkin et al. [76]:

- The open porosity of the network allows the permeation of molecular species and,
- The Si-O-Si bridges network provides defects (as lattice vacancies) through which structural self-diffusion occurs with breaking and reforming of the bonds.

For the two diffusion processes, similar activation energies for dry oxidation were determined (with a value of 113 kJ/mol for molecular permeation of oxygen [44] and values between 85 and 121 kJ/mol for oxygen self-diffusion through amorphous scale). Likely, the oxygen transport mechanism is similar in both cases, and this conclusion can be applied for diffusion of water through silica. In fact, activation energies for water permeation are lower than the ones for oxygen, as it is reported in the table below. Doremus showed that this was due to high diffusivity of water in the network which is linked to its high solubility and high reactivity with silanol bonds.

For the parabolic regime, the activation energy is related to the energy needed for oxidant diffusion. As high activation energy was recorded, Doremus stated that, because the diffusion mechanism could proceed via the breaking of Si-O or Si-OH bonds, the diffusion mechanism is defect-based. Contrarily to other authors, he proposed that O_2 dissociates into atomic oxygen, which implies the diffusion of charged species into the silica scale [82]. According to Narushima et al., O_2^- could be the diffusing species, whereas Singhal hypothesized that hydroxyl ions are the reason for high parabolic activation energy recorded under water vapor. However, for lower values, molecular permeation is the most probable diffusion mechanism [76].

Regarding the linear regime of SiC oxidation, the surface-reaction mechanism is thought to proceed via bond breaking at the SiC/SiO_2 interface. Some authors proposed two possible pathways: one via the breaking of Si-C bonds, which required an energy of 290 kJ/mol [70]; or the other via the Si-Si bonds breaking, which required an energy of 177 kJ/mol, according to Pauling [73] (see Table 6 above).

Table 6. Activation energy for water and oxygen molecules through silica and silicon-containing bond energies.

Material/Silicon Bonding	Activation Energy in the Parabolic Regime Which Is Limited by Diffusion of Oxidant Species through Silica				Linear Regime Limited by the Interface Reaction
	Conditions (°C)	Oxygen Permeation (kJ/mol)	Conditions (°C)	Water Permeation (kJ/mol)	Breaking Energy (kJ/mol)
Fused silica	950–1100	113 [44]	300–1100	70 [81]	
Cristobalite	1000–1400 on SiC substrate	430 [87]		/	
Tridymite	1070–1280	195 [88]		/	
β-Quartz	870–1180	195 [88]	600–800 + 100 MPA H_2O	142 (//to c) [89]	
Si-Si					177 [73]
Si-C					290 [73]
Si-O or Si-OH					377 [81]

3.5. Conclusions

The wet (air) oxidation process was expressed, and different parameters were identified as highly influent on the oxidation kinetics of SiC and Si materials. A new model for wet oxidation kinetics is described as the paralinear model, which takes into account the volatilization of the silica scale at high pressures and high gas velocities. Finally, the main conclusions (i.e., effect of impurities and temperature) made for dry oxidation appear to be valid for the wet oxidation. However, the water dramatically increases the oxidation rate of SiC materials, even for small percentages of H_2O. In fact, water is the primary oxidant in wet oxidation and has the capacity to enhance the oxidation rates by loosening the silica network. As a consequence, impurities and oxygen can diffuse faster in the silica network and enhance the oxidation rates. In fact, the impurities have the capacity to increase the local viscosity of the amorphous layer (i.e., decrease the stress) and limit the growth of cristobalite crystals. In opposition to that, water accelerates the silica devitrification (i.e., the crystallization rate), and the oxidation rates decrease as the layer becomes crystalline. The competition between these two antagonist phenomena could explain the instability of the oxide layer grown during wet oxidation.

Finally, Doremus proposed a diffusion–reaction model to show that diffusion coefficient of water through oxide scales is modified as reaction with the network occurs [80,82]. This model agrees well with the data from the literature, and it shows a time dependence when local equilibrium has not been established. It also shows that diffusion coefficients depend on a variety of factors, but it does not depend on the activation energy of water diffusion. Fortier and Gilletti [86] were able to correlate the ionic porosity with the activation energy of water permeation. A linear relation was obtained, showing that water diffusion increases with the openness of the oxide structure. In this model, water reacts with silanol groups to diffuse until it reaches the SiC interface, which gives rise to hydrothermal oxidation. Therefore, the degradation of SiC materials is caused by water reaction and belongs to the "Hydrothermal Corrosion" classification.

4. Hydrothermal Corrosion of Silicon Carbides Materials

Under wet atmosphere, the SiC material is highly damaged, and its mechanical properties decrease. Since the hydrothermal oxidation reaction leads to the degradation of the material, it is classified as a chemical corrosion. In Figure 23, the corrosion behavior classification is given.

First, electrochemical corrosion is more developed for metals, as ceramics can be insulators or semiconductors and do not give up electrons easily. Then, mechanochemical corrosion occurs mostly for structural ceramics under thermal or mechanical loading, whereas tribochemical interactions can happen between ceramics and water [90].

Finally, chemical corrosion is the most studied case of corrosion in aqueous environments. In fact, the lack of corrosion resistance in water and water vapor of SiC is due to the formation of stable silicon hydroxides. As the oxides cannot act as a protective barrier, the chemical corrosion leads to rapid consumption of the material by a typical grain dissolution mechanism.

According to Kim et al., silicon carbides contain grain boundary layers, which drives their corrosion resistance [91]. Thus, the corrosion behavior of ceramics depends not only on composition but on microstructure as well. Under hydrothermal conditions, the chemical corrosion occurs as follows (Figure 24):

1. Reaction of grain-boundary phases,
2. Water transport along grain boundaries into the bulk of ceramics,
3. Reaction of ceramic grains.

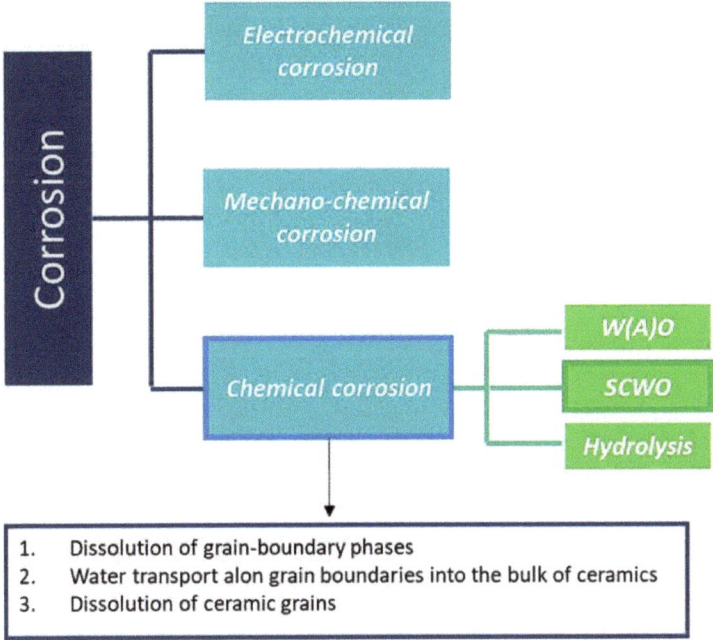

Figure 23. Ordering of the corrosion behavior of ceramic oxide materials. W(A)O stands for wet (air) oxidation and SCWO for supercritical water oxidation.

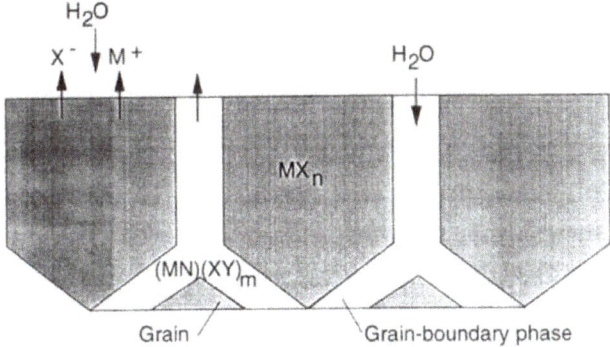

Figure 24. Schematic sketch showing dissolution of grains and grain-boundary phases (mostly silicates) and water transport along grain boundaries into the bulk of ceramics. Reprinted with permission from Reference [90]. Copyright 1994 Materials Research Society.

In order to better understand this corrosion mechanism, several reaction models were proposed to explain the SiC/water interaction.

4.1. Reaction Model for Chemical Corrosion
4.1.1. Yoshimura's Model for Hydrothermal Oxidation of SiC in Supercritical Water

From previous studies [35–37], the hydrothermal oxidation of SiC was thought to release CO_2 species, as follows (41):

$$SiC + 4H_2O \rightarrow SiO_2 + CO_2 + 4H_2 \tag{41}$$

However, Yoshimura et al. discovered the formation of a large amount of CH_4 in comparison to CO_2 [92] and concluded that the oxidation reaction was (42) [58]:

$$SiC + 2H_2O \rightarrow SiO_2 + CH_4 \qquad (42)$$

In this model, no reactions occurred below 500 °C even at 100 MPa H_2O, whereas above, a weight gain was observed. Thus, SiC material is transformed into amorphous silica, which crystallizes to form cristobalite and tridymite above 700 °C.

The following reaction model is proposed for SiC oxidation in H_2O, as it is represented on Figure 25 below. After the amorphous silica layer is formed on the surface of SiC particles, H_2O and CH_4 diffuse, respectively, inward and outward.

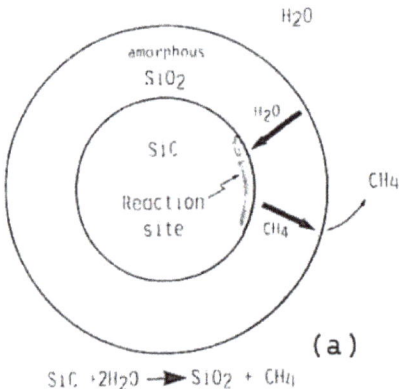

Figure 25. Reaction model of the hydrothermal oxidation of SiC powder. Reprinted with permission from Reference [92]. Copyright 1986 Materials Research Society.

It is probable that non-polar species encounter difficulties to diffuse through the amorphous silica. Thus, the oxidation rate of SiC in H_2O seems to be controlled by the outward diffusion of CH_4, or by the separate diffusion of carbon and hydrogen, if CH_4 has dissociated in the amorphous silica. The hydrothermal oxidation of SiC could proceed by two simultaneous reactions (43) and (44):

$$SiC + 2H_2O \rightarrow SiO_2 + CH_4 \qquad (43)$$

$$CH_4 + 2O_2 \rightarrow CO_2 + 2H_2O \qquad (44)$$

At a high temperature, reaction (22) is more stable than the reaction (45), and water could oxidize H_2 (46):

$$SiC + 4H_2O \rightarrow SiO_2 + CO_2 + 4H_2 \qquad (45)$$

$$4H_2 + 2O_2 \rightarrow 4H_2O \qquad (46)$$

Thus, the oxidation of SiC in hydrothermal medium proceeds via a two-step mechanism instead of a one-step reaction (47):

$$SiC + 2O_2 \rightarrow SiO_2 + CO_2 \qquad (47)$$

Yoshimura et al. noticed that the activation energy under 10 MPa (194 kJ/mol) was higher than the one calculated at 100 MPa (167 kJ/mol). Indeed, the oxidation rates were accelerated when high water pressure is used, and these two values were slightly smaller than for dry oxidation (above 200 kJ/mol). Therefore, it was concluded that, firstly, the water pressure did not affect the oxidation mechanism and, secondly, that the diffusing species might differ under dry oxygen and pressurized water.

The reason for this is that H$_2$O diffuses faster than oxygen in the silica layer, so the kinetics of reactions (43) plus (44) or (45) plus (46) would be higher than for the reaction (47). This could be the key element for higher oxidation of SiC under hydrothermal conditions. However, Hirayama et al. proposed that it might be due to the formation of hydrosoluble silica scale.

4.1.2. Hirayama's Model for SiC Corrosion in Water Vapor

In 1989, Hirayama et al. [93] investigated the corrosion behavior of silicon carbide ceramics after immersion in 290 °C water solutions with different pH. After a 72-h exposure, the exposed α-SiC materials demonstrate a higher dependency on pH for weight loss in oxygenated water than in deoxygenated water. Figure 26 relates their results:

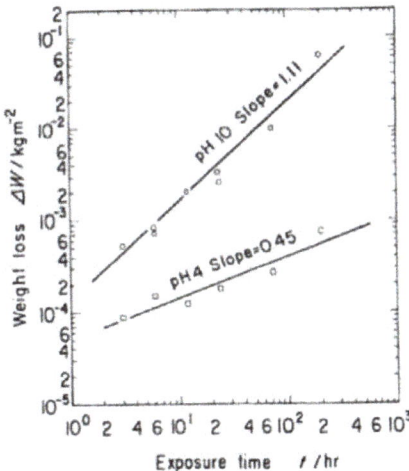

Figure 26. Weight change–time diagram for SiC ceramics immersed in an oxygenated solution (dissolved [O$_2$] is 32 ppm) at 290 °C. Reprinted with permission from Reference [93]. Copyright 1989 The American Ceramic Society/Wiley.

From the model of SiC dissolution presented by Yoshimura et al., the following reaction is expected to occur [92]:

$$\text{SiC} + 2\text{H}_2\text{O} \rightarrow \text{SiO}_2 + \text{CH}_4 \tag{48}$$

This reaction contributes to the weight gain because solid SiC changes into solid SiO$_2$. However, Hirayama et al. found out that the weight of the materials decreased with the exposure time, as silica easily dissolves in alkaline water, following the below-mentioned reaction (46):

$$\text{SiO}_2 + \text{H}_2\text{O} = \text{H}_2\text{SiO}_3 = \text{HSiO}_3^- + \text{H+} = \text{SIO}_3^{2-} + 2\text{H}^+ \tag{49}$$

Thus, two protons are generated, so the reaction is pH-dependent. The higher the pH values, the more the reaction can be accelerated, and the larger the weight loss. Indeed, a high dissolution rate of SiO$_2$ films is recorded for high pH values. Moreover, when oxygen is dissolved in the solution, reactions (50), (51 and (52) seem to participate in the SiC oxidation:

$$\text{SiC} + 2\text{O}_2 \rightarrow \text{SiO}_2 + \text{CO}_2 \tag{50}$$

$$\text{CH}_4 + 2\text{O}_2 \rightarrow \text{CO}_2 + 2\text{H}_2\text{O} \tag{51}$$

$$\text{CH}_4 + 2\text{O}_2 \rightarrow 2\text{H}^+ + \text{CO}_3^{2-} + \text{H}_2\text{O} \tag{52}$$

The reaction (50) and (48) contribute to weight gain. The reaction (52) can shift the equilibrium of reaction (48) to the right and then contributes to weight gain rather than losses. However, the authors are not able to explain the important weight loss generated by oxygenated water in comparison with deoxygenated water. As SiO_2 is not identified on the surface (SEM, X-ray or AES), no production of SiO_2 is assumed, and another dissolution model is proposed, in which a poorly adherent layer, such as $Si(OH)_4$, is produced on the surface:

$$SiC + 4H_2O \rightarrow Si(OH)_4 + CH_4 \tag{53}$$

with the dissolution reaction (53) as follows:

$$Si(OH)_4 \rightarrow H_3SiO_4^- + H^+ \rightarrow H_2SiO_4^{2-} + 2H^+ \tag{54}$$

In the case of oxygenated solution, the oxidation of SiC is as follows:

$$SiC + 2O_2 + 2H_2O \rightarrow Si(OH)_4 + CO_2 \tag{55}$$

Again, the CO_2 produced dissolves in water as follows:

$$Si(OH)_4 + CO_2 + H_2O \rightarrow H_3SiO_4^- + HCO_3^- + 2H^+ \rightarrow H_3SiO_4^- + CO_3^{2-} + 4H \tag{56}$$

This model demonstrates that reaction (53) is linked to the equilibrium of reaction (54) in deaerated solution, and the reaction (56) dominates the equilibrium of reaction (55), in aerated solution. Therefore, the reactions (54) and (56) contribute to weight loss because of the dissolution of the substrate. As the pH increases, the equilibrium of reactions (53) and (55) shifts to the right, so more proton ions (H^+) are released. In the case of oxygenated water, the oxidation reaction (56) produces four H^+, whereas for deoxygenated water, only two are released by the reaction (54). This results in higher weight loss for oxygenated solution.

The new model is illustrated in Figure 27:

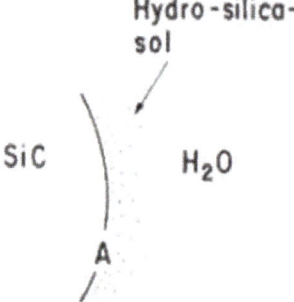

Figure 27. Dissolution model, assuming that a hydro–silica–sol is produced. Reprinted with permission from Reference [93]. Copyright 1989 The American Ceramic Society/Wiley.

The overall hydrothermal oxidation reactions with SiC materials in alkaline medium are as follows:

— For a deoxygenated solution:

$$SiC + 4H_2O = H_2SiO_4^{2-} + 2H^+ + CH_4 \tag{57}$$

— For an oxygenated solution:

$$SiC + 2O_2 + 2H_2O = H_3SiO_4^- + CO_3^{2-} + 4H^+ \tag{58}$$

This model shows that the hydrothermal oxidation of SiC materials is not able to produce a protective silica layer to prevent further oxidation. Furthermore, multiple oxidation reactions occur with oxygen as the primary oxidant species.

The conclusion is that the production of a non-protective hydrosoluble silica layer is the key element which allows for the understanding of the high oxidation rate of SiC in water. Furthermore, the multiple roles of oxygen are proof of high synergy between water and oxygen, which leads to higher oxidation rates. Finally, as these hydrothermal reactions are pH-dependent, they are not considered as oxidation reactions but instead related to hydrolysis. This point is discussed in the following section on Allongue's model.

4.1.3. Allongue's Model for Dissolution of Silicon in Liquid Water

An in situ scanning tunneling microscopy (STM) is used to investigate the etching of Si (111) in alkaline solutions [94]. During the dissolution process, the silicon surface is covered with Si-H bonds followed by the formation of etch pits. The etch rates are deduced from the weight loss of the material and a model of chemical dissolution of Si-crystals is designed [95].

In this model (Figure 28), the dissolution occurs at a kink site, (structure A on Figure 28). A kink site is a defect in the crystal where one Si atom cannot be involved within four bonds with other Si atoms. As this site is located at the edge, the kink Si atom has only two bonds linked to the crystal lattice while the two other bonds are involved in Si-H bonds. The structure A evolves into C, and then D and A' by successive hydrolysis. At the end, the last product is decomposed to give $Si(OH)_4$.

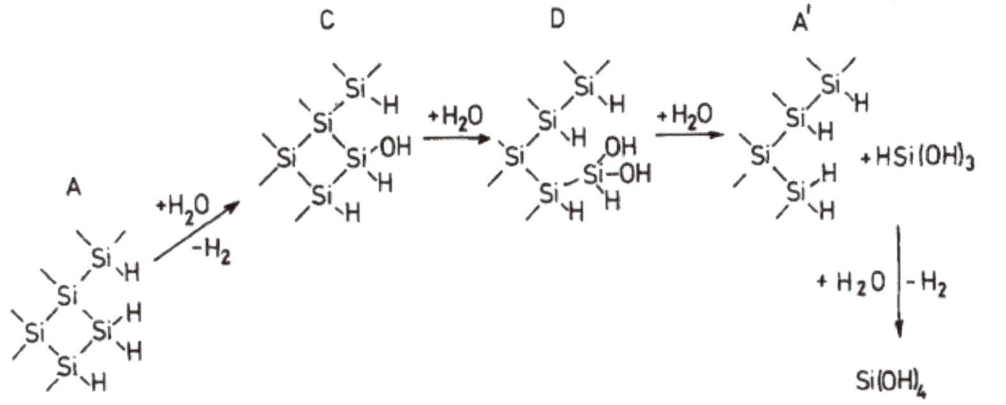

Figure 28. Main reaction path for the chemical dissolution of Si. Reprinted with permission from Reference [94] Copyright 1993 IOP.

These successive steps are described in detail:
First, the direct Si-Si backbones hydrolysis occurs (59):

$$Si\text{-}Si + H_2O \rightarrow Si\text{-}H + Si\text{-}OH \tag{59}$$

The first step of Si-H bonds hydrolysis released H_2 as follows (60):

$$Si\text{-}H + H_2O \rightarrow Si\text{-}OH + H_2 \tag{60}$$

The third step of hydrolysis produces $Si(OH)_2$ (61):

$$Si\text{-}OH + H_2O \rightarrow Si(OH)_2 \tag{61}$$

Then, the secondary product HSi(OH)$_3$ is obtained (62):

$$Si(OH)_2 + H_2O \rightarrow HSi(OH)_3 \tag{62}$$

A second H$_2$ molecule is produced by the decomposition of the primary product HSi(OH)$_3$ in solution (63):

$$HSi(OH)_3 + H_2O \rightarrow Si(OH)_4 + H_2 \tag{63}$$

The final product entering the solution is silicic acid, Si(OH)$_4$, or its equivalent dissolution product in alkaline solution (64):

$$Si(OH)_4 + H_2O \rightarrow [SiO_4H_2]^{2-} \tag{64}$$

So, the overall reaction produces the silicate oxyanions and releases hydrogen (62):

$$Si + 2HO^- + 2H_2O \rightarrow [SiO_4H_2]^{2-} + 2H_2 \tag{65}$$

Regarding the kinetics of dissolution, the limiting step could be the A→C or A→D chemical paths.

The key element for all these reactions to occur is that the surface is exhibiting H terminations (see Figure 29). Therefore, a positive charge is induced by the polarization of the Si bond with OH or H ligands. This leads to the attachment of the OH ligand on the kink atom site, and the H ligand on the Si atom is underneath when the Si-Si backbone is hydrolyzed.

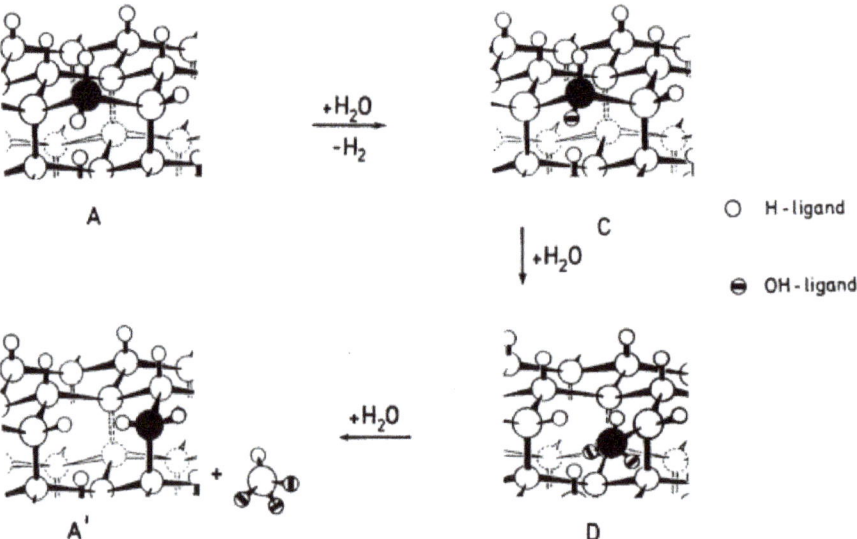

Figure 29. Three-dimensional model for the chemical dissolution of a kink Si atom (black), from the structure A to A'. Reprinted with permission from Reference [95]. Copyright 1993 IOP.

During the consecutive hydrolysis steps of silicon, the addition of OH ligand induces the polarization of the Si-OH bond, which dislocates the kink site atoms. Due to this, is the O-H ligand quickly removed from the lattice.

Finally, the OH$^-$ ions catalyze the hydrolysis of Si-Si and Si-H bonds (59) and (60) and allow for the dissolution of hydrolysis products. Hydroxyl ions then tend to enhance the hydrolysis of Si.

This model demonstrates good agreement with the Hirayama's model of SiC dissolution. Indeed, if silicon is substituted by SiC in Equation (65), the hydrolysis reactions will lead to the release of methane gas instead of hydrogen (66):

$$SiC + 2HO^- + 2H_2O \rightarrow [SiO_4H_2]^{2-} + CH_4 \tag{66}$$

If only water molecules participate to the hydrolysis, Equation (67) is obtained:

$$SiC + 4H_2O = [SiO_4H_2]^{2-} + 2H^+ + CH_4 \tag{67}$$

Now, one can see that this equation is equivalent to the hydrothermal oxidation expressed by Hirayama's model for deoxygenated solutions. Therefore, its validity is confirmed.

4.1.4. Conclusions

According to the Yoshimura's model, a silica scale is formed during SiC corrosion through which oxidant species and volatile by-products diffuse. He concluded that water diffused faster than oxygen due its diffusion reaction into oxide scales, according to the Doremus model.

In opposition to that, Hirayama's model does not consider the formation of a silica scale on the SiC surface. Several reactions were expressed for which hydrosoluble silica is produced but neither protective nor non-protective silica were identified.

Allongue's model suits Hirayama's model well. Indeed, water is found to hydrolyze silicon by a five-step mechanism with the production of silicic acid. Still, what can be considered about a hydrolysis mechanism that is valid for SiC interface?

Indeed, a microstructure study can reveal if the corrosion mechanism is the same for silicon and silicon-containing materials. In the next section, the corroded surface of such materials is compared.

4.2. Hydrothermal Corrosion of the Surface

Allongue et al. conducted in situ STM etching of Si (111) face in alkaline solution [94] and reported the etching mechanism of the Si double layer steps. In fact, the Si (111) face consists of smooth terraces separated by a single double layer of 3.14 Å height, at the edge of which kink sites are found.

The first observation is that silicon can be hydrolyzed by water and this reaction is catalyzed by hydroxyl ions. At high pH values, the hydrolysis becomes dominant over the electrochemical mechanism responsible for the Si corrosion.

The second observation is that at negative bias, the Si etching proceeds by shrinkage of the terraces and an increase of the surface roughness. This so-called "step-flow" mechanism competes with the nucleation of etch pits at cathodic bias. The corrosion evolution of the Si surfaces is reported on the Figures below, as in Figure 30. The picture on the left side shows the smooth terraces of the starting Si materials. As the corrosion occurs, pit formation and growth are observed on the terraces, as it is seen in the center image. After the coalescence of the etch pits, the corroded surface reveals sharp edges. The surface topography on the rightmost image is consistent with corrosion etching via a step-flow mechanism and etch pit formation.

Corrosion pitting is also highly dependent on the chemical environment, along with the homogeneity and morphology of materials. Smialek et al. showed that the hot corrosion of α-SiC proceeded via preferential attack at structural discontinuities, whereas excessive corrosion of pressures sintered β-SiC produced bubbles and resulted in high-roughness surfaces due to high carbon content [96]. Indeed, Henager et al. claimed that the preferential formation of bubbles or localized corrosion along grain boundaries or aligned with grains is due to higher dissolution rates at boundaries [97].

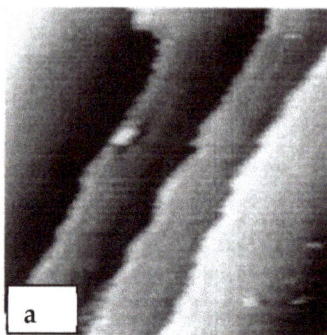 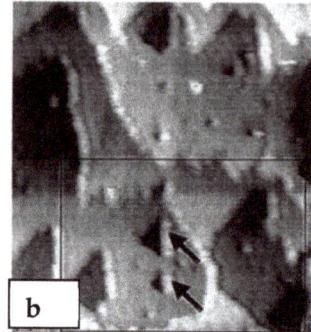 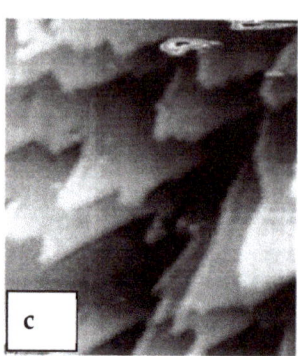

Figure 30. In situ STM observation of the surface of n-Si(111) in 2 M NaOH solution under relatively strong hydrogen evolution of 200 µA/cm^2 (**a**) Image (380 × 380 Å^2) of the starting material. The following tunneling conditions: $U_S = -0.63$ V/Pd-H; $I_S = -150$ µa/cm^2; tip: $U_T = +300$ mV/Pd; $IT = 0.2$ nA (**b**) Image (1076 × 1193 Å^2) of the growth of triangular etch pits which is followed by their coalescence. The following tunneling conditions: $U_S = -0.63$ V/Pd-H; $I_S = -150$ µa/cm^2; tip: $UT = +300$ mV/Pd; $IT = 0.2$ nA (**c**) Image (1280 × 1470 Å^2) at the end of corrosion under the same tunneling conditions as (**b**) and terraces go upwards from the bottom left to the upper right. Reprinted with permission from Reference [94]. Copyright 1992 Elsevier.

After exposure of CVD-SiC materials to deoxygenated water at 300 °C and 10 MPa the surfaces revealed both embryonic and large pits, after initial exposure for 4000 h. The small pits could have been produced by the formation of volatile carbon species, such as CO, CO_2 or CH_4 that act to disrupt the surface silica layer. Locally, the CO and CO_2 would increase the acidity and accelerate the corrosion process, resulting in the formation of local pits due to the agglomeration of small pits on the surface.

Therefore, pitting degradation affects the mechanical properties of SiC materials, and there is a need to identify which combustion conditions lead to the highest strength reductions

Hirayama et al. carried out the first microstructural studies of SiC materials revealing that corrosion occurred at grain boundaries and suggested a preferential intergranular attack [93]. Kim et al. confirmed this observation by performing corrosion experiments on sintered SiC (SSiC) and chemically deposited SiC (CVD-SiC) ceramics, in distilled water at 360 °C and over 10 days [90]. As the CVD-SiC has higher purity than SSiC and clean grain boundaries, the sample showed a stronger corrosion resistance.

In fact, in SiC samples, boron and carbon atoms are known to be segregated at grain boundaries, and during corrosion, these impurities are rapidly attacked. As a result, the disintegration of grains into water occurred due to the weakening of grain boundaries at extended corrosion times. On the contrary, the CVD-SiC specimen underwent low weight loss and the corroded surface showed feature like the columnar growth pattern of the SiC deposits, it can be seen on Figure 31, below:

No SiO_2 or $Si(OH)_4$ amorphous layer was observed because of the very thin thickness of the film or due to its instability. According to Hirayama, Kim et al. suggested that the oxidation of SiC material produces a poorly adherent film: $Si(OH)_4$ instead of SiO_2, in high-temperature water. However, Barringer et al. [98] recorded lower linear mass loss for CVD-SiC exposure to deoxygenated water at 360 °C and assumed that the control of oxygen helped to reduce corrosion rates. Finally, the corrosion behavior of reaction-bonded silicon carbide (RBSC) in pure water at 360 °C showed an increase of the weight loss with the amount of free Si atoms in the material [99].

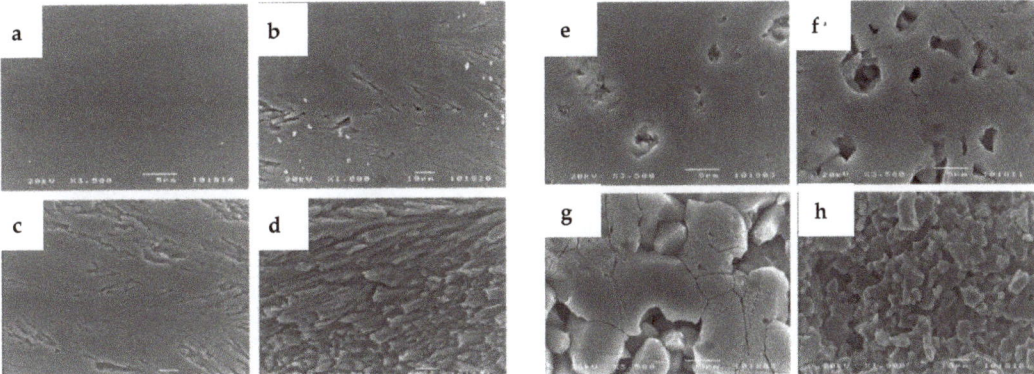

Figure 31. Surface microstructures of the (**Left**)—(**a**) as-polished and corroded SSiC specimens after corrosion testing for (**b**) 1, (**c**) 5 and (**d**) 7 days in water at 360 °C and (**Right**)—(**e**) as-polished and corroded CVD SiC specimens after corrosion testing for (**f**) 3, (**g**) 7 and (**h**) 10 days in water at 360 °C. Reprinted with permission from Reference [91]. Copyright 2003 Springer Nature.

In Figure 32 below the light and dark gray phases represent the Si phase and SiC particles, respectively. During corrosion, the Si-free phase was preferentially corroded in high-temperature water, which led to the formation of large voids in the material.

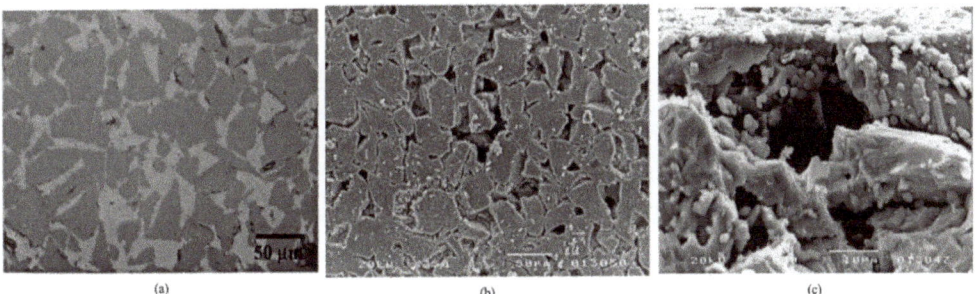

Figure 32. Optical micrograph of the RBSC specimen before corrosion test (**a**) and SEM micrographs of the surface (**b**) and the cross-section (**c**) of the same specimen after corrosion at 360 °C for 7 days. Reprinted with permission from Reference [99]. Copyright 2002 Springer Nature.

Recently, Park et al. [100] carried out long-term corrosion tests on high-purity CVD-SiC specimens in water and steam up to 400 °C. They showed that the dissolved oxygen content was the most dominant factor controlling the corrosion of SiC in water. The grain boundary dissolves preferentially during the early stage of corrosion. The grains detach from the surface when the grain boundaries become thinner, thereby leading to an acceleration of the weight loss.

In Figure 33, above, the top view shows significant dissolution of the SiC throughout the surface. In the side view, the shapes of many grains are exposed because of the preferential dissolution at the grain boundary during the corrosion process. These observations were confirmed by Tan et al. [101] who exposed CVD-SiC materials to 500 °C water under 25 MPa (Figure 34). The image of strain distribution and Image Quality revealed that corrosion primarily occurred at regions with high intensity of strains associated with small grains.

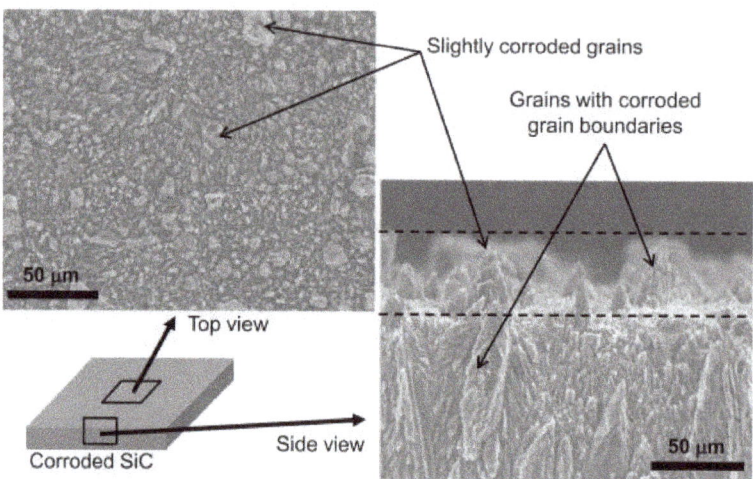

Figure 33. Scanning electron micrographs of the surface morphology of the CVD Si corroded in the 360 °C static water autoclave for 90 days. Reprinted with permission from Reference [100]. Copyright 2013 Elsevier.

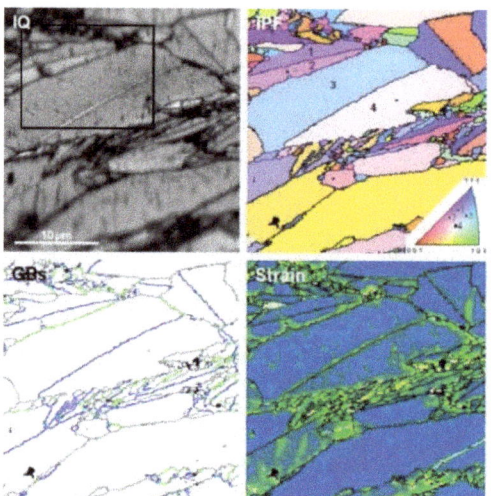

Figure 34. Electron Back-Scatter Diffraction (EBSD) analysis of the cross-section surface (28 × 28 μm) of the sample exposed to the SCW with 10 ppb oxygen for 333 h. GBs denote grain boundaries with blue, green and gray lines denoting three different network, and general boundaries. Reprinted with permission from Reference [101]. Copyright 2009 Elsevier.

It was concluded that the pre-oxide was not able to protect the SiC material from dissolving into the water, but it helped to mitigate the dissolution of the SiC.

To conclude on the microstructure study of SiC corrosion, Barringer et al. [96] believed that corrosion occurs via hydrolysis to hydrated silica species at the surface, which are then rapidly dissolved into the water. This conclusion is based on no analytical methods revealing the presence of a measurable oxide scale. In addition to that, a reduction of the amount of oxygen present at the surface was recorded as the exposure time increased. This may be consistent with the removal of Si atoms to form $Si(OH)_4$ via water hydrolysis, which allows the Allongue's model to be extended to SiC materials. Finally, Henager et al. showed that pits resulted from the formation of a local galvanic cell, which was due to hydrolysis reactions involving the dissolution of silica and the formation of carbon [97]. However, tribocorrosion process can lead to such observations—the next section details the tribochemical corrosion of SiC surfaces.

4.3. Reaction Model for Tribochemical Corrosion

Under dynamic conditions, the exposure of non-oxide ceramics to oxidizing atmosphere leads to the formation of an oxide film and it is referenced as "tribo-oxidation" whereas the formation of a chemical layer under hydrothermal conditions is referenced as "tribochemistry". Tribo-oxidation leads to the modification of SiC physical properties, and, according to the wear oxidation model of Quinn, tribo-oxidation is a consequence of a local temperature rise when the sample undergoes friction [102,103].

Both action of friction and temperature (because of fast kinetics) influence the tribochemical SiC degradation. Some studies reveal that the higher the humidity, the lower the coefficient of friction and wear rates [104,105] and Zum et al. demonstrated that when SiC is submitted to water, tribochemical polishing of its surface is expected due to boundary lubrication [106].

Therefore, water seems to affect the tribochemical behavior of SiC. However, the effect of water, including its temperature and pressure, is not clear. Hence, Presser et al. studied the hydrothermal behavior of SiC materials and its wear properties evolution [107] by conducting degradation experiments under static (at 500 °C up to 700 MPa) and dynamic (tribological) conditions in water. They showed that the smoothing process of SiC was responsible for the decrease of the surface roughness and the increase of the sliding behavior, which resulted in the decrease of the friction coefficient. As well, silicon leaching and amorphization of the surface occurs under tribological tests in water medium (without pressure).

However, the hydrothermal degradation leads to pitting formation and an increase of the roughness with bubbles and silica precipitates which agrees with the observations of Barringer et al. [98]. Finally, a tribochemical wear model is designed into three steps (see Figure 35):

1. Interaction between SiC and water: The bulk reaction leads to the formation of OH groups and to the saturation of dangling bonds. Moreover, weak hydrogen bridges are created.
2. Amorphization: Initially, mechanical stress causes the superficial amorphization of SiC. Therefore, disordered layers and strained Si-C bonds might form with higher susceptibility to be attacked by water. However, neither silica nor oxycarbidic phases were identified knowing that the detection limit of XRD and Raman spectroscopy is about 100 nm.
3. Tribochemical corrosion: Simultaneously, silica dissolves in water, and, for low water-to-SiC ratios, it precipitates. Likewise, a cavitation-like wear phenomenon created by the release of gaseous compounds can cause the delamination of the layer.

In conclusion, this is another corrosion mechanism can lead to the degradation of SiC materials under hydrothermal conditions. This corrosion, tribocorrosion, is enhanced by water, as the dissolution of amorphous SiC is possible by the formation of hydrosoluble silica and by the release of volatile carbonaceous species. Simultaneous hydrothermal corrosion also occurs at grain boundaries.

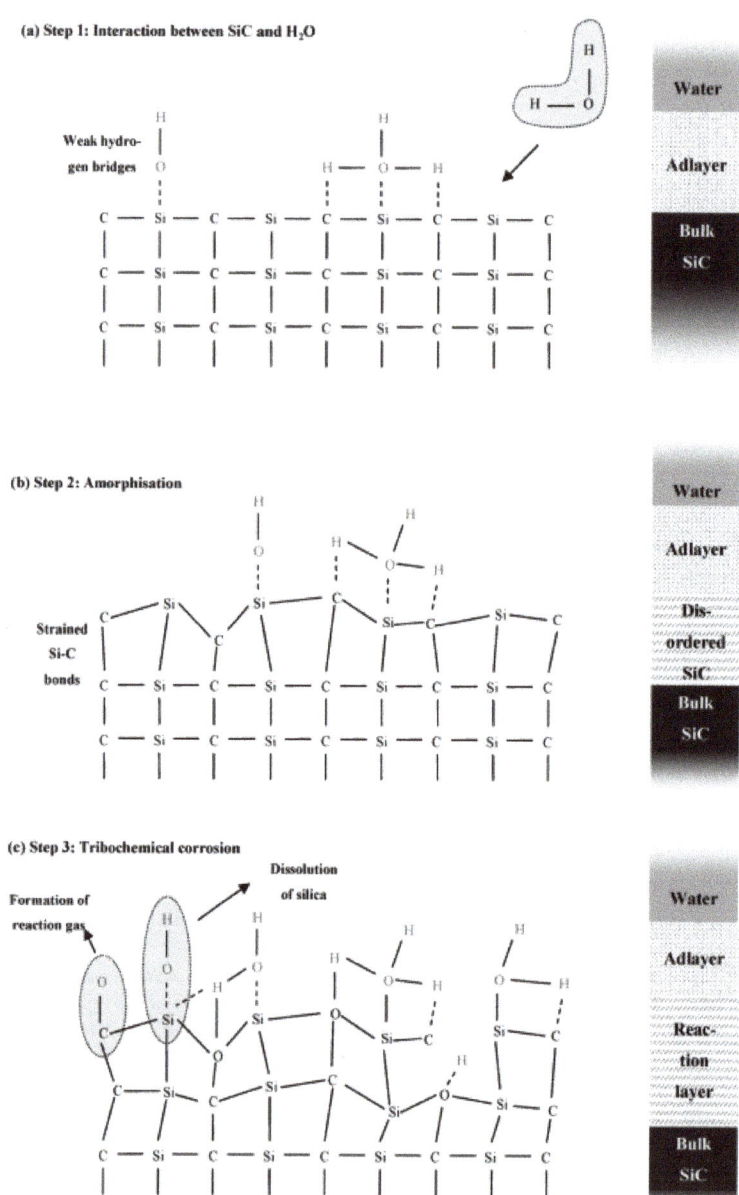

Figure 35. Three-step tribomodel of SiC corrosion. Reprinted with permission from Reference [107]. Copyright 2009 Elsevier.

4.4. Conclusions

From the microstructural studies, the etching of Si materials seems similar to that of SiC materials. Indeed, the degradation of silicon-containing materials occurs via pit formation and growth on the surface, which results in high weight losses. The corroded surface shows a high degree of pitting corrosion via preferential attack at the grain boundary, which highly depends on the chemical environment and the morphology of SiC samples. For CVD-SiC, typical columnar corrosion shape is observed, and for SSiC, the etching of carbon and boron atoms occurs locally on its surface. Finally, the degradation process depends on the sample purity: CVD-SiC samples show high corrosion resistance, whereas the presence of Si-free atoms in the RBSC sample leads to a high degradation rate. Moreover, the etching of Si-free atoms is thermodynamically favored, as the corrosion products are stable (silica easily dissolves in water); however, several authors were not able to detect the dissolved silica [98–100]. Nevertheless, these authors agreed with the Hirayama's model of SiC dissolution by water [93] but not with the one of Yoshimura which suggests the formation of a silica scale. These findings also agree with the Allongue's model for the dissolution of silicon crystals. Indeed, it appears that successive hydrolysis of the Si-Si, or Si-C bonds in the case of SiC materials, causes the formation of OH groups and leads to the etching of Si atoms.

In conclusion, the corrosion behavior of SiC materials seems complex, as several chemical reactions compete, and different mechanisms are taking part in the degradation of the materials, as is suggested by the tribochemical corrosion model.

In the final part of the review paper, the water corrosion of these materials under sub- and supercritical conditions is examined. Several chemical reactions are in competition depending on the temperature and pressure conditions along with the molar ratio of $H_2O:SiC$. Indeed, hydrothermal oxidation leads to a silica layer formation whereas hydrolysis reaction leads to a carbon film formation on top of SiC substrates (powders, fibers and substrates).

5. Supercritical Water Corrosion of Silicon Carbide Materials

5.1. Supercritical Water Characteristics

High temperature and high-pressure medium needs to be defined to have an insight of the interesting properties that a fluid can demonstrate under these particular conditions. Critical coordinates are defined for all pure components and represent the beginning of the supercritical domain. By increasing the temperature and the pressure, the equilibrium state of water evolves as it is shown on the diagram below:

As seen on Figure 36, the three states of water are in equilibrium at the triple point defined for $T_{TP} = 0.01\ °C$ and $P_{PT} = 612$ Pa.

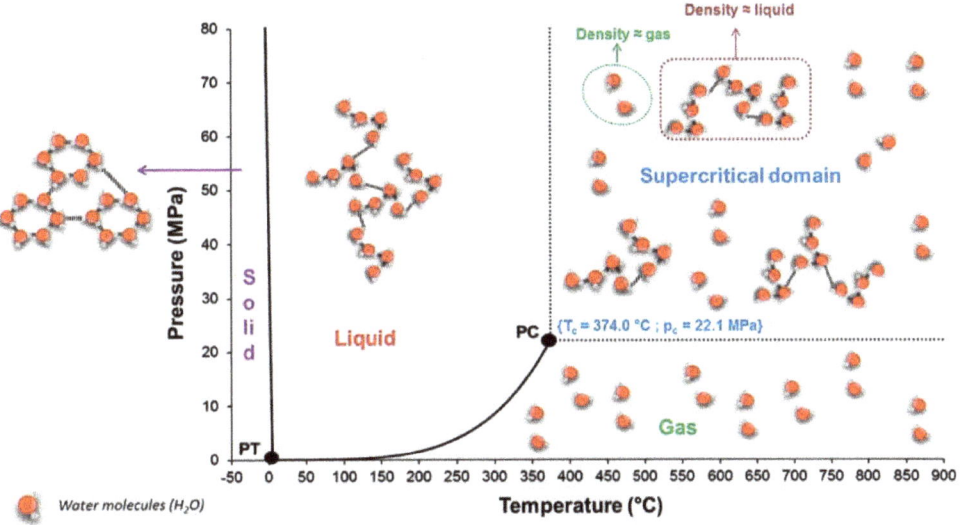

Figure 36. Pressure–temperature phase diagram of pure water. TP is the triple point ($T_{TP} = 0.01\ °C$, $P_{PT} = 612$ Pa) and CP is the critical point ($T_{CP} = 374.1\ °C$, $P_{CP} = 22.1$ MPa). Water molecules are draw schematically (planar view) for each state—solid, liquid and gas—and in the case of the supercritical conditions. Reprinted with permission from Reference [108]. Copyright 2010 Wiley.

At high temperature and high pressure, the liquid-vapor equilibrium curve ends when the supercritical point of water is reached (T_{CP} = 374 °C and P_{CP} = 22.1 MPa). From there, the matter is called fluid as it demonstrates physicochemical properties in between those of liquid and gas phases.

Indeed, the main characteristics of supercritical fluids (SCF) are low viscosity and no surface tension, which gives rise to high diffusion properties. The SCF region shows discontinuities, as some micro-domains demonstrating liquid-like properties coexist with others which exhibit gas-like properties. Finally, the fluid density evolves linearly with the increase of temperature and pressure.

Under high pressure and high temperature conditions, the dielectric constant and the density of water sharply decrease. This can be seen with the increase of temperature at a constant pressure of 24 MPa in Figure 37. However, the ionic product reaches a maximum at 300 °C before falling, similar to other parameters.

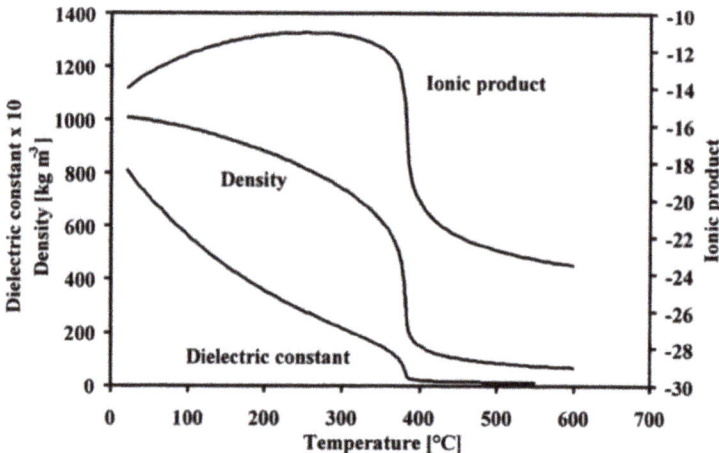

Figure 37. Course of ionic product, density, and dielectric constant of water versus temperature at a pressure of 24 MPa. Reprinted with permission from Reference [109]. Copyright 1999 Elsevier.

In other terms, salt dissolution ability of water decreases with the density due to the increase of temperature. The distance between water molecules increases, which weakens the hydrogen bonds and dipole electrostatic interactions. Therefore, low-density water acts as a non-polar solvent, in which salts precipitate, in the supercritical region [110]. However, for a constant temperature of 400 °C, the density, ionic and dielectric constants increase with the pressure. High-density water demonstrates high solvency for inorganic compounds but keeps its solvent-like properties [108]. These observations evidence the high tunability of water properties by playing with the temperature and/or pressure parameters.

Table 7 highlights the existence of another region called the sub- or near critical region which defines water in the liquid state for a temperature range of 250 °C < T < 450 °C and P < Pc [111].

Table 7. Physicochemical properties of water as a function of temperature and pressure. Reprinted with permission from Reference [111]. Copyright 1999 Wiley.

	"Normal Water"	"Subcritical Water"	"Supercritical Water"		Superheated Steam
T [°C]	25	250	400	400	400
P[MPa]	0.1	5	25	50	0.1
ρ[g·cm^{-3}]	0.997	0.80	0.17	0.58	0.0003
ε	78.5	27.1	5.9	10.5	1
pK$_w$	14.0	11.2	19.4	11.9	/
C$_p$[kJ·kg^{-1}·K^{-1}]	4.22	4.86	13	6.8	2.1
η[mPa.s]	0.89	0.11	0.03	0.07	0.02
λ[mW·m^{-1}·K^{-1}]	608	620	160	438	55

In the subcritical region, water has infinite compressibility, and, near 250–300 °C, the ionic constant reaches its maximum. Consequently, water acts as an acid or base catalyst and ionic reactions are favored over radical ones. Thus, the electrochemical corrosion is enhanced. Moreover, high-density water and high temperature allows fast reaction kinetics which makes the sub- and supercritical region a high corrosive medium. Kritzer illustrated the corrosion-determining factors and their interdependences below (Figure 38) [109]:

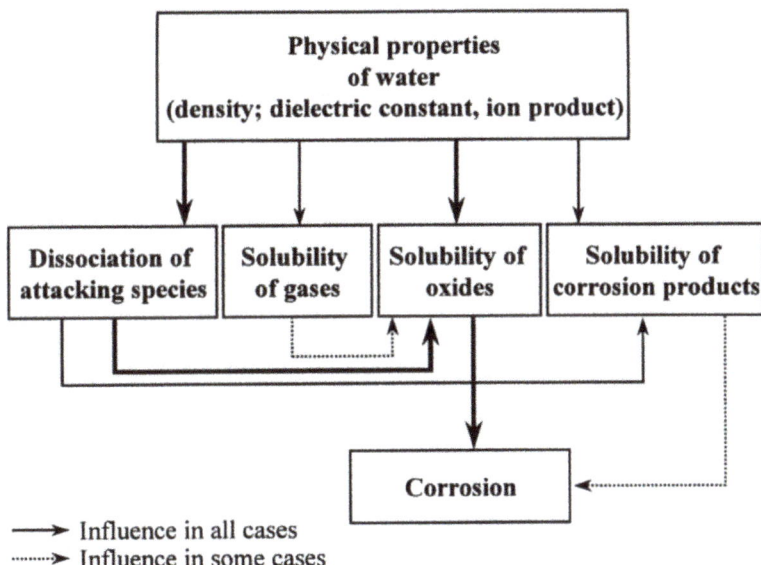

Figure 38. Interdependences of the corrosion-determining factors in high-temperature aqueous solutions. Reprinted with permission from Reference [109]. Copyright 1999 Elsevier.

The most used supercritical fluids are water, alcohol and carbon dioxide but the interest for supercritical ammonia and halogenated gases is increasing [112]. Supercritical fluids fill applications to address environmental challenges for wastewater treatment [113] chemical recycling [114] and extraction process [115]. SCFs are involved in organic reactions, for example, in catalyst [116,117], and polymer processing [118]. Finally, SCFs take part in material engineering for synthesis [119,120], surface treatment [121], microfluidics [122] and recently, in nanotechnology [123] and nanomedicine [124].

To conclude, the chemical versatility of sub- and supercritical medium is nearly infinite, as the water properties can be tuned by varying the temperature and pressure conditions. Hence, water can demonstrate high solvent power for organic, salts and gases, along with varying properties from liquid-like to gas-like. This section aims to highlight the different roles of water and its interaction with the matter when using high temperature and high-pressure conditions. More particularly, the interaction between silicon carbide and high pressure/high temperature water is discussed.

5.2. Reaction Model of the SiC Hydrothermal Oxidation

The work of Gogotsi and Yoshimura consists of the application of hydrothermal technology to SiC fibers from the studies of Hirano et al. [125] and Yoshimura et al. [92].

Hirano shown that the hydrothermal treatment of SiC at 900 °C and 1000 MPa water pressure enhanced graphitization of carbonaceous materials by the formation of intermediates and gaseous compounds.

Yoshimura et al. studied the interaction of SiC materials with H_2O under hydrothermal conditions between 400 and 800 °C, at pressures up to 100 MPa and for 72 h and found that the effects of water vapor on the oxidation of SiC powders was of interest, but no model was hypothesized on the reaction of H_2O and SiC [92].

Surprisingly, Gogotsi and Yoshimura demonstrated that the corrosion of SiC materials (α-SiC single crystals, α-SiC platelets, α and β-SiC powders and β-SiC whiskers) leads to the formation of carbon film above 500 °C through the following relation (68) [126,127]:

$$SiC + 2H_2O \rightarrow SiO_2 + C + 2H_2 \tag{68}$$

Gogotsi and Yoshimura worked on the oxidation and hydrothermal corrosion of SiC by first investigating the hydrothermal corrosion of SiC-based fibers at high temperature and high-pressure water [127]. The experiments were performed on Tyranno fibers under 100 MPa in distilled water at 300–600 °C for 25 h. It was observed that above 300–450 °C, a smooth and uniform carbon film was grown via the following reaction (69):

$$SiC_xO_y + nH_2O \rightarrow SiO_2 + xC + nH_2 \tag{69}$$

In this reaction, the free carbon is likely responsible for the formation of carbon films. CVD is traditionally used to prepare carbon films and resulted in a decrease of strength. However, this new and inexpensive method allows for the transformation of the surface layer of carbides into carbon instead of depositing a film from a solution. The two major advantages of this technique is that the carbon films can be grown on top of different kind of SiC materials (powders, platelets, fibers and single-crystals) and its thickness can vary from 10–20 nm up to 1–2 μm, depending on the experimental conditions. In fact, the following parameters strongly influence the hydrothermal treatment of SiC:

- The temperature and time of treatment can affect the composition of the carbon film from amorphous to graphitic carbon,
- Above a certain temperature and reaction time, the yield of carbon reaches a maximum value,
- Above a certain temperature, the carbon film is oxidized,
- The influence of H_2O:SiC is not well understood.

Indeed, thermodynamic calculations showed that the formation of a carbon deposit on the surface of SiC follows three regimes depending on the H_2O:SiC molar ratio [128]:

1. At low H_2O:SiC molar ratios (1:10), both carbon and silica were deposited,
2. At intermediate H_2O:SiC molar ratios (2:1), both carbon and silica were produced, but silica is dissolved in the water as follows (70):

$$SiO_2 + H_2O = H_2SiO_3 = HSiO_3^- + H^+ = SIO_3^{2-} + 2H^+ \tag{70}$$

A dissolution rate of approximately 0.8 μm was recorded for glassy silica in distilled water at 285 °C under 100 MPa pressure [129], meaning that the equilibrium of reactions (68) and (69) can be shifted to the right. Thus, a carbon-rich layer is created on top of the SiC materials.

3. At higher H_2O:SiC molar ratios (10:1), neither carbon nor silica was identified on the surface of SiC (for a nanoscale detection limit) as the carbon reacts with water to form CO/CO_2 and silica dissolves in water.

Indeed, due to corrosion under supercritical conditions, water is able to oxidize the carbon coating according to the following reactions:

$$C + H_2O \rightarrow CO + H_2 \text{ (water-gas reaction)} \quad (71)$$

$$2C + 2H_2O \rightarrow CO_2 + CH_4 \quad (72)$$

$$C + 2H_2 \rightarrow CO_2 + 2H_2 \quad (73)$$

$$3C + 2H_2O \rightarrow 2CO + CH_4 \quad (74)$$

Generally, for long corrosion times and intermediate H_2O:SiC molar ratios, the carbon film at the surface is oxidized, and the oxidation is followed by silica deposition [130].

Gogotsi and Yoshimura investigated the behavior of SiC materials under dry and wet conditions. First, low-temperature oxidation (850 °C) was performed on SiC (Tyranno) fibers for up to 300 h [131]. The Tyranno fiber consisted of amorphous Si-Ti-C-O material with different oxygen contents (12 or 18 wt%), diameters (8.5 or 11 µm) and mechanical properties. This fiber contained silicon carbide crystals embedded in an amorphous matrix (Si-C-O-Ti). The oxidation started above 500 °C and led to a mass gain and an increase of the fiber diameter. Above 1200 °C, the mass gain increased sharply due to high diffusion rates of oxygen and carbon oxides through the silica layer.

In Figure 39 the SEM micrographs of S fibers oxidized in air for 300 h at 800 °C are represented.

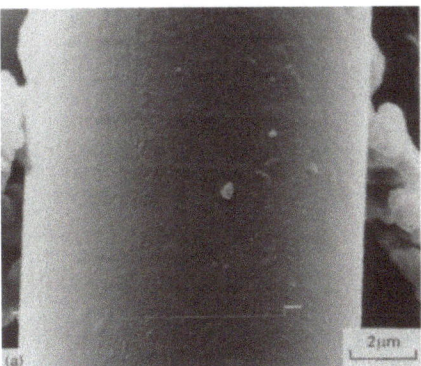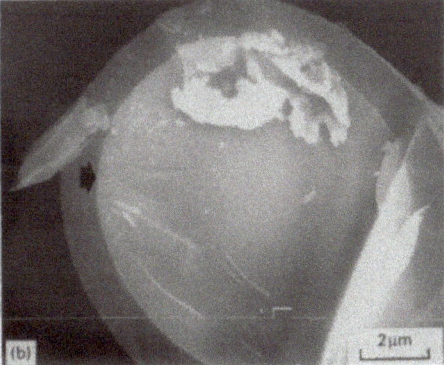

Figure 39. SEM micrographs of the surface (**a**) and fracture surface (**b**) of the S fibers oxidized for 300 h in air at 800 °C. The arrow shows the fracture origin. Reprinted with permission from Reference [131]. Copyright 1994 Springer Nature.

The thin, uniform (Figure 38a), glassy oxide layer formed (Figure 40b, in dark) was not dense and/or thick enough to limit the flaws on the surface. Moreover, the layer had low mechanical strength and showed cracking after mechanical tests (Figure 40b). As a result, the strength of the fibers decreased with increasing oxidation temperature/thickness of the oxide layer. This could be due to the decrease of the effective fiber diameter, which resulted from the oxide formation, and/or due to internal stress in the oxide layer.

As the oxide film on the surface was microscopically uniform, the oxidation is thought to be simultaneous for all constituents of the fiber [132]. Indeed, AES demonstrated the presence of a transition region at the SiC/SiO_2 interphase, where silicon oxycarbide and free carbon could be produced, in accordance with other research [133].

Secondly, Gogotsi and Yoshimura observed the degradation of SiC (Tyranno) fibers in high-temperature and high-pressure water [134] for which amorphous or microcrystalline carbon constituted the carbon rich layer. The reaction leading to the formation of carbon (75):

$$SiC + 2H_2O \rightarrow SiO_2 + C + 2H_2 \quad (75)$$

dominates over the reactions which produces silica (76), (77) and (78).

$$SiC + 2H_2O \rightarrow SiO_2 + CH_4 \quad (76)$$

$$SiC + 4H_2O \rightarrow SiO_2 + CO_2 + 4H_2 \quad (77)$$

$$SiC + 3H_2O \rightarrow SiO_2 + CO + 3H_2 \quad (78)$$

The interaction of SiC fiber with water was studied by Kraft et al. [135], between 400 and 700 °C, under 200 MPa water. However, the experiments were conducted at very low H_2O: SiC molar ratio (1:10) and the formation of silica dominates in accordance with Yoshimura et al. [92]. However, at higher molar ratios, hydrosoluble silica is formed and the experiments agree well with the Hirayama's model [92]. Finally, Gogotsi and Yoshimura demonstrated that oxidation and hydrothermal corrosion competition depends mostly on the temperature.

At 300 °C, the dissolution of the SiC_xO_y and silica phases start, as AES depth profiles revealed no oxygen atoms at the surface of the fiber. The low consumption of SiC materials does not affect its strength properties [132].

Then, at 400 °C, poor low protective carbon film is produced by the following reaction (79):

$$SiC_xO_y + nH_2O \rightarrow SiO_2 + xC + nH_2 \quad (79)$$

Thus, the oxycarbidic phase of the Tyranno fibers starts to corrode, so the tensile strength and Young's modulus decrease with the effective diameter of the fiber.

At 500 °C, the hydrothermal corrosion of crystalline SiC grains leads to the formation of thick carbon coatings (80):

$$SiC + 2H_2O \rightarrow SiO_2 + C + 2H_2 \quad (80)$$

The layer demonstrates high protective and good adhesion properties. Above 800 °C, silica starts to crystallize into cristobalite and quartz, which have lower solubility in water. Therefore, the carbon film growth is limited, and hydrothermal treatment at high temperature and longer times engender the full consumption of the fiber.

Futatsuki et al. [136] studied the oxidation of SiC surface at low temperature and high-pressure water. They explored SCW (400 °C, 25 MPa) and SubCW (350 °C, 16.5 MPa) conditions to set up a high–efficiency method of oxidation.

The density of high-pressure and high-temperature water is in the range of 100–600 kg/m³, which is greater than the oxygen density under typical thermal oxidation conditions of 1200 °C, at atmospheric pressure, which is 0.003 kg/m³. The SC medium has a high oxygen solubility, and, due to its low viscosity and surface tension, its high diffusivity contributes to a strong oxidation potential.

They found out that the oxide thickness increases with the oxidation time, according to the following reactions:

$$SiC + 3/2\, O_2 \rightarrow SiO_2 + CO \quad (81)$$

$$SiC + 2O_2 \rightarrow SiO_2 + CO_2 \quad (82)$$

$$SiC + H_2O \rightarrow SiO_2 + 3H_2 + CO \quad (83)$$

$$SiC + H_2O \rightarrow SiO_2 + 4H_2 + CO_2 \quad (84)$$

However, the oxidation of SiC was not possible without adding oxygen to water, so reactions (83) and (84) did not happen. When oxygen was added, they remarked that, for the NCW oxidation, the oxide film was very thin. Thus, the temperature might be too low to promote SiC oxidation. For SCW, oxidation occurs according to reactions (81) and (82).

Finally, the conclusions on this matter are as follows:

→ SCW can facilitate the transfer of carbon by-products (CO and CO_2) because of its high diffusivity. Therefore, the transfer of carbon by-products is one of the key factors promoting SiC oxidation.
→ For NCW, the temperature is too low for the release of carbon by-products, so no SiC oxidation occurs.

However, NCW demonstrates high ionic constant which promotes hydrolysis reaction via Si atom etching. Recently, Yoko et al. [136] recycled silicon from silicon sludge by using SCW in a semi-batch reactor (400 °C, 25 MPa). It was noticed that silicon undergoes oxidation and/or degradation via Si dissolution into water. Indeed, between 260 and 380 °C, the amount of dissolved silicon increased with the temperature, whereas between 400 °C and 500 °C, it decreased. For temperatures higher than 1000 °C, the oxidation rate of silicon dominates. This behavior may be due to the low ion product of water between 400 and 500 °C under a pressure of 25 MPa.

According to Yoko et al. [137] and Futatsuki et al. [136], it seems how oxidation and hydrolysis compete depends not only on the temperature, but also on the ionic product. Indeed, NCW between 250 and 350 °C allows the etching of Si atoms, which dominates over oxidation.

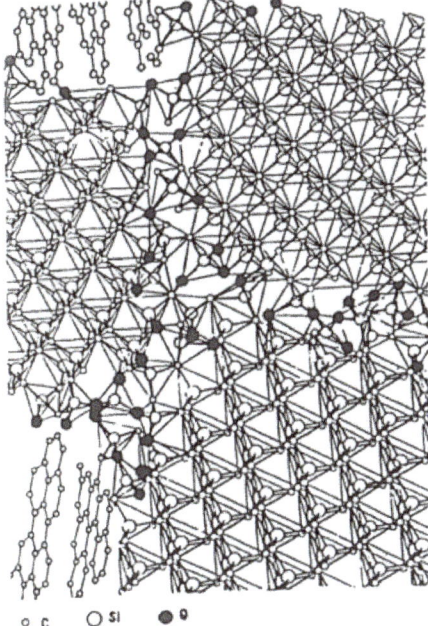

Figure 40. Texture model of the Nicalon NLM-202 fibers. Reprinted with permission from Reference [138]. Copyright 1999 Elsevier.

5.3. Nanoporous Carbon Film Formation

As SiC materials are available in a continuous-fiber shape, they are promising to integrate into ceramic matrices for thermomechanical reinforcement properties. Therefore, it is necessary not only to improve the SiC corrosion resistance, but also to tune it in order for SiC surface to demonstrate new properties.

Two different fibers made of SiC are commercially available. The Nicalon fiber is obtained by polymerization of organosilane monomers followed by pyrolysis via the Yajima's method [2]. At the end of the process, a microcomposite material is obtained which is made of a crystalline β-SiC phase (55 wt%), an intergranular oxycarbidic phase (40 wt%) and free carbon (5 wt%) [138] (Figure 40):

The other commercially available fiber is the Tyranno fiber which is made by pyrolysis of an organometallic polymer precursor synthetized from polycarbosilane and titanium tetrabutoxide [139]. At the end of the process, the continuous fiber consists of SiC crystals separated by an amorphous Si-Ti-C-O phase. Then, the Nicalon and Tyranno share the same texture model, which is described in Figure 40.

To have good reinforcement properties, the fiber should demonstrate good compatibility with the matrix. For that, a carbon coating at the fiber/matrix interface is suitable in order for the composite to have a non-fragile behavior when submitted to stress. Gogotsi et al have studied the hydrothermal treatment of carbon-based surfaces to form such a nanostructure carbon coating, as schematized below [140]. Under supercritical conditions, the selective etching of metals from carbides leads to the formation of a carbon-rich layer with tribological properties. This allows the fiber-reinforced materials to maintain high thermomechanical properties and therefore have the potential to be used in the aerospace field. The principle is represented in Figure 41.

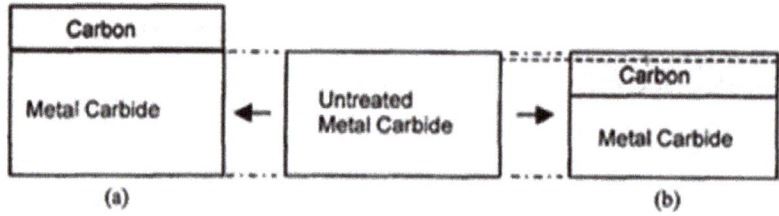

Figure 41. Carbon coating via vapor deposition (**a**) and selective leaching of carbides (**b**). Reprinted with permission from Reference [140]. Copyright 2000 Springer Nature.

Similar to the chemical or physical vapor deposition method, the carbon coating is grown on the surface, thus avoiding any adhesion issues.

Gogotsi et al. showed that sp^3-bonded carbon could be produced on the surface of SiC materials under hydrothermal conditions [141,142]. Various allotropes of carbon were obtained, including diamond, in the temperature range of 300–800 °C and at pressures below 500 MPa. Raman, infrared (IR), electron diffraction (EDS) and Auger Electron Spectroscopy (AES) were performed in addition to X-Ray Diffraction (XRD) to characterize treated and untreated materials. Depending on the shifting or the splitting of the G and D bands, their width and position, the Raman spectroscopy suggests the presence of disordered diamond and other non-graphitic carbon phases. The Raman spectra of SiC fibers before and after the hydrothermal treatment are given in Figure 42 [126]:

Graphite exhibits one Raman peak at 1580 cm^{-1} referred to as G mode, which is characteristic of the C=C stretching vibrations, whereas another peak appears at 1350 cm^{-1}, which is referred to as D mode, when defects are present. This mode also appears for nanocrystalline graphite, and its intensity increases with the hydrothermal treatment temperature. The bands can broaden if strained bonds and defects are present in the materials. For the D band down-shift, Gogotsi at al. concluded that bond angle disorder came from strained sp^3 carbon atoms. The up-shift of the G band towards the high-frequency edge, however, arose from small size crystallites, which are dominated by sp^2-hydridized carbon rather than sp^3 [143].

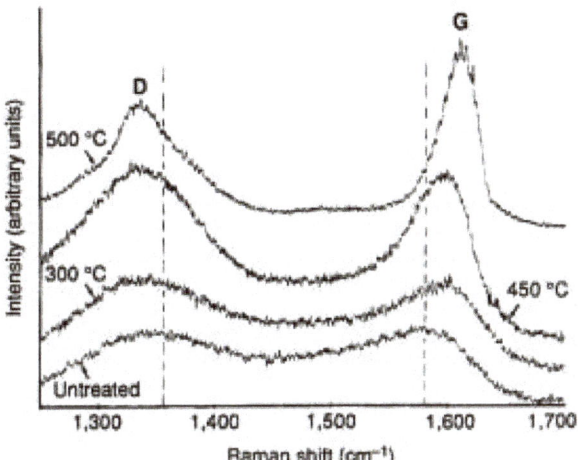

Figure 42. Raman spectra of the LoxM-garde Tyranno fibers before and after hydrothermal treatments at various temperatures. Ar laser radiation at a wavelength of 488 nm was used. Vertical dashed lines show the predicted positions of the nanocrystalline graphite line (G) at ~1580 cm^{-1} and its disorder-induced line (D) at ~1355 cm^{-1}. The shift of the observed lines indicates the presence of bond-angle disorder. Reprinted with permission from Reference [126]. Copyright 1994 Springer Nature.

The IR spectroscopy assumes that H-bonded and free sp^3-hybridized carbons were produced. The reflections at 2.06 Å was assigned to diamond on XRD patterns, whereas the narrow diamond Raman peak was missing, and the authors could not claim the presence of well-ordered cubic diamond. Indeed, the diamond sharp peak at 1332 cm^{-1} [144] is difficult to determine by conventional visible wavelength Raman spectroscopy when it is mixed with graphite. To have better sensitivity, UV excitation Raman spectroscopy could be used [145].

Finally, several key points could explain the formation of sp^3-bonded carbons instead of simple amorphous or graphitic carbon network [142]:

(1) The presence of a good substrate: by acting as a template, the cubic structure of β-SiC could allow the diamond growth,
(2) The formation of hydrogen during hydrothermal treatment of SiC suggests that diamond was produced as hydrogen plays a role in the nucleation during diamond growth,
(3) Tetrahedral carbon in SiC is believed to be transformed into diamond and not into graphite for energetical reason,
(4) Preferential oxidation of sp^2-bonded carbon by water seems to lead to the formation of carbon nuclei if the reaction (82) is replaced by the two following ones (85) and (86):

$$SiC + 2H_2O \rightarrow SiO_2 + C + 2H_2 \tag{85}$$

$$SiC + 2H_2O \rightarrow SiO_2 + C_{(Si)} + 4H^\bullet \tag{86}$$

$$C_{(Si)} + 4H^\bullet \rightarrow C_{(diamond)} + 2H_2 \tag{87}$$

where $C_{(Si)}$ is a carbon atom that has at least one bond with a Si atom which could act as a nucleus for diamond growth following the mechanism below, on Figure 43.

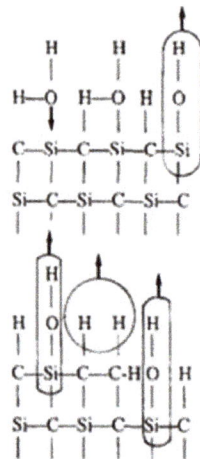

The process includes three steps:

i) Interaction between H₂O molecules and the SiC surface,
ii) Stabilization of the dangling bonds by hydrogen,
iii) Release of hydrated silica and H₂ gas followed by the formation of sp³-bonded carbon.

Figure 43. Two-dimensional sketch of the reaction mechanism for producing diamond. Reprinted with permission from Reference [142]. Copyright 1996 Elsevier.

By comparison with the model of the SiC etching via chlorine gas, suggested by Peng et al. [146], it seems that both mechanisms are similar. However, under high pressure conditions, it seems that the carbon-rich layer has a higher chance to evolve into sp^3-hybridized structure, which is diamond [142] whereas at ambient pressure, it gives rise to few layers of graphene.

Moreover, carbon formation could be a consequence of the three following processes:

○ From the free carbon (USB) located around the SiC crystals, in Nicalon and Tyranno fibers,
○ From a C-H-O-(Si) fluid which is created under hydrothermal conditions [129],
○ The growth of the diamond particles could be due to the reaction of carbonaceous by-products (88) in a H_2O-dominated fluid [147]:

$$CO_2 + CH_4 \rightarrow 2C + 2H_2O \tag{88}$$

As CH_4 and CO_2 gas are not stable at high temperature, the reaction is thermodynamically possible [148].

To conclude, multiple parameters, such as the composition of the hydrothermal media and the temperature and pressure conditions, allow for the control of the nature of the carbon layer, along with its properties.

Likewise, it was shown that the molar ratio and time of treatment is an important factor during the hydrothermal treatment as hydrolysis or oxidation processes can be favored for thermodynamic and/or kinetic reasons. As a result, the SCW treatment is efficient for producing either carbon films or silica films, and, sometimes, neither film.

5.4. The Use of Silicon Carbide in the Nuclear Field

Recently, silicon carbide materials have attracted attention in the nuclear field. Since the Fukushima Daishi incident in Japan in 2011, replacement of Zircaloy cladding in Pressurized Water Reactors (PWR) has been an important issue. SiC materials constitute an alternative due to their excellent thermomechanical properties, accident tolerance and resistance to irradiation [149]. Moreover, the development of TRISO particles, as alternative fuel material is of great interest [150]. Therefore, information and comprehension about the behavior of SiC materials in high temperature and high pressure water is necessary. More precisely, factors influencing the material's resistance to high pressure water oxidation have been investigated, as well as solutions to improve this resistance.

Terrani et al. [151] studied the corrosion of CVD-SIC and SiC composites materials in conditions similar to Light Water Reactor (LWR) operating conditions ($290 \leq T \leq 330\ °C$). From a chemical-mechanism point of view, they did not observe any silica on the surface, suggesting that it was dissolved by the water media. In accordance with Hirayama's model, the recession of SiC was limited by surface reaction, i.e., silica formation on the surface of SiC.

$$SiC + O_2 \rightarrow SiO_2 + CO_2 \qquad (89)$$

Moreover, oxygen activity in the system is believed to play a major role, as it is the chemical species responsible for SiC oxidation. Indeed, increasing oxygen activity (from 2.0×10^{-36} to 1.7×10^{-6}) leads to an important increase of reaction rate (from 7.3×10^{-9} to 4.43×10^{-7}). It is important to note that oxygen activity is directly related to irradiation of the media, so irradiation is directly responsible for important corrosion of the silicon carbide phase [152].

On the contrary, dissolved hydrogen is believed to have the opposite effect [153], as it reduces the chemical affinity of the chemical reaction forming SiO_2. More precisely, H_2 is believed to act as a scavenger of O_2, which supposedly decreases oxygen activity in the media.

Doyle et al. [154] established a model to determine SiC recession rate according to roughness, amount of oxygen dissolved and temperature:

$$\frac{0.1458}{1+SA}T\quad (1.09(1-10^{-3}T)[O_2]e^{\frac{-1.275*10^4}{T}} + 7.91 * 10^{-6}e^{-\frac{7.39*10^3}{T}}) \qquad (90)$$

However, this model does not take into account a certain number of parameters or mechanisms, such as grain fallout or morphology of the sample (fibers, for example).

Shin et al. [155] gave another important parameter regarding the recession of SiC, which is electrical conductivity of the sample. Indeed, higher electrical conductivity of the sample leads to higher exchange current density and promotes electrochemical reactions. Because the formation of silica (or silicon hydroxide) is electrochemical in nature, a higher electrical conductivity of the sample induces an increase in the recession rate.

The behavior of SiC materials also obviously depends on their microstructure and their method of elaboration.

Manufacturing of dense silicon carbide monoliths often requires the use of sintering additives that can have a strong influence on the corrosion behavior of the material.

Parish et al. [156] studied the influence of different sintering additives for the NITE process (liquid phase sintering using nanopowder of SiC) (Figure 44).

Results of the study show that sintering additives greatly decrease the SiC material's resistance toward oxidation. Additives used in this study include alumina, ceria and zirconia. Alumina, which is the most used sintering additive in NITE process, display such high recession rate that the material disappeared after two days of exposure to reactive media. In fact, the behavior of the sintered material depends on the affinity of the additive with water. Alumina is rapidly dissolved by water, thus leaving the grain boundaries of the material open, which is catastrophic for corrosion resistance. On the contrary, yttria(Y_2O_3) used with alumina [157] forms a eutectic compound ($Y_3Al_5O_{12}$) which acts as a protective barrier for corrosion. Similarly, adding chromium leads to the formation of chromium oxide, another protective oxide [158]. However, a precise balance between sintering additives and protective oxides needs to be considered to make the sintering possible while maintaining an acceptable corrosion behavior.

More precisely, Suyama et al. [159] precisely described the hydrothermal corrosion of the constituents of a SiC_f/SiC composite.

As expected, silicon carbide fibers, whatever purity or crystallinity, were affected by hydrothermal corrosion. No hypothesis was made about the mechanism, but it is highly probable that silica was formed and dissolved by the media. Likewise, the SiC matrix did undergo corrosion, which was mainly affected by the surface state of the sample (Figure 45).

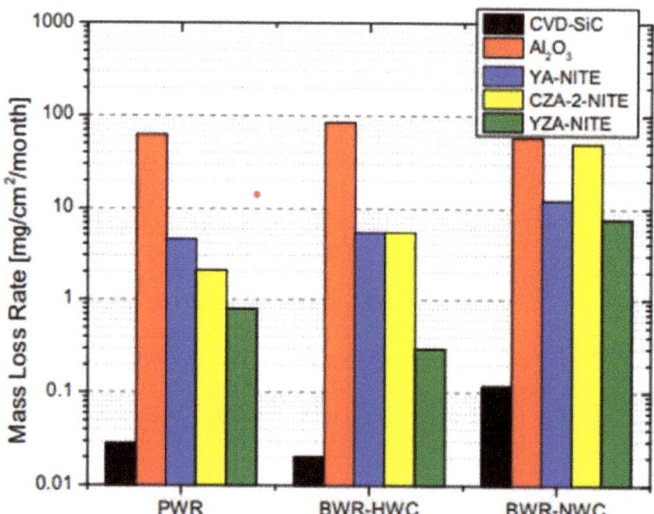

Figure 44. Evolution of mass loss for different SiC materials with different additives. Reprinted with permission from Reference [156]. Copyright 2017 Elsevier.

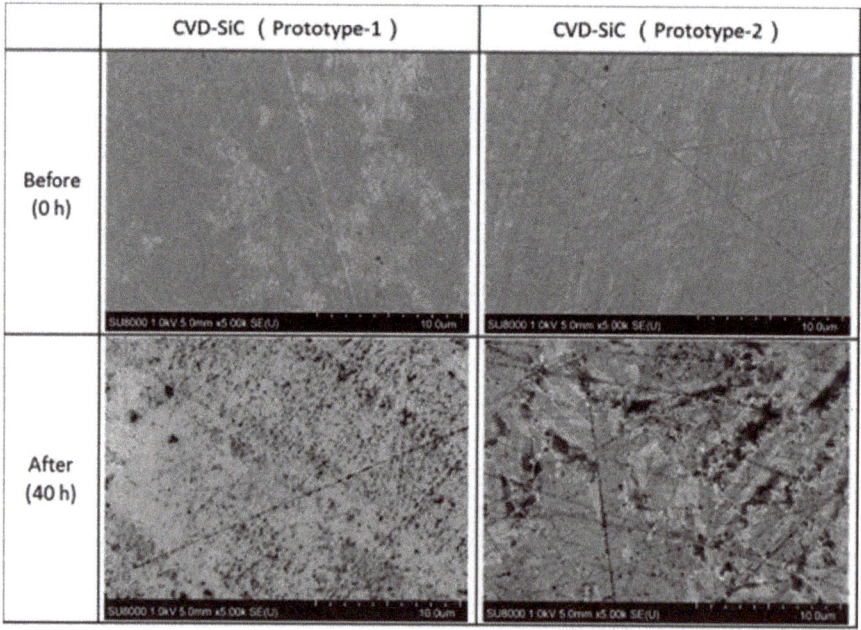

Figure 45. SEM images of CVD-SiC before and after corrosion testing. Reprinted with permission from Reference [159]. Copyright 2019 MDPI AG/Creative Commons Attribution License 4.0.

To conclude, recent studies on hydrothermal corrosion of silicon carbide focused on the use of SiC materials in nuclear environments. The mechanism of hydrothermal corrosion has been widely studied in the literature, and very accurate models can be used. However, the effect of the chemical structure of the candidate materials needs to be elucidated in order to optimize the materials elaboration and maximize the corrosion resistance.

5.5. Conclusions

Hydrothermal corrosion via hydrolysis leads to the formation of a carbon-rich layer by Si etching whereas the hydrothermal oxidation leads to the formation of silica. These two processes compete depending both on the temperature and the H_2O:SiC molar ratios, as follows:

As shown, for high H_2O:SiC molar ratios, silicon carbide is hydrolyzed by water without carbon formation. However, by increasing the temperature, the hydrothermal oxidation is favored, and silica and carbon are formed. For low molar ratios, no hydrolysis is observed, whereas above 500 °C, the oxidation of SiC starts. Thus, carbon and silica are recovered on the SiC surface, as silica does not dissolve. In this case, increasing the temperature does not affect the carbon film, whereas for higher molar ratios, carbon is oxidized by water above 600 °C.

Carbon coating can find applications in many fields as it improves ceramics' sinterability and allows for control over electrical conductivity. Indeed, graphitic carbon can decrease friction coefficients for lubricating applications, whereas diamond coating demonstrates high hardness for abrasive applications.

Finally, the improvement of the fiber/matrix compatibility is essential for ceramic composites to have a ductile behavior when submitted to stress. For maximizing the compatibility, the coating should be tailored in term of composition, structure, porosity and thickness. [160].

Understanding the numerous chemical phenomena involved in hydrothermal corrosion of SiC has proven to be useful in the design of materials for harsh environments, such as a nuclear reactor. Most recent studies therefore focus on the behavior of SiC or SiC composite materials in hydrothermal conditions typical for the new generation of reactors, studying the effect of experimental conditions and microstructure of the materials (summarized in Table 8).

Table 8. Competition between the hydrothermal oxidation and hydrolysis of crystalline silicon carbide, oxycarbidic and silica phases by water depending on the temperature and the H_2O: SiC molar ratios. Extracted from References [92,93,126].

P = 10–100 MPa	Low (H_2O:SiC) Molar Ratios (1: 10)	Intermediates (H_2O:SiC) Molar Ratios (2: 1)	High (H_2O:SiC) Molar Ratios (10: 1)
Observations	Deposition of carbon and silica, according to Yoshimura	Formation of a carbon layer and dissolution of silica, according to Gogotsi	Oxidation of carbon and dissolution of silica, according to Hirayama
300 °C	No reactions	$SiC_xO_y + nH_2O \rightarrow SiO_2 + xC + nH_2$ $SiO_2 + H_2O \rightarrow SiO_3^{2-} + 2H^+$	$SiC + 4H_2O \rightarrow Si(OH)_4 + CH_4$ $SiC_xO_y + 4H_2O \rightarrow Si(OH)_4 + xCH_4$
400 °C			
500 °C	$SiC + 2H_2O \rightarrow SiO_2 + C + 2H_2$ $SiC + 2H_2O \rightarrow SiO_2 + CH_4$ $O_2 + CH_4 \rightarrow CO_2 + 2H_2O$	$SiC + 2H_2O \rightarrow SiO_2 + C + 2H_2$ $SiC_xO_y + nH_2O \rightarrow SiO_2 + xC + nH_2$ $SiO_2 + H_2O \rightarrow SiO_3^{2-} + 2H^+$	$SiC + 2H_2O \rightarrow SiO_2 + C + 2H_2$ $SiC_xO_y + nH_2O \rightarrow SiO_2 + xC + nH_2$ $SiO_2 + H_2O \rightarrow SiO_3^{2-} + 2H^+$
600 °C	$SiC + 2H_2O \rightarrow SiO_2 + C + 2H_2$ $SiC + 3H_2O \rightarrow SiO_2 + CO + 3H_2$ $SiC + 4H_2O \rightarrow SiO_2 + CO_2 + 4H_2$	$C + H_2O \rightarrow CO + H_2$ $2C + 2H_2O \rightarrow CO_2 + CH_4$ $3C + 2H_2O \rightarrow 2CO + CH_4$	$C + H_2O \rightarrow CO + H_2$ $2C + 2H_2O \rightarrow CO_2 + CH_4$ $3C + 2H_2O \rightarrow 2CO + CH_4$

6. Conclusions

First, Deal and Grove established a kinetic model for the silicon oxidation under wet and dry atmosphere [11] and parabolic and linear regimes were defined for expressing the growth of silica. The parabolic regime expresses a diffusion-controlled mechanism whereas the linear one expresses a surface-controlled mechanism. As this model fits the oxidation behavior of silicon and silicon carbide materials, it was widely used for expressing the silica growth on SiC materials under dry and wet conditions.

It should be mentioned that recently, Hijitaka et al. [18] developed and proved the effectiveness of the Si an C emission model to explain the SiC oxidation for the whole thickness of oxide.

For both processes, dry and wet (air) oxidation, the kinetics show time-dependence. Indeed, for short oxidation time, a thin amorphous oxide film is created, and the growth of silica is linear, whereas at longer time, a thick film is created for which the kinetics follow the parabolic regime. The parabolic regime shows a temperature-dependence, as the amorphous silica crystallizes with the temperature. Indeed, tridymite, cristobalite and quartz phases show lower permeability to oxidant species and make the diffusion process slower. This, in turn, slows the oxidation rates.

From a microstructural point of view, differences between pure dense SiC and SiC within other microstructures should be emphasized. As SiC oxidation is occurring at grained boundaries first, different behaviors can be observed, depending on the microstructure of the material.

When adding water to the system of C-O-Si, oxidation and subsequent degradation of SiC can be drastically enhanced, even with small amounts of H_2O. Indeed, water has the capacity to loosen the oxide network due to its high reactivity with silanol bonds, which allows a faster solution and diffusion of the molecule through the scale. In fact, the primary oxidant during wet air oxidation is water, but oxygen molecules diffuse as well. Therefore, a kind of synergy is developed as water facilitates the transport of oxidant species through the scale and oxygen demonstrates high reactivity with the Si-C bonds. That is why a new oxidation regime is defined, the paralinear regime, for which porous and non-protective scales are obtained due to high oxide volatilization.

When high temperature and high-pressure water interacts with SiC materials, chemical corrosion occurs. Three models were proposed, and the microstructure study of the corroded SiC surface shows agreement with the Allongue's model [93]. Indeed, he proposed that successive hydrolysis of the Si-Si and Si-H bonds was involved in the corrosion process of silicon crystals. As no silica scale was found at the surface, Hirayama proposed the formation of hydrosoluble silica, which is pH-dependent [92]. From that aspect, presence of alkaline ions (Na for example) can also dramatically increase corrosion by promoting dissolution of silica scale.

The oxidation behavior of SiC seems complex, as several reactions compete along with different corrosion processes, as is suggested by the tribochemical corrosion model of Presser et al. [106].

However, the corrosion of SiC can be controlled using SCW due to its high tunability. By varying the temperature, SCW demonstrates high solvent power for organic, salts or gases and from liquid-like to gas-like properties. Gogotsi and Yoshimura [125] created a carbon coating on top of SiC materials by optimizing the properties of high temperature and high-pressure water. Its high reactivity allows the etching of Si atoms which leads to the carbon enrichment of the SiC surface; however, high temperature leads to the oxidation of carbon. As SCW shows high diffusivity, the thickness of the layer evolves linearly with time. Finally, the H_2O:SiC molar ratio has great influence, as it favors either hydrolysis or oxidation reactions. Indeed, low ratios lead to the oxidation of SiC and silica and carbon are created. However, high ratios lead to the oxidation of SiC with creation of carbon. The hydrothermal corrosion of SiC has therefore been the subject of an extensive study in the literature, and mechanisms have been elucidated. That is why recent studies on the supercritical oxidation of SiC are focused on the influence of the structure of the material (composition, architecture) and on its corrosion by high pressure, high temperature water.

Author Contributions: N.B. was involved in the literature reviewing and the original draft preparation. L.H. was involved in conceptualization, reviewing of literature and writing of the manuscript. T.A. and M.Y. were involved in reviewing manuscript after first draft preparation. C.A. was involved in supervision of N.B. and L.H., as well as conceptualization and reviewing of the manuscript after first draft preparation. All authors have read and agreed to the published version of the manuscript.

Funding: This research was founded by "Région Nouvelle Aquitaine" and "Direction Générale de l'Armement" (French ministry of defense).

Institutional Review Board Statement: Not applicable to this work.

Informed Consent Statement: Not applicable to this work.

Data Availability Statement: Not applicable to this work.

Acknowledgments: The authors express their gratitude to Monica Keszler for English editing and correction of the manuscript.

Conflicts of Interest: The authors declare no conflict of interest.

References

1. Riedel, R. *Handbook of Ceramic Hard Materials*; Wiley-VCH: Hoboken, NJ, USA, 2000; ISBN 9783527299720.
2. Yajima, S.; Hayashi, J.; Omori, M. Continuous silicon fibre of high tensile strength. *Chem. Lett.* **1975**, *9*, 931–934. [CrossRef]
3. Roewer, G.; Herzog, U.; Trommer, K.; Muller, E.; Fruhauf, S. Silicon Carbide–A Survey of Synthetic Approaches, Properties and Applications. In *Structure and Bonding*; Jansen, M., Ed.; Springer: Berlin/Heidelberg, Germany, 2002; Volume 12, pp. 59–127.
4. Gulbransen, E.A.; Jansson, S.A. The high-temperature oxidation, reduction, and volatilization reactions of silicon and silicon carbide. *Oxid. Met.* **1972**, *4*, 181–201. [CrossRef]
5. Rosenqvist, T. *Principles of Extractive Metallurgy*; McGraw-Hill: New York, NY, USA, 1974.
6. Hinze, J.W.; Tripp., W.C.; Graham, H.C. *Mass Transport Phenomena in Ceramics*; Cooper, A.R., Heuer, A.H., Eds.; Plenum Press: New York, NY, USA; London, UK,, 1975; p. 383.
7. Jacobson, N.S. Corrosion of silicon-based ceramics in combustion environments. *J. Am. Ceram. Soc.* **1993**, *76*, 3–28. [CrossRef]
8. Bhatt, R.T. Oxidation effects on the mechanical properties of SiC fiber-reinforced reaction-bonded silicon nitride matrix composites. *J. Am. Ceram. Soc.* **1992**, *75*, 406–412. [CrossRef]
9. Zheng, Z.; Tressler, R.E.; Spear, K.E. Oxidation of Single-Crystal Silicon Carbide. Part 1990, I. Experimental Studies. *J. Electrochem. Soc.* **1990**, *137*, 854–858. [CrossRef]
10. Park, D.J.; Jung, Y.I.; Kim, H.G.; Park, J.Y.; Koo, Y.H. Oxidation behavior of silicon carbide at 1200 °C in both air and water-vapor-rich environments. *Corros. Sci.* **2014**, *88*, 416–422. [CrossRef]
11. Deal, B.E.; Grove, A.S. General Relationship for the Thermal Oxidation of Silicon. *J. Appl. Phys.* **1965**, *36*, 3770–3778. [CrossRef]
12. Harris, R.C.A. Oxidation of 6H- Silicon Carbide Platelets. *J. Am. Ceram. Soc.* **1975**, *58*, 7–9. [CrossRef]
13. Massoud, H.Z. Thermal Oxidation of Silicon in Dry Oxygen Growth-rate Enchancement in the Thin Regime. I. Experimental Results. *J. Electrochem. Soc. Solid State Sci. Technol.* **1985**, *132*, 2685.
14. Massoud, H.Z. Thermal Oxidation of Silicon in Dry Oxygen Growth-rate Enchancement in the Thin Regime. II. Physical Mechanisms. *J. Electrochem. Soc. Solid State Sci. Technol.* **1985**, *132*, 2693.
15. Goto, D.; Hijikata, Y. Unified theory of silicon carbide oxidation based on the Si and C emission model. *J. Phys. D Appl. Phys.* **2016**, *49*, 225103. [CrossRef]
16. Kageshima, H.; Shiraishi, K.; Uematsu, M. Universal theory of Si oxidation rate and importance of interfacial Si emission. *Jpn. J. Appl. Phys.* **1999**, *38*, L971. [CrossRef]
17. Kouda, K.; Hijikaya, Y.; Yagi, S.; Yaguchi, H.; Yoshida, S. Oxygen partial pressure dependence of the SiC oxidation process studied by in-situ spectroscopic ellipsometry. *J. Appl. Phys.* **2012**, *112*, 24502. [CrossRef]
18. Hijikata, Y. Macroscopic simulations of the SiC thermal oxidation process based on the Si and C emission model. *Diam. Relat. Mater.* **2019**, *92*, 253–258. [CrossRef]
19. Hasunuma, R. Interfacial transition layer in thermally grown SiO_2 film on 4H-SiC. In Proceedings of the IEEE International Conference on IC Design and Technology (ICICDT), Austin, TX, USA, 23–25 May 2017; pp. 1–4.
20. Dartora, G.H.S.; Pitthan, E.; Stedile, F.C. Unraveling the mechanisms responsible for the interfacial region formation in 4H-SiC dry thermal oxidation. *J. Appl. Phys.* **2017**, *122*, 215301. [CrossRef]
21. Akiyama, T.; Hori, S.; Nakamura, K.; Ito, T.; Kageshima, H.; Uematsu, M.; Shiraishi, K. Reaction mechanisms at 4H-SiC/SiO_2 interface during wet SiC oxidation. *Jpn. J. Appl. Phys.* **2018**, *57*, 04FR08. [CrossRef]
22. Gao, H.; Wang, H.; Niu, M.; Su, L.; Fan, X.; Wen, J.; Wei, Y. Oxidation simulation study of silicon carbide nanowires: A carbon-rich interface state. *Appl. Surf. Sci.* **2019**, *493*, 882–888. [CrossRef]
23. Park, T.; Park, C.; Jung, J.; Yun, G.J. Investigation of silicon carbide oxidation mechanism using ReaxFF molecular dynamics simulation. *J. Spacecr. Rocket.* **2020**, *57*, 1328–1334. [CrossRef]
24. Shimizu, T.; Akiyama, T.; Pradipto, A.-M.; Nakamura, K.; Ito, T.; Kageshima, H.; Uematsu, M.; Shiraishi, K. Ab initio calculations for the effect of wet oxidation condition on the reaction mechanism at 4H-SiC/SiO_2 interface. *Jpn. J. Appl. Phys.* **2020**, *59*, SMMD01. [CrossRef]
25. Ervin, G. Oxidation Behavior of Silicon Carbide. *J. Am. Ceram. Soc.* **1958**, *41*, 347–352. [CrossRef]
26. Lu, W.J.; Steckl, A.J.; Chow, T.P.; Katz, W. Thermal oxidation of sputtered silicon carbide thin Films. *J. Electrochem. Soc. Solid State Sci. Technol.* **1984**, *131*, 1907–1914. [CrossRef]

27. Ramberg, C.E.; Cruciani, G.; Spear, K.E.; Tressler, R.E. Passive-oxidation kinetics of high-purity silicon carbide from 800° to 1100 °C. *J. Am. Ceram. Soc.* **1996**, *79*, 2897–2911. [CrossRef]
28. Singhal, S.C. Oxidation kinetics of hot-pressed silicon carbide. *J. Mater. Sci.* **1976**, *11*, 1246–1253. [CrossRef]
29. Costello, J.A.; Tressler, R.E. Oxidation kinetics of hot-pressed and sintered alpha-SiC. *J. Am. Ceram. Soc.* **1981**, *64*, 327–331. [CrossRef]
30. Costello, J.A.; Tressler, R.E. Oxidation kinetics of silicon carbide crystals and ceramics: I, In dry oxygen. *J. Am. Ceram. Soc.* **1986**, *69*, 674–681. [CrossRef]
31. Narushima, T.; Goto, T.; Hirai, T. High-temperature passive oxidation of chemically vapor deposited silicon carbide. *J. Am. Ceram. Soc.* **1989**, *72*, 1386–1390. [CrossRef]
32. Opila, E.J. Oxidation Kinetics of Chemically Vapor-Deposited Silicon Carbide in Wet Oxygen. *J. Am. Ceram. Soc.* **1994**, *77*, 730–736. [CrossRef]
33. Ogbuji, L.U.J.T.; Opila, E.J. A comparison of the oxidation kinetics of SiC and Si3N. *J. Electrochem. Soc.* **1995**, *142*, 925–930. [CrossRef]
34. Goto, T.; Homma, H.; Hirai, T. Effect of oxygen partial pressure on the high-temperature oxidation of CVD SiC. *Corros. Sci.* **2002**, *44*, 359–370. [CrossRef]
35. Nakatogawa, T. Silicon carbide non-ohmic resistors. II. Oxidation rates of silicon carbide. *J. Soc. Chem. Ind. Jpn.* **1954**, *57*, 348–350.
36. Adamsky, R.F. Oxidation of silicon carbide in the temperature range 1200 to 1500 °C. *J. Phys. Chem.* **1953**, *63*, 305–307. [CrossRef]
37. Jorgensen, P.J.; Wadsworth, M.E.; Cutler, I.B. Oxidation of Silicon Carbide. *J. Am. Ceram. Soc.* **1959**, *4*, 613–616. [CrossRef]
38. Jorgensen, P.J.; Wadsworth, M.E.; Cutler, I.B. Effects of Oxygen Partial Pressure on the Oxidation of Silicon Carbide. *J. Mater. Chem.* **1960**, *43*, 209–212. [CrossRef]
39. Ainger, F.W. The formation and devitrification of oxides on silicon. *J. Mater. Sci.* **1966**, *1*, 1–13. [CrossRef]
40. Pultz, W.W. Temperature and oxygen pressure dependence of silicon carbide oxidation. *J. Phys. Chem.* **1967**, *71*, 4556–4558. [CrossRef]
41. Rosner, D.E.; Allendorf, D.H. High temperature kinetics of the oxidation and nitridation of pyrolytic silicon carbide in dissociated gases. *J. Phys. Chem.* **1970**, *74*, 1829–1839. [CrossRef]
42. Antill, J.E.; Warburton, J.B. Oxidation of silicon and silicon carbide in gaseous atmospheres at 1000–1300 °C. In Proceedings of the AGARD Conference Proceedings, Paris, France, 1 January 1970.
43. Norton, F.J. Permeation of gaseous oxygen through vitreous silica. *Nature* **1961**, *191*, 701. [CrossRef]
44. Motzfeldt, K. On the rate of oxidation of silicon and of silicon carbide in oxygen, and correlation with the permability of silica glass. *Acta Chem. Scand.* **1964**, *18*, 1596–1606. [CrossRef]
45. Jorgensen, P.J.; Wadsworth, M.E.; Cutler, I.B. Effects of water vapour on oxidation of silicon carbide. *J. Am. Ceram. Soc.* **1960**, *44*, 258–261. [CrossRef]
46. Presser, V.; Loges, A.; Hemberger, Y.; Nickel, K.G. Microstructural evolution of silica on single-crystal silicon carbide. Part I Devitrification and oxidation rates. *J. Am. Ceram. Soc.* **2009**, *92*, 724–731. [CrossRef]
47. Wagstaff, F.E. Crystallization kinetics of internally nucleated vitreous silica. *J. Mater. Chem.* **1968**, *51*, 449–453. [CrossRef]
48. Wei, W.-C.; Halloran, J.W. Phase transformation of diphasic aluminosilicate gels. *J. Am. Ceram. Soc.* **1988**, *71*, 166–172. [CrossRef]
49. Lewis, E.A.; Irene, E.A. The Effect of Surface Orientation on Silicon Oxidation Kinetics. *J. Electrochem. Soc.* **1987**, *134*, 2332–2339. [CrossRef]
50. Suzuki, A.; Ashida, H.; Furui, N.; Mameno, K.; Matsunami, H. Thermal oxidation of SiC and elelectric properties of Al-SiO$_2$-SiC MOS structure. *Jpn. J. Appl. Phys. Jpn.* **1982**, *21*, 579–585. [CrossRef]
51. Onneby, C.; Pantano, C.G. Silicon oxycarbide formation on SiC surfaces and at the SiC/SiO2 interface. *J. Vac. Sci. Technol. A Vac. Surf. Film.* **1997**, *15*, 1597–1602. [CrossRef]
52. Vickridge, I.; Ganem, J.; Hoshino, Y.; Trimaille, I. Growth of SiO2 on SiC by dry thermal oxidation: Mechanisms. *J. Phys. D Appl. Phys.* **2007**, *40*, 6254. [CrossRef]
53. Gavrikov, A.; Knizhnik, A.; Safonov, A.; Scherbinin, A.; Bagatur'yants, A.; Potapkin, B.; Chatterjee, A.; Matocha, K. First-principles-based investigation of kinetic mechanism of SiC 0001 dry oxidation including defect generation and passivation. *J. Appl. Phys.* **2008**, *104*, 093508. [CrossRef]
54. Chang, K.C.; Nuhfer, N.T.; Porter, L.M.; Wahab, Q. High-carbon concentrations at the silicon dioxide-silicon carbide interface identified by electron energy loss spectroscopy. *Appl. Phys. Lett.* **2000**, *77*, 2186–2188. [CrossRef]
55. Suzuki, H. A study of the oxidation of pure silicon carbide powders. *Yogyo Kyokaishi* **1966**, *65*, 88. [CrossRef]
56. Chaudhry, M.I. A study of native oxides of B-SiC using Auger electron spectroscopy. *J. Mater. Res.* **1989**, *4*, 404–407. [CrossRef]
57. Fitzer, E.; Ebi, R. *Kinetic Studies on the Oxidation of Silicon Carbide*; Marshall, R.C., Faust, J.W., Jr., Ryan, C.E., Eds.; University of South Carolina Press: Colombia, SC, USA, 1974; pp. 320–328.
58. Singhal, S.C. Effect of water vapor on the oxidation of hot-pressed silicon nitride and silicon carbide. *J. Am. Ceram. Soc.* **1976**, *59*, 81–82. [CrossRef]
59. Tedmon, C.S. The effect of oxide volatilization on the oxidation kinetics of Cr and Fe-Cr alloys. *J. Electrochem. Soc.* **1966**, *113*, 766. [CrossRef]
60. Irene, E.; Ghez, R. Silicon oxidation studies: The role of H$_2$O. *J. Electrochem. Soc. Solid State Sci. Technol.* **1977**, *124*, 1757–1761.
61. Opila, E.J.; Hann, R.E. Paralinear oxidation of CVD SiC in water vapor. *J. Am. Ceram. Soc.* **1997**, *80*, 197–205. [CrossRef]

62. Opila, E.J.; Fox, D.S.; Jacobson, N. Mass spectrometric identification of Si-O-H(g) species from the reaction of silica with water vapor at atmospheric pressure. *J. Am. Ceram. Soc.* **1997**, *80*, 1009–1012. [CrossRef]
63. Opila, E.J.; Robinson, R.C. The oxidation rate of SiC in high pressure water vapor environments. In *High-temperature Corrosion Materials Chemistry*; McNallan, M.J., Opila, E.J., Maruyama, T., Narita, T., Eds.; The Electrochemical Soc.: Pennington, NJ, USA, 1999; pp. 398–406.
64. Opila, E.J. Oxidation and volatilization of silica formers in water vapor. *J. Mater. Chem.* **2003**, *86*, 1238–1248. [CrossRef]
65. Opila, E.J.; Robinson, R.C.; Cuy, M.D. High temperature corrosion of silicon carbide and silicon nitride in water vapor. In *Advances in Science and Technology*; Vincenzini, P., Ed.; NASA Glenn Researcher Center: Cleveland, OH, USA, 2003; pp. 243–254.
66. Opila, E.J. Variation of the oxidation rate of silicon carbide with water-vapor pressure. *J. Am. Ceram. Soc.* **1999**, *82*, 625–636. [CrossRef]
67. Tortorelli, P.F.; More, K.L. Effects of high water-vapor pressure on oxidation of silicon carbide at 1200 °C. *J. Am. Chem. Soc.* **2003**, *86*, 1249–1255. [CrossRef]
68. Haycock, E.W. Transitions from Parabolic to Linear Kinetics in Scaling of Metals. *J. Electrochem. Soc.* **1959**, *106*, 771–775. [CrossRef]
69. Palmour, J.W.; Kim, H.J.; Davis, R.F. WET and Dry Oxidation of Single Crystal β-SiC: Kinetics and Interface Characteristics. *MRS Online Proc. Libr.* **1985**, *54*, 553–559. [CrossRef]
70. Cappelen, H.; Johansen, K.H.; Motzfeldt, K. Oxidation of silicon carbide in oxygen and in water vapour at 1500 °C. *Acta Chem. Scand.* **1981**, *35*, 247–254. [CrossRef]
71. Fung, C.D.; Kopanski, J.J. Thermal oxidation of 3C silicon carbide single-crystal layers on silicon. *Appl. Phys. Lett.* **1984**, *45*, 757–759. [CrossRef]
72. Narushima, T.; Goto, T.; Iguchi, Y.; Hirai, T. High-temperature oxidation of chemically vapor-deposited silicon carbide in wet oxygen at 1823 to 1923 K. *J. Am. Ceram. Soc.* **1990**, *73*, 1580–1584. [CrossRef]
73. Pauling, L. *The Nature of the Chemical Bond*; Ithaca, N., Ed.; Cornell University Press: Ithaca, NY, USA, 1960.
74. Maeda, M.; Nakamura, K.; Ohkubo, T. Oxidation of silicon carbide in a wet atmosphere. *J. Mater. Sci.* **1988**, *23*, 3933–3938. [CrossRef]
75. Sosman, R.B. The properties of silica: An introduction to the properties of substances in the solid non conducting state. *Chem. Cat. Co.* **1927**, *37*, 856.
76. Lamkin, M.A.; Riley, F.L. Oxygen mobility in silicon dioxide and silicate glasses: A review. *J. Eur. Ceram. Soc.* **1992**, *10*, 347–367. [CrossRef]
77. Wagner, C. Passivity during the Oxidation of Silicon at Elevated Temperatures. *J. Appl. Phys.* **1985**, *29*, 1295–1297. [CrossRef]
78. Moulson, A.J.; Roberts, J.P. Water in silica glass. *Trans. Faraday Soc.* **1961**, *57*, 1208–1216. [CrossRef]
79. Hetherington, G.; Jack, K.H. Water in Vitreous Silica. *J. Phys. Chem. Glasses* **1962**, *3*, 129–133.
80. Brückner, J. Properties and structure of vitreous silica. *J. Non Cryst. Solids* **1971**, *5*, 177. [CrossRef]
81. Doremus, R.H. The diffusion of water in fused silica. In *Reactivity of Solids*; Mitchell, J.W., Devries, R.C., Eds.; Wiley: New York, NY, USA, 1969; pp. 667–673.
82. Doremus, R.H. Diffusion of water in crystalline; glass oxides: Diffusion-reaction-model. *J. Mater. Res.* **1999**, *14*, 3754–3758. [CrossRef]
83. Shewmon, P.G. *Diffusion in Solids*; McGraw-Hill Book Company: New York, NY, USA, 1963.
84. Roberts, G.J.; Roberts, J.P. An oxygen tracer investigation of the diffusion of 'water' in silica glass. *Phys. Chem. Glasses* **1966**, *7*, 82–89.
85. Wakabayashi, H.; Tomozawa, M. Diffusion of Water into Silica Glass at Low Temperature. *J. Am. Ceram. Soc.* **1989**, *72*, 1850–1855. [CrossRef]
86. Fortier, S.M.; Giletti, B.J. An empirical model for predicting diffusion coefficients in silicate minerals. *Science* **1989**, *245*, 1481–1484. [CrossRef] [PubMed]
87. Bremen, W.; Naoumidis, A.; Nickel, H. Oxidationsverhalten des pyrolytisch abgeschiedenen B-SiC unter einer atmosphäre aus CO-CO2-gasgemischen. *J. Nucl. Mater.* **1977**, *71*, 56–64. [CrossRef]
88. Schachtner, R.; Sockel, H. Study of Oxygen Diffusion in Quartz by Activation Analysis. In *Reactivity of Solids*; Wood, J., Lindqvist, O., Helgesson, C., Vannerberg, N.-G., Eds.; Springer: Berlin/Heidelberg, Germany, 1977; pp. 605–609.
89. Giletti, B.J.; Semet, M.P.; Yund, R.A. Studies in diffusion: III. Oxidation in feldspars: An ion microprobe determination. *Geochem. Cosmochim. Acta* **1978**, *42*, 45–57. [CrossRef]
90. Gogotsi, Y.; Yoshimura, M. Water effects on corrosion behavior of SiC ceramics. *MRS Bull.* **1994**, *XIX*, 39–45. [CrossRef]
91. Kim, W.-J.; Hwang, H.S.; Park, J.Y.; Ryu, W.-S. Corrosion behaviors of sintered; chemically vapor deposited silicon carbide ceramics in water at 360 °C. *J. Mater. Sci. Lett.* **2003**, *22*, 581–584. [CrossRef]
92. Yoshimura, M.; Kase, J.; Somiya, S. Oxidation of SiC powder by high-temperature, high-pressure H_2O. *J. Mater. Res.* **1986**, *1*, 100–103. [CrossRef]
93. Hirayama, H.; Kawakubo, T.; Goto, A. Corrosion behavior of silicon carbide in 290 °C water. *J. Am. Ceram. Soc.* **1989**, *72*, 2049–2053. [CrossRef]
94. Allongue, P.; Brune, H.; Gerischer, H. In situ STM observations of the etching of n-Si (111) in NaOH solutions. *Surf. Sci.* **1992**, *275*, 414–423. [CrossRef]

95. Allongue, P.; Costa-Kieling, V.; Gerischer, H. Etching of Silicon in NaOH Solutions. Part II. Electrochemical studies of n-Si (111) and (100) and mechanism of the dissolution. *J. Electrochem. Soc.* **1993**, *140*, 1018–1026. [CrossRef]
96. Smialek, J.L.; Jacobson, N.S. Mechanism of Strength Degradation for Hot Corrosion of alpha-SiC. *J. Am. Ceram. Soc.* **1986**, *69*, 741–752. [CrossRef]
97. Henager, C.H.; Schemer-Kohrn, A.L.; Pitman, S.G.; Senor, D.J.; Geelhood, K.J.; Painter, C.L. Pitting corrosion in CVD SiC at 300 °C in deoxygenated high-purity water. *J. Nucl. Mater.* **2008**, *378*, 9–16. [CrossRef]
98. Barringer, E.; Faiztompkins, Z.; Feinroth, H.; Allen, T.; Lance, M.; Meyer, H.; Gog Lara-Curzio, E. Corrosion of CVD Silicon Carbide in 500 °C Supercritical Water. *J. Am. Ceram. Soc.* **2007**, *90*, 315–318. [CrossRef]
99. Kim, W.-J.; Hwang, H.S.; Park, J.Y. Corrosion behavior of reaction-bonded silicon carbide ceramics in high-temperature water. *J. Mater. Sci. Lett.* **2002**, *21*, 733–735. [CrossRef]
100. Park, J.-Y.; Kim, I.-H.; Jung, Y.-I.; Kim, H.-G.; Park, D.-J.; Kim, W.-J. Long-term corrosion behavior of CVD SiC in 360 °C water and 400 °C steam. *J. Nucl. Mater.* **2013**, *443*, 603–607. [CrossRef]
101. Tan, L.; Allen, T.R.; Barringer, E. Effect of microstructure on the corrosion of CVD-SiC exposed to supercritical water. *J. Nucl. Mater.* **2009**, *394*, 95–101. [CrossRef]
102. Quinn, T. Oxidation wear modeling: I. *Wear* **1992**, *153*, 179–200. [CrossRef]
103. Quinn, T. Oxidational wear modelling: Part II. The general theory of oxidational wear. *Wear* **1994**, *175*, 199–208. [CrossRef]
104. Boch, P.; Platon, F.; Kapelski, G. Tribological and interfacial phenomena in Al_2O_3/SiC and SiC/SiC couples at high temperature. *J. Eur. Ceram. Soc.* **1989**, *5*, 223–228. [CrossRef]
105. Sasaki, S. The effect of the surrounding atmosphere on the friction and wear of alumina, zirconia, silicon-carbide and silicon-nitride. *Wear* **1989**, *134*, 185–200. [CrossRef]
106. Zum Gahr, K.-H.; Blattner, R.; Hwang, D.-H.; Pöhlmann, K. Micro- and macro-tribological properties of SiC ceramics in sliding contact. *Wear* **2001**, *250*, 299–310. [CrossRef]
107. Presser, V.; Krummhauer, O.; Nickel, K.; Kailer, A.; Berthold, C.; Raisch, C. Tribological and hydrothermal behaviour of silicon carbide under water lubrication. *Wear* **2009**, *266*, 771–781. [CrossRef]
108. Loppinet-Serani, A.; Aymonier, C.; Cansell, F. Supercritical water for environmental technologies. *J. Chem. Technol. Biotechnol.* **2010**, *85*, 583–589. [CrossRef]
109. Kritzer, P.; Boukis, N.; Dinjus, E. Factors controlling corrosion in high-temperature aqueous solutions: A contribution to the dissociation and solubility data influencing corrosion processes. *J. Supercrit. Fluids* **1999**, *15*, 205–227. [CrossRef]
110. Duverger-Nédellec, E.; Voisin, T.; Erriguible, A.; Aymonier, C. Unveiling the complexity of salt(s) in water under transcritical conditions. *J. Supercrit. Fluids* **2020**, *165*, 104977. [CrossRef]
111. Bröll, D.; Kaul, C.; Krämer, A.; Krammer, P.; Richter, T.; Jung, M.; Vogel, H.; Zehner, P. Chemistry in Supercritical Water. *Angew. Chem. Int. Ed.* **1999**, *38*, 2998–3014. [CrossRef]
112. Aymonier, C.; Philippot, G.; Erriguible, A.; Marre, S. Playing with chemistry in supercritical solvents and the associated technologies for advanced materials by design. *J. Supercrit. Fluids* **2018**, *134*, 184–196. [CrossRef]
113. Loppinet-Serani, A.; Aymonier, C.; Cansell, F. Current and foreseeable applications of supercritical water for energy and the environment. *ChemSusChem* **2008**, *1*, 486–503. [CrossRef]
114. Morin, C.; Loppinet-Serani, A.; Cansell, F.; Aymonier, C. Near- and supercritical solvolysis of carbon fibre reinforced polymers (CFRPs) for recycling carbon fibers as a valuable resource: State of the art. *J. Supercrit. Fluids* **2012**, *66*, 232–240. [CrossRef]
115. Adachi, Y.; Lu, B.C.-Y. Supercritical fluid extraction with carbon dioxide and ethylene. *Fluid Phase Equilibria* **1983**, *14*, 147–156. [CrossRef]
116. Jessop, P.G.; Ikariya, T.; Noyori, R. Homogeneous catalysis in supercritical fluid. *Chem. Rev.* **1999**, *99*, 475–493. [CrossRef] [PubMed]
117. Savage, P.E. A perspective on catalysis in sub and supercritical water. *J. Supercrit. Fluids* **2009**, *47*, 407–414. [CrossRef]
118. Kazarian, S. Polymer processing with supercritical fluids. *Polym. Sci. Ser. C* **2000**, *42*, 78–101.
119. Aymonier, C.; Loppinet-Serani, A.; Reveron, H.; Garrabos, Y.; Cansell, F. Review of supercritical fluids in inorganic materials science. *J. Supercrit. Fluids* **2006**, *38*, 242–251. [CrossRef]
120. Adschiri, T.; Kanazawa, K.; Arai, K. Rapid and continuous hydrothermal crystallization of metal oxide particles in supercritical water. *J. Am. Chem. Soc.* **1992**, *75*, 1019–1022. [CrossRef]
121. Cansell, F.; Aymonier, C.; Loppinet-Serani, A. Review on materials science and supercritical fluids. *Curr. Opin. Solid State Mater. Sci.* **2003**, *7*, 331–340. [CrossRef]
122. Marre, S.; Jensen, K.F. Synthesis of micro and nanostructures in microfluidic systems. *Chem. Soc. Rev.* **2010**, *39*, 1183–1202. [CrossRef]
123. Byrappa, K.; Adschiri, T. Hydrothermal technology for nanotechnology. *Progress Cryst. Growth Charact. Mater.* **2007**, *53*, 117–166. [CrossRef]
124. Campardelli, R.; Baldino, L.; Reverchon, E. Supercritical fluids applications in nanomedicine. *J. Supercrit. Fluids* **2015**, *101*, 193–214. [CrossRef]
125. Hirano, S.-I.; Nakamura, K.; Sōmiya, S. Graphitization of carbon in presence of calcium compounds under hydrothermal condition by use of high gas pressure apparatus. In *Hydrothermal Reactions for Materials Science and Engineering*; Sōmiya, S., Ed.; Springer: Dordrecht, The Netherlands, 1989; pp. 331–336.

26. Gogotsi, Y.; Yoshimura, M. Formation of carbon films on carbides under hydrothermal conditions. *Nature* **1994**, *367*, 630–638. [CrossRef]
27. Gogotsi, Y.; Yoshimura, M.; Kakihanna, M.; Kanno, Y.; Shibuya, M. Hydrothermal synthesis of carbon films on SiC fibers and particles. *Ceram. Trans.* **1995**, *51*, 243–247.
28. Jacobson, N.; Gogotsi, Y.; Yoshimura, M. Thermodynamic and experimental study of carbon formation on carbides under hydrothermal conditions. *J. Mater. Chem.* **1995**, *5*, 595–601. [CrossRef]
29. Ito, S.; Tomozawa, M. Stress corrosion of silica glass. *J. Am. Ceram. Soc.* **1981**, *64*, C160. [CrossRef]
30. Kase, J. Master's Thesis. Tokyo Institute of Technology, Yokohama, Japan, 1987; pp. 189–190.
31. Gogotsi, Y.; Yoshimura, M. Oxidation and properties degradation of SiC fibres below 850 °C. *J. Mater. Sci. Lett.* **1994**, *13*, 680–683. [CrossRef]
32. Gogotsi, Y.G.; Yoshimura, M. Low-temperature oxidation, hydrothermal corrosion, and their effects on properties of SiC (tyranno) fibers. *J. Am. Ceram. Soc.* **1995**, *78*, 1439–1450. [CrossRef]
33. Li, J.; Eveno, P.; Huntz, A.M. Oxidation of silicon carbide. *Werkst. Korros.* **1987**, *41*, 716–725. [CrossRef]
34. Gogotsi, Y.; Yoshimura, M. Degradation of SiC-based fibres in high-temperature, high pressure. *J. Mater. Sci. Lett.* **1994**, *13*, 395–399. [CrossRef]
35. Kraft, T.; Nickel, K.G.; Gogotsi, Y. Hydrothermal degradation of chemical vapour deposited SiC fibres. *J. Mater. Sci.* **1998**, *33*, 4357–4364. [CrossRef]
36. Futatsuki, T.; Oe, T.; Aoki, H.; Kimura, C.; Komatsu, N.; Sugino, T. Low-temperature oxidation of SiC surfaces by supercritical water oxidation. *Appl. Surf. Sci.* **2010**, *256*, 6512–6517. [CrossRef]
37. Yoko, A.; Oshima, Y. Recovery of silicon from silicon sludge using supercritical water. *J. Supercrit. Fluids* **2013**, *75*, 1–5. [CrossRef]
38. Le Coustumer, P.; Monthioux, M.; Oberlin, A. Understanding Nicalon Fibre. *J. Eur. Ceram. Soc.* **1993**, *11*, 95–103. [CrossRef]
39. Yajima, S.; Iwai, T.; Yamamura, T.; Okamura, K.; Hasegawa, Y. Synthesis of a polytitanocarbosilane and its conversion into inorganic compounds. *J. Mater. Res.* **1981**, *16*, 1349–1355. [CrossRef]
40. Gogotsi, Y. Nanostructure Carbon Coatings. NATO Advanced Research Workshop on Nanostructured Films and Coating. *Ser. 3 High Technol.* **2000**, *78*, 25–40.
41. Gogotsi, Y.; Kofstad, P.; Yoshimura, M.; Nickel, K.G. Formation of sp3-bonded carbon upon hydrothermal treatment of SiC. *Diam. Relat. Mater.* **1996**, *5*, 151–162. [CrossRef]
42. Gogotsi, Y.; Nickel, K.G.; Bahloul-Hourlier, D.; Merle-Mejean, T.; Khomenko, G.E.; Skjerlie, K.P. Structure of carbon produced hydrothermal treatment of β-SiC powder. *J. Mater. Chem.* **1996**, *6*, 595–604. [CrossRef]
43. Wang, W.; Wang, T.; Chen, B. Primary study on the irradiation effects of high energy C^+ and H^+ on diamond-like carbon films. *J. Appl. Phys.* **1992**, *72*, 69–72. [CrossRef]
44. Badzian, A.; Badzian, T. Diamond homoepitaxy by Chemical Vapor Deposition. *Diam. Relat. Mater.* **1993**, *2*, 147–157. [CrossRef]
45. Prawer, S.; Nugent, K.W.; Jamieson, D.N. The Raman Spectrum of Amorphous Diamond. *Diam. Relat. Mater.* **1998**, *7*, 106–110. [CrossRef]
46. Peng, T.; Lv, H.; He, D.; Pan, M.; Mu, S. Direct transformation of amorphous silicon carbide into graphene under low temperature and ambient pressure. *Sci. Rep.* **2013**, *3*, 1148. [CrossRef] [PubMed]
47. DeVries, R.C. Hydrothermal carbon: A review from carbon in Herkimer diamonds to that in real diamonds. *Adv. Ceram.* **1990**, *3*, 181–205.
48. Rumble, D., III; Hoering, T.C. Carbon isotope geochemistry of graphite vein deposits from New Hampshire, U.S.A. *Geochim. Cosmochim. Acta* **1986**, *50*, 1239–1247. [CrossRef]
49. Pint, B.A.; Terrani, K.A.; Brady, M.P.; Cheng, T.; Keiser, J.R. High temperature oxidation of fuel cladding candidate materials in steam-hydrogen environments. *J. Nucl. Mater.* **2013**, *440*, 420–427. [CrossRef]
50. Seibert, R.L.; Jolly, B.C.; Balooch, M.; Schappel, D.P.; Terrani, K.A. Production and characterization of TRISO fuel particles with multilayered SiC. *J. Nucl. Mater.* **2019**, *515*, 215–226. [CrossRef]
51. Terrani, K.A.; Yang, Y.; Kim, Y.-J.; Rebak, R.; Meyer, H.M., III; Gerczak, T.J. Hydrothermal corrosion of SiC in LWR coolant environments in the absence of irradiation. *J. Nucl. Mater.* **2015**, *465*, 488–498. [CrossRef]
52. Kondo, S.; Lee, M.; Hinoki, T.; Hyodo, Y.; Kano, F. Effect of irradiation damage on hydrothermal corrosion of SiC. *J. Nucl. Mater.* **2015**, *464*, 36–42. [CrossRef]
53. Kim, D.; Lee, H.-G.; Park, J.Y.; Park, J.-Y.; Kim, W.-J. Effect of dissolved hydrogen on the corrosion behavior of chemically vapor deposited SiC in a simulated pressurized water reactor environment. *Corros. Sci.* **2015**, *98*, 304–309. [CrossRef]
54. Doyle, P.J.; Zinkle, S.; Raiman, S.S. Hydrothermal corrosion behavior of CVD SiC in high temperature water. *J. Nucl. Mater.* **2020**, *539*, 152241. [CrossRef]
55. Shin, J.H.; Kim, D.; Lee, H.-G.; Park, J.Y.; Kim, W.-J. Factors affecting the hydrothermal corrosion behavior of chemically vapor deposited silicon carbides. *J. Nucl. Mater.* **2019**, *518*, 35–356. [CrossRef]
56. Parish, C.M.; Terrani, K.A.; Kim, Y.-J.; Koyanagi, T.; Katoh, Y. Microstructure and hydrothermal corrosion behavior of NITE-SiC with various sintering additives in LWR coolant environments. *J. Eur. Ceram. Soc.* **2017**, *37*, 1261–1279. [CrossRef]
57. Xie, X.; Liu, B.; Liu, R.; Zhao, X.; Ni, N.; Xiao, P. Comparison of hydrothermal corrosion behavior of SiC with Al_2O_3 and Al_2O_3 + Y_2O_3 sintering additives. *J. Am. Ceram. Soc.* **2020**, *103*, 2024–2034.

158. Lobach, K.V.; Sayenko, S.Y.; Shkuropatenko, V.A.; Voyevodin, V.M.; Zykova, H.V.; Zuyok, V.A.; Bykov, A.O.; Tovazhnyans'kyy, L.L.; Chunyaev, O.M. Corrosion resistance of ceramics based on SiC under hydrothermal conditions. *Mater. Sci.* **2020**, *55*, 672–682. [CrossRef]
159. Suyama, S.; Ukai, M.; Akimoto, M.; Nishimura, T.; Tajima, S. Hydrothermal corrosion behaviors of constituent materials of SiC/SiC composites for LWR applications. *Ceramics* **2019**, *2*, 47. [CrossRef]
160. Henry, L.; Biscay, N.; Huguet, C.; Loison, S.; Aymonier, C. A water-based process for the surface functionalization of ceramic fibres. *Green Chem.* **2020**, *22*, 8308–8315. [CrossRef]

MDPI
St. Alban-Anlage 66
4052 Basel
Switzerland
Tel. +41 61 683 77 34
Fax +41 61 302 89 18
www.mdpi.com

Nanomaterials Editorial Office
E-mail: nanomaterials@mdpi.com
www.mdpi.com/journal/nanomaterials

www.ingramcontent.com/pod-product-compliance
Lightning Source LLC
LaVergne TN
LVHW070156120526
838202LV00013BA/1259